Essentials of Thermal Processing

Essentials of Thermal Processing

Second Edition

Gary Tucker
Campden BRI, Gloucestershire, UK

Susan Featherstone
Safe Food Consulting, Cape Town, South Africa

WILEY Blackwell

This second edition first published 2021

Edition History
Wiley-Blackwell (1e, 2010)

Registered Office(s)
John Wiley & Sons, Inc., 111 River Street, Hoboken, NJ 07030, USA
John Wiley & Sons Ltd, The Atrium, Southern Gate, Chichester, West Sussex, PO19 8SQ, UK

Editorial Office
The Atrium, Southern Gate, Chichester, West Sussex, PO19 8SQ, UK

For details of our global editorial offices, customer services, and more information about Wiley products, visit us at www.wiley.com.

Wiley also publishes its books in a variety of electronic formats and by print-on-demand. Some content that appears in standard print versions of this book may not be available in other formats.

Library of Congress Cataloging-in-Publication Data
Names: Tucker, Gary, author. | Featherstone, Susan, author.
Title: Essentials of thermal processing / Gary Tucker, Campden BRI, Gloucestershire, UK, Susan Featherstone, Safe Food Consulting Cape Town, South Africa.
Description: Second edition. | Hoboken, NJ : Wiley-Blackwell, [2021] | Includes bibliographical references and index.
Identifiers: LCCN 2020056297 (print) | LCCN 2020056298 (ebook) | ISBN 9781119470373 (cloth) | ISBN 9781119470359 (adobe pdf) | ISBN 9781119470328 (epub)
Subjects: LCSH: Food–Effect of heat on. | Food–Preservation. | Food–Microbiology.
Classification: LCC TP371 .T83 2021 (print) | LCC TP371 (ebook) | DDC 664/.028–dc23
LC record available at https://lccn.loc.gov/2020056297
LC ebook record available at https://lccn.loc.gov/2020056298

Cover Design: Wiley
Cover Image: © H G Molenaar & Co. (Pty) Ltd

Set in 11/13pt Times by SPi Global, Pondicherry, India
Printed and bound by CPI Group (UK) Ltd, Croydon, CR0 4YY

C9781119470373_270421

Contents

Preface

Essentials of Thermal Processing is written by two authors with many years' experience of thermal processing practice, one gained in the northern hemisphere and one in the southern hemisphere. This gives the book a unique appeal. It covers all aspects of thermal processing from its beginnings in 1795 with Nicolas Appert through to modern day computer-controlled processing systems and electronic data capture. The intention was to write a book of practical use to students studying food science and technology as well as for their lecturers, individuals in companies, and research centres that have a need to understand thermal processing principles.

In this second edition, we have tried to expand on the information given so that this book is the one stop, single text required to assist with all thermal processing queries. The second edition starts with the history of thermal processing, from the prize-winning invention in the Napoleonic era, and describes many of the developments that resulted in the precise science that we know today. The next chapter covers basic microbiological principles that govern microbial growth and death. Detail about microorganisms of particular interest to thermal processing specialists is given as well as information about solving spoilage problems. The third chapter is dedicated to the hurdles to the growth of microorganisms so that the reader gains a thorough understanding of how to ensure a safe product.

The next chapters take the reader through the different food categories that present their unique challenges for thermal processing. This includes the traditional sector of low acid foods in which the familiar $F_0 3$ concept was derived from heat resistance studies originally undertaken by Esty and Meyer in 1922. Low acid foods are a group of foods that do not contain any preservation hurdles to microorganism growth and rely on the heat process to control microorganism numbers in the food and a hermetically sealed package to prevent recontamination. Chapters 5 and 6 are on acid and acidified foods. The chapters are separated to describe the different principles and the types of thermal processes that are applied. In the acid foods chapter, information is included about other products that can be pasteurized. The last of the chapters that deals with a specific food group is on heat preserved chilled foods, which is one of the most rapidly growing sectors in Europe. This includes the ready meal concept and ready-to-eat meat, fish, and poultry products. Shelf-life requires refrigerated storage and is typically up to 10 days when a very mild process is used or it can be extended beyond this with more severe thermal treatments.

Chapter 8 describes processing methods that can be used for manufacturing heat preserved foods. It is divided into in-pack or in-line systems, often known as retorts and continuous systems. The focus of this chapter is the operating principles of the different designs. The main types of retort are described, for example steam, steam-air, water immersion, raining, and sprayed water, also included are the in-pack continuous retorts such as hydrostatic and reel and spiral cooker–cooler systems. Equipment choice for a continuous or in-line system depends mostly on the food viscosity and whether the flow behaviour is laminar or turbulent.

Thermal processing of foods has a dual purpose, which is to commercially sterilize the product and to cook it to an acceptable level. The chapter on cooking and process optimization addresses the challenges of maximizing a specific quality attribute without damaging the thermal processing effect on microbial reduction. Examples are given on the different quality parameters appropriate to heat preserved foods and how these can be mathematically analysed with the aim of adjusting processes so that maximum quantities of a nutrient, such as a vitamin, are retained.

Techniques for establishing and validating thermal processes are described in two chapters. Temperature and heat transfer are explained and process calculation methods are discussed. Measurement techniques for in-pack processes include temperature sensors of various types and log reduction methods that can be either microbiological or biochemical. Process calculation techniques are introduced, which allow process conditions to be calculated from the temperature measurements and analysis of deviations to be carried out. Differences in the techniques required for batch (in-pack) and continuous (in-flow) heat processes are described.

Chapter 12 discusses cooling water; almost all types of microorganisms can be found in water, and water is most often the vehicle of transmission for these organisms. Poor quality cooling water or inadequate cooling is arguably the most significant causes of public health issues from canned foods in the last century. Examples of spoilage incidents caused by cooling issues are dealt with in several chapters, and so, this chapter solely describes the different methods of disinfecting cooling water.

Despite the best efforts to control all aspects of a thermal processing operation, all factories will experience process deviations from time to time. Chapter 13 is dedicated to process deviations and describes methods to assess whether a process deviation has made a critical impact on the commercial sterility or safety of a product. Methods to assess a deviation are many and include calculation methods as well as experimentally recreating the deviation using retort simulators.

Packaging of the product also demanded its own chapter. Understanding packaging options and the way they work together with the product is critical to the success of any food packaging system. The functions of packaging are to contain, protect, preserve, portion, inform, promote, and make foods portable. Packaging options for thermally processed foods are many and include metal (mainly tinplate, but some aluminium), glass, cartons, and specific plastics (laminates and composites). Primary packaging must be able to hold a hermetic seal, withstand the process temperature, provide a physical barrier, withstand the physical stresses during processing, transport, storage, and distribution, and must not react adversely with the food.

Incubation of processed packs is commonplace in the food industry; however, a variety of incubation practices occurs, which suggests that there is no standardized approach. Chapter 15 describes the objectives of incubation together with some recommendations for times and temperatures that different types of microorganism require. It includes suggestions for sample size and methods for interpretation of results.

The penultimate chapter is about the critical control factors (CCP's) involved in thermal processing of food. HACCP (Hazard Analysis Critical Control Point) is a tool to assess hazards and establish control systems that focus on prevention rather than relying mainly on end-product testing. Examples are given of CCPs of relevance to thermal processing.

The final chapter presents a different focus for thermal processing in that it deals with environmental aspects and presents a positive case for thermal processing. As a technology for preserving food, it can be environmentally beneficial because the goods are stored ambient, thus avoiding the need for refrigeration, and allows agricultural products to be processed close to where they are grown, thus avoiding transportation emissions and forcing crops to grow out of season. An example of a carbon footprint assessment is given in this chapter for a bottled apple juice.

Gary Tucker
Susan Featherstone

The authors during a practical training course session in Cape Town.

Glossary of Terms

Acid	A substance which increases the concentration of hydrogen ions (H+) in water, and reacts with a base to form a salt.
Acid Foods	A food that has a natural pH of 4.6 or below (in Europe it is common to use pH 4.5 as the upper limit).
Acidified Low-Acid Food	A food which has been treated so as to attain an equilibrium pH of 4.6 (or 4.5 in Europe) or lower after heat processing.
Acidified Food	A low-acid food to which acid(s) or acid food(s) are added and which has a finished equilibrium pH of 4.6 or below and a water activity (a_w) greater than 0.85.
Acidulant	An acidifying agent, such as acetic acid or vinegar.
Aciduric	Micro-organisms that can grow in high acid foods, i.e., with a pH value below 3.0. Generally, are of low heat resistance.
Aerobes	Micro-organisms that need oxygen for growth. Obligate aerobes cannot survive in the absence of oxygen.
Aerobic	Living or active only in the presence of free oxygen.

Anaerobes	Micro-organisms that grow in the absence of oxygen. Obligate anaerobes cannot survive in the presence of oxygen. Facultative anaerobes normally grow in oxygen, but can also grow in its absence.
Anaerobic	Living or active in the absence of free oxygen.
Anti-Microbial	A compound which inhibits the growth of a microbe.
Antioxidants	Substances that retard the oxidative rancidity of fats, or the oxidation of other substances.
Aseptic Processing and Packaging	The filling of a commercially sterile product into sterilised containers followed by hermetical sealing with a sterilised closure in an atmosphere free from micro-organisms.
Asepsis	Clean and free of micro-organisms.
Autoclave	A vessel in which high temperatures can be reached by using high steam pressure. Bacteria are destroyed more readily at elevated temperatures, and autoclaves are used to sterilize food, for example in cans.
Bacillus	A rod-shaped bacterium. Some *Bacillus* produce spores.
Bacillus Cereus	Spore-forming, rod-shaped bacterium, aerobic to facultative aerobic, proteolytic. It produces gastroenteritis caused by the release of an exo-enterotoxin during lysis of *B. cereus* in the intestinal tract.
Bactericide	Any substance that destroys bacteria, although not necessarily the spores of bacteria.
Bentonite	Colloidal clay used as an absorbent. Also used in model systems for determining rate of heat penetration.

Biodegradability	Susceptibility of a chemical compound to depolymerization by the action of biological agents.
Biological Oxygen Demand (BOD)	Micro-organisms consume oxygen in their respiration. The BOD test determines uptake of oxygen by a contaminated material, e.g., sewage, water, etc., as a measure of microbial activity.
Blanching	Heating by direct contact with hot water or live steam. It softens the tissues, eliminates air from the tissues, destroys enzymes, and washes away raw flavours.
Body Hook	That flange portion of the can body that is turned back for the formation of the double seam.
Body Maker	A machine for automatic forming of a cylindrical metal can or drums body from a body blank. In the manufacture of tin cans, the body maker may also automatically weld the side seam.
Boiler Scale	Deposit left inside boilers caused by evaporation of water and precipitation of water-soluble and insoluble substances.
Botulism	A poisoning caused by substances formed by the bacterium *Clostridium botulinum* under conditions of improper processing and storage of food. The spores of this bacterium are often found in soil and are likely to be present on soil-contaminated food.
Break-Point Chlorination	Addition of chlorine to water beyond the point where chloramines are oxidized, and where further increases in the dosage of chlorine will result in a proportional increase of chlorine residual.
Brines	Salt solutions used in canning and pickling.

Brix	The measure of density of a solution, more particularly a solution containing sucrose, as determined by a hydrometer. Degrees Brix equal Percent sucrose in water solution at 20°C.
Broken Heating Curve	A heating curve which shows a distinct change in the rate of heat transfer such that the curve may be represented by two or more distinct straight lines.
Broth Medium	A liquid medium for growth of microorganisms.
Browning Reaction	A reaction in foods, usually deteriorative, involving amino (e.g., from amino acids or proteins) and carbonyl (e.g., from glucose) groups; this reaction often leads to a brown discolouration and sometimes to off-flavours and changes in texture.
Buffer	Any substance in a fluid which tends to resist the change in pH (hydrogen-ion concentration) when acid or alkali is added.
Canned Food	Commercially sterile food in hermetically sealed containers.
Carbohydrates	Carbohydrates are essential food nutrients that can get broken down into glucose to be used as energy. They are comprised of C, H and O. Examples include starches, sugars and fibre.
Carcinogen	A cancer causing agent.
Cation	Positively charged ion such as K^+, NH_4^+.

Cationic Surfactants	Ionic surface active agents in which the portion that associates with the internal phase is the cation. They include simple amine salts, quaternary ammonium salts, amino imides and imidazolines. Cationic surfactants often have germicidal, anticorrosive, and anti-static properties.
Celsius (°C)	Temperature on a scale of 100° between the freezing point (0°) and the boiling point (100°) of water.
Chemical Oxygen Demand (COD)	An indirect measure of the biochemical load exerted on the oxygen content of a body of water when organic wastes are introduced into the water. When the wastes contain only readily available organic bacterial food and no toxic matter, the COD values can be correlated with BOD values obtained from the same wastes.
Chlorination	Building up the chlorine content (as hypochlorous acid) to process or sanitize water supplies.
Clostridia	Genus of spore forming bacteria. *Clostridium botulinum* is the most heat resistant of the food-poisoning organisms; its growth is inhibited at pH 4.6 and below, thus it is only a problem in low-acid foods. Produces an endotoxin, highly toxic in minute doses, but destroyed by heat. Destruction of this organism is generally accepted as the minimum standard of processing for low-acid and medium-acid canned food, although other *Clostridia* are more heat-resistant.
Closure	The joint or seal which is made in attaching the cover to the glass container. Also, the type of closure, such as friction, lug, screw top, etc.
Coccus	Shape of bacteria (plural cocci). A round / spherical cell.

Codex Codex Alimentarius (Latin for "Book of Food") is a collection of internationally recognized standards, codes of practice, guidelines and other recommendations relating to foods, food production and food safety. The texts are developed and maintained by the Codex Alimentarius Commission, a body that established by the Food and Agriculture Organization of the United Nations (FAO), and later was joined by the World Health Organization (WHO).

Coliform Bacteria Group of aerobic bacteria of which *Escherichia coli* is the most important member. Many coliforms are not harmful, but as they arise from faeces, they are useful as a test of contamination, particularly as a test for water pollution

Colony A microscopically visible growth of micro-organisms on a solid culture medium.

Come-Up Time (CUT) The time which elapses between the introduction of steam into the closed retort and the time when the retort reaches the required processing temperature.

Commercial Sterility The condition achieved by application of heat which renders such food free of viable forms of micro-organisms having public health significance, as well as any micro-organisms of non-health significance capable of reproducing in the food under normal non-refrigerated conditions of storage and distribution. "Commercial sterility of equipment and containers used for aseptic processing and packaging of food means the condition achieved by application of heat, chemical sterilant(s), or other appropriate treatment which render such equipment and containers free of viable forms of micro-organisms having public health significance, as well as any micro-organisms of non-health significance capable of reproducing in the food under normal non-refrigerated conditions of storage and distribution.

Consistometer One of the several types of instruments used to measure the consistency of foods.

Contamination Entry of undesirable organisms into some material or container.

Controlled Atmosphere (CA) Storage Storage of foods in a hermetic warehouse where the concentrations of O_2, CO_2, and N_2 are controlled at specific levels.

Convection Natural or forced motion in a fluid induced by heat or the action of gravity.

Corrosion Chemical action of dissolving the surface of a metal (e.g. tin in food medium).

Corrosion Accelerator Chemical species with the ability to accept electrons, which will increase the rate of a corrosion reaction.

Countersink Depth The measurement from the top edge of the double seam to the end panel adjacent to the chuck wall.

Critical Factor Any property, characteristic, condition, aspect, or other parameter, variation of which may affect the scheduled process delivered and thus the commercial sterility of the product. This does not include factors which are controlled by the processor solely for purposes of product appearance, quality, and other reasons which are not of public health significance.

Culture A population of micro-organisms cultivated in a medium; pure culture-single kind of micro-organism, mixed culture-two or more kinds of micro-organisms growing together.

Culture Medium Any substance or preparation suitable for and used for the growth and cultivation of micro-organisms.

D-value (decimal reduction time) Time in minutes at a specified temperature required to destroy 90% of the micro-organisms in a population.

Deaeration Removal of oxygen from produce juices to prevent adverse effects on juice properties.

Dehydration A food processing unit operation resulting in the removal of water from the food generally to the point where spoilage is prevented.

Detergent Surface-active material or combination of surfactants designed for removal of unwanted contamination from the surface of an article.

Deterioration A non-biological, physical, or chemical change in food which adversely affects quality.

Detinning The process of corrosion, where the internally plain tin coating is slowly dissolved by the food medium; rapid detinning refers to abnormally fast tin dissolution, caused by the presence of corrosion accelerators.

Disinfectant An agent that frees from infection by killing the vegetative cells of micro-organisms.

Disinfection The reduction, without adversely affecting the food, by means of hygienically satisfactory chemical agents and /or physical methods, of the number of micro-organisms to a level that will not lead to harmful contamination of food.

Double Seam To attach an end to a can body by a method in which five (5) thicknesses of plate are interlocked or folded and pressed firmly together. A joint formed by interlocking the edges of both the end and body of a can.

Drained Weight Weight of the solid portion of the product after draining the covering liquid for a specified time with the appropriate sieve.

Effluent Waste water or other liquid, partially or completely treated or untreated, flowing out of a process operation, processing plant, or treatment plant.

Emulsion System consisting of two incompletely miscible liquids, one being dispersed as finite globules in the other. A small amount of a third substance may render the dispersion stable. The liquid broken up into globules is the dispersed (discontinuous) phase; the surrounding liquid is the external (continuous) phase.

Endotoxin A toxin produced with an organism liberated only when the organism disintegrates.

Enterotoxin A toxin specific for cells of the intestine. Gives rise to symptoms of food poisoning.

Enzymatic Browning The darkening of plant tissues or products produced by enzymatic reactions.

Enzyme A compound of biological origin which accelerates a specific chemical reaction.

Equilibrium pH The pH of the macerated (thoroughly blended) contents of the product container.

Equilibrium Relative Humidity	The relative humidity of the ambient atmosphere surrounding a substance when the substance neither gains nor loses moisture.
Escherichia Coli	The strains of *E. coli* that produce an entero-pathogenic food poisoning syndrome.
Exotoxin	A toxin excreted by a micro-organism into the surrounding medium.
F-value	The number of minutes required to destroy a stated number of micro-organisms at a defined temperature, usually 121.1°C (250°F), and when the z-value is 10°C (18°F). "F" value is a common term employed in the canning industry to express the lethality or sterilizing value of a sterilization process. See also z-value.
Facultative Bacteria	Bacteria which can exist and reproduce under either aerobic or anaerobic conditions.
Fahrenheit	A temperature scale in which 32°F represents the freezing point of water and 212°F represents its boiling point.
Fermentation	The action of micro-organisms upon foods. Anaerobic respiration. Usually fermentation is undesirable, but sometimes it is produced intentionally, such as in the manufacture of vinegar from apple cider.
Fill Weight	The weight of the product before processing. It does not include the weight of the container or covering liquid.
Flash Pasteurization	Process in which the material is held at a much higher temperature than in normal pasteurization, but for a considerably shorter period.

Flat sours	Thermophilic and thermoduric bacteria, facultative anaerobes that attack carbohydrates with the production of acids, but without gas formation. Flat-sour spoiled canned foods therefore show no swelling of the ends.
Flexible Container	A container where the shape or contours of the filled, sealed container are affected by the enclosed product.
Flipper	A can having both ends flat, but with insufficient vacuum to hold the ends in place, thus a sharp blow will cause the end to become convex, but both ends may be pressed to their normal position.
Food Infection	An illness caused by an infection produced by invasion, growth and damage to the tissue of the host due to the ingestion of viable pathogenic micro-organisms associated with the food.
Food Intoxication	An illness resulting from the ingestion of bacterial toxin with or without viable cells. The illness does not require actual growth of cells in the intestinal tract.
Food Poisoning	A general term applied to all stomach or intestinal disturbances due to food contaminated with certain micro-organisms or their toxins.
Fortified	Food to which specific nutrients have been added "enriched".
Fungicide	Any substance that destroys fungi or inhibits the growth of spores or hyphae. Legally, sometimes the term is interpreted as also including yeasts and bacteria.
Fungistatic Agent	Prevents growth of fungi (moulds) without necessarily killing the existing cells.

GRAS Generally Recognized as Safe.

HACCP Hazard Analysis of Critical Control Points. An inspectional approach that determines what points in the process are critical for the safety of the product and how well the firm controls these points.

Halophilic Can grow or survive in a medium with a relatively high salt concentration.

Hard Swell Spoilage in which can ends are swelled too hard to be readily depressed by applying thumb pressure.

Headspace, Gross The vertical distance between the level of the product (generally the liquid surface) and the inside surface of the lid in an upright rigid container (the top of the double seam of a can or the top edge of a glass jar).

Headspace The vertical distance between the level of the product (generally the liquid surface) and the inside surface of the lid in an upright, rigid container having a double seam, such as a can.

Heat Exchanger Equipment for heating or cooling liquids rapidly by providing a large surface area and turbulence for the rapid and efficient transfer of heat.

Heat, Latent Heat absorbed or liberated in a change of physical state such as evaporation, condensation, freezing or sublimation. Expressed as BTU per lb., kCal per kg or joules per kg.

Heating Curve A graphical representation of the rate of temperature change in the food throughout the heat process; this is usually plotted on semi-log graph paper so that the temperature on an inverted log scale is plotted against time on a linear scale.

Heating factor (f_h) Time taken for the heating or cooling curve plotted on semi-log coordinates to traverse one log cycle.

HMF Hydroxymethylfurfural. Is an organic compound derived from dehydration of certain sugars. HMF is practically absent in fresh food, but it is naturally generated in sugar-containing food during heat-treatments like drying or heat processing. HMF can be used as an indicator for excess heat-treatment.

Incubation Holding cultures of micro-organisms under conditions favourable to their growth. Also, the holding of a sample at a specified period of time before examination.

Incubation Time The time period during which micro-organisms inoculated into a medium are allowed to grow.

Initial Temperature (IT) The average temperature of the contents of the coldest container to be processed at the time the sterilizing cycle begins, as determined after thorough stirring or shaking of the filled and sealed container.

Inoculate The artificial introduction of micro-organisms into a system.

Inoculation	The artificial introduction of micro-organisms into a growth medium. This can refer to the introduction of test organisms to food, to the accidental introduction of organism to food.
Inoculum	The material containing micro-organisms used for inoculation.
Intoxication	The adverse physiological effects of an organism consuming a toxic material.
Lag factor (j)	The lag in achieving a logarithmic heating rate for the heating or cooling curve plotted on semi-log coordinates.
Latent Heat	The quantity of heat, measured in BTU's or calories, necessary to change the physical state of a substance without changing its temperature.
Lethal	Capable of causing death.
Lethality Value (F-value)	This degree of sterility is referred to as commercial sterility. The F-value is introduced as a standard on which to base the sterilization of food products. The F-value is defined as the number of minutes which it takes to reduce the initial spore count of a certain micro-organism to a desired safety level at a defined lethal reference temperature.
Low-acid Foods	Any food (other than alcoholic beverages) with a finished equilibrium pH greater than 4.6 (4.5 in Europe) and a water activity greater than 0.85.
Maillard Reaction	A group of organic reactions, especially between amino acids and reducing sugars, producing brown colour and flavour changes in many foods. Also known as non-enzymatic browning.

Moulds	Micro-organisms that belong to the fungi. Moulds are much less heat resistant than bacteria.
Mycotoxins	Toxins produced by moulds or fungi.
Nesting	Containers that fit within one another when stacked.
Net Weight, Minimum	The weight of all the product in the container, including brine or sauce, but not including the weight of the container.
Non-Newtonian	Materials whose resistance to flow changes with a change in rate of shear.
Nutrients	Compounds that promote biological growth.
Osmophilic	Can grow or survive in a medium very low in humidity or of low water activity.
Osmosis	A process by which molecules in a solvent tend to pass through a semi-permeable membrane from a less concentrated solution into a more concentrated one.
Oxidation	The act of oxidizing which is brought about by increasing the number of positive charges on an atom or the loss of negative charges.
Oxidative Rancidity	The deterioration of fats and oils due to oxidation.
Package	Any container or wrapping in which a consumer commodity is enclosed for delivery or display to retail purchasers.
Panelling	Distortion (side wall collapse) of a container caused by development of a reduced pressure (too high vacuum) inside the container.

Pasteurization	A heat treatment of food usually below 100°C.
Pathogen	Disease producing micro-organism.
Pathogenic	Capable of producing disease.
Patulin	A mycotoxin.
ppb	Parts per billion.
pH	A figure expressing the acidity or alkalinity of a solution on a logarithmic scale on which 7 is neutral, lower values are more acid and higher values more alkaline.
Potable Water	Drinkable water / fit for human consumption.
ppm	Parts per million.
Preservation	Any physical or chemical process which prevents or delays decomposition of foods.
Preservatives	Any substance capable of retarding or arresting food spoilage or deterioration.
Primary Spoilage	That spoilage due to bacterial or chemical action of product packed within the can. See also Secondary Spoilage.
Process Authority	The person or organization that scientifically establishes thermal processes for low-acid canned foods or processing requirements for acidified foods.
Process, Scheduled	The process selected by the processor as adequate under the conditions of manufacture for a given product to achieve commercial sterility. This process is in excess of that necessary to ensure destruction of micro-organisms of public health significance.

Process Temperature	The calculated temperature at a particular time (process time) for which a specific can size and food product need to be heated in order to achieve commercial sterility.
Process Time	The calculated time at a particular temperature (process temperature) for which a specific can size and food product need to be heated in order to achieve commercial sterility.
PSIG	Pounds per square inch gauge pressure.
Psychrometer	An instrument for measuring the humidity (water-vapour) content of air by means of two thermometers, one dry and one wet.
Putrefaction	Decomposition of proteins by microorganisms, producing disagreeable odours.
Refractometer	Optical instrument that measures the percent of soluble solids in solution by the extent to which a beam of light is bent (refracted). Soluble solids scale is based on sugar concentration in a pure sucrose solution (Brix).
Relative Humidity	The ratio of actual humidity to the maximum humidity which air can retain without precipitation at a given temperature and pressure. Expressed as percent of saturation at a specified temperature.
Retort	Any closed vessel or other equipment used for the thermal sterilization of foods.
Retort (Retortable) Pouch	A flexible container in which food is placed to be heated to commercial sterility in a retort or other sterilization system.

Rigid Container A container whereby the shape or contours of the filled and sealed container are neither affected by the enclosed product nor deformed by an external mechanical pressure of up to 0.7 kg/cm^2 (10 psig), (i.e. normal firm finger pressure).

Salmonella A genus of bacteria that can cause infections in man that are characteristically gastrointestinal. Destroyed by adequate heating as in the canning process.

Salmonellosis Infectious disease caused by bacteria of the genus *Salmonella.*

Sanitizer A chemical agent that reduces the number of microbial contaminants on food contact surfaces to safe levels from the standpoint of public health requirements.

Secondary Spoilage Rusted, corroded and perforated cans (from the outside towards the inside) as a result of product leaking onto them from other spoilt containers. May occur during warehousing or transport.

Semi-rigid Container A container whereby the shape or contours of the filled, sealed container are not affected by the enclosed product under normal atmospheric temperature and pressure but can be deformed by an external mechanical pressure of less than 0.7 kg/cm^2 (10 psig), (i.e. normal firm finger pressure).

Shelf-Life The length of time that a container, or material in a container, will maintain market acceptability under specified conditions of storage.

Spoilage	A process whereby food is rendered unacceptable through microbial or chemical action. See also Primary Spoilage and Secondary Spoilage.
Spores	Some rod forms of bacteria produce spores. In the spore state, bacteria can survive extremes of cold, heat, drying, and other unfavourable conditions for long periods of time; and when the environment is again favourable, the spores germinate, and the organisms start another cycle of growth. Growing cells are called "vegetative" cells.
Springer	Swelled can with only one end remaining out; on pressing this end it will return to normal, but the other end will bulge out.
Stack Burn	Condition resulting from placing cased cans in piles insufficiently cooled. Food may vary from over cooked to definitely burned flavour and colour. Also, excessive corrosion of interior of container may result.
Staphylococci	A genus of spherical bacteria (cocci) occurring in pairs, tetrads or irregular, grape-like clusters.
Staphylococcus aureus	Species of bacteria that are important as a cause of human infections and of food poisoning.
Sterile	Free of living organisms.
Sterilization	Any process, chemical or physical, which will destroy all living organisms.
Sterilization Process	The time-temperature treatment necessary to render canned foods commercially sterile.

Sterilization Temperature	The temperature maintained throughout the thermal process as specified in the scheduled process.
Sterilization Time	The time that lapses between the moment a retort reaches sterilization temperature, until steam is cut off.
Sterilization Value (F_0)	The number of minutes at a reference temperature of 121.1°C (250°F) required to kill a known population of micro-organisms with a z-value of 10°C (18°F).
Still Retort	A retort for the sterilization of canned foods in batch amounts, with no agitation.
Streptococci	A genus of bacteria that are spherical or oval shaped and that divide in such a way that pairs or chains are formed.
Superheated Steam	Saturated steam that has been heated at constant pressure above its saturation temperature.
Swell	(1) (Noun) A container with either one or both ends bulged by moderate or severe internal pressure. (2) (Verb) To bulge out by internal pressure, as by gases caused by biological or chemical action.
Swell, Hard	A can of food which has spoiled to the point where both ends are bulged out and show no appreciable yield to thumb pressure.
Thermal Process	The application of heat to food, either before or after sealing in a hermetically sealed container, for a period of time and at a temperature scientifically determined to achieve a condition of commercial sterility.

Thermocouple A bi-metallic device to measure temperatures electrically.

Thermoduric Micro-organisms that have the ability to withstand high temperatures, i.e. are highly heat resistant.

Thermolabile Fairly easily destroyed by heat.

Tin Plate Sheet steel, usually of special formula and temper, coated on both sides with a controlled thickness of pure tin.

Toxin An organic poison, a product of the growth of an organism. Some toxins are given off as waste products of a micro-organism, and are called "exotoxins". Others are contained within the cells, and are liberated only when the cell dies and disintegrates. These are called "endotoxins".

UHT Ultra High Temperature. Term used in reference to pasteurization of commercial sterilization of food by heating it for an extremely short period, around 1–2 seconds, at a temperature exceeding 135°C (275°F).

Ultraviolet Irradiation A disinfection method that uses ultraviolet (UV) light at sufficiently short wavelength to kill micro-organisms. It destroys the nucleic acids in these organisms so that their DNA is disrupted.

Vacuum Pack Products packed with little or no brine or water, which are sealed under a high mechanical vacuum, and which require maintenance of high vacuum to assure process adequacy.

Vegetative Cells Stage of active growth of the micro-organism, as opposed to the bacterial spore.

Venting	Eliminating air from a retort prior to sterilizing canned foods.
Vents	Openings controlled by gate, plug, cock, or other adequate valves used for the elimination of air during the venting period.
Viscosity	The internal friction or resistance to flow of a liquid. The constant ratio of shear stress to shear rate. In liquids for which this ratio is a function of stress, the term "apparent viscosity" is defined as this ratio.
Water Activity (a_w)	A measure of water availability in food for microbial growth. The ratio of water vapour pressure of a food to the vapour pressure of pure water under identical conditions of temperature and pressure.
Yeasts	Micro-organisms classified in the kingdom Fungi. They are spherical or more or less elongated cells. Most yeasts break down sugars to carbon dioxide and alcohol. That process is called fermentation.
Z-value	Z-value is a term used in thermal death time calculations. It is the temperature required for one log reduction in the D-value for a specific bacteria.

1 History of Thermal Processing

The application of high temperatures is one of the most commonly used methods to kill or control the numbers of micro-organisms present within foods and on packaging surfaces. Evidence for its importance in today's food industry can be seen in the percentage of shelf space taken up in any retailer by foods preserved by heat. Along with low temperature storage (e.g. refrigeration and freezing) and dried foods, thermal processing remains of key importance to the worldwide food industry. There have been many attempts to introduce foods preserved by new technologies (e.g. ultra high pressure, pulsed electric fields) into retail food markets, but their significance is still relatively minor in comparison with thermal processing.

Thermal processing is a general term that describes all forms of heat treatments in which micro-organism numbers are controlled by heat. This includes heat processed container types such as metal cans, plastic trays, pouches, glass jars, and even cartons. It also includes continuous thermal processes that take place outside of the package and are usually linked with aseptic or hot filling. Nowadays, the terms 'canning' and 'thermal processing' are often used interchangeably; however, as most of the original work was done in cans, thermally processed foods that are in other packaging formats are still often referred to as 'canned'.

1.1 A BRIEF HISTORY OF THE SCIENCE AND TECHNOLOGY OF THERMAL PROCESSING

The early Greeks believed in 'spontaneous generation' (that living things could originate from non-living matter). Although Aristotle discarded this notion, he still believed that animals could arise

Essentials of Thermal Processing, Second Edition. Gary Tucker and Susan Featherstone.

spontaneously from other unlike organisms or from soil. His influence regarding this concept of spontaneous generation was still felt as late as the seventeenth century.

The sixteenth and seventeenth centuries were a time of great scientific advancement in areas of chemistry, mathematics, and physics. This was known as the Scientific Revolution which laid the foundations for the Age of Enlightenment in the eighteenth century, a period where science became popular with the ordinary person, and an increasingly literate population was hungry for knowledge, information, and to learn. Evening Science lectures, with demonstrations, were very popular as a form of entertainment for the working class.

In addition at that time, the Industrial Revolution was taking place in Europe in the eighteenth and nineteenth centuries. Great strides were made in the areas of textiles, steam generation, and metallurgy. Steam generation was much more efficient; fossil fuels were used for the first time, instead of wood based fuel, resulting in a much more efficient source of energy. Advances in mining techniques and metal-working, especially iron founding, resulted in many new uses for metals like iron, copper, and tin (http://www.wikipedia.org).

Towards the end of the seventeenth century, a chain of observations, experiments, and arguments began that disproved the belief that life could be generated from non-life. Microbiology, as a science, can be said to have begun with the development of the microscope. In the mid-seventeenth century, although not the inventor of the microscope, Antonie van Leeuwenhoek (1632–1723) (Fig. 1.1), a Dutch draper, ground his own lenses and made small microscopes that could magnify up to 500 times. He had exceptional attention to detail and was the first to provide proper descriptions of his observations, which included protozoans from the guts of animals and bacteria from teeth scrapings. His descriptions and drawings were excellent, and he conveyed his findings in a series of letters to the British Royal Society during the mid-1670s. Although the observation of his 'animalcules' stimulated much interest, it remained as an oddity until the eighteenth century.

In France, the French Revolution (1789–1799) took place, largely as a result of growing dissatisfaction due to poverty and a shortage of food and increasing malnutrition. The Napoleonic Wars (1803–1815) further influenced the economy and developments in Europe.

It was in this time that Nicolas Appert was working as a confectioner. He was born on 17 November 1749 at Charlon-sur-Marne.

Fig. 1.1 Antonie van Leeuwenhoek, painting by Jan Verkolje.

His family were not wealthy, and young Nicolas acquired an education largely through personal effort. He was trained as a chef and worked and experimented with different kinds of food processes (fermenting, distilling, and preserving) for his own interest all of his life (http://www.appert-aina.com).

The French Army was busy with many military campaigns, and a major concern to the French Government was that they were losing more troops to diseases like scurvy and malnutrition, than to battle casualties. In 1795, they offered a prize of 12 000 Francs (a lot of money at that time) to anyone who could find a way to safely preserve food. This offer caught the attention of Nicolas Appert. He had noticed that the sugar syrups that he used for his confectionary, kept almost indefinitely when heated, and stored in stoppered glass bottles. He began experimenting with preserving other types of food, also by heating them and storing them in stoppered jars. He worked scientifically and had a keen attention to detail. He started with using champagne bottles, but soon had them modified with a wider mouth, so that he could fill them more easily.

By 1804, he felt confident enough to test some product on the French Navy. The test was a huge success. In 1806, more tests were conducted. His invention was assessed, and early in 1810, he was informed that he could claim the award of 12 000 Francs, but had to publish the exact details of his discovery, which he did (Goldblith 1971b).

Appert is known as the Father of Canning. Heat sterilization is also known as Appertization. Although his first products were in glass bottles, he soon started using metal cans at the cannery he built with the prize money he won for his invention. He had to deliver 200 copies of his processing methods (printed at his own expense) to the French Government before they would give him his award. The book was very detailed and described the canning process much as it is still today (Appert 1810). Appert described the process as follows (Featherstone 2012):

- Enclose the foods to be preserved in bottles.
- Cork the bottles carefully.
- Boiling the bottles in water for various lengths of time (depending on the foods).
- Remove the bottles and cool them.

In 1811, he has a second edition printed in French as well as English and Swedish, and in 1823 a third, and in 1831 a fourth. The later editions were translated into many other languages (Bitting 1937).

Although Appert's methods worked, no one at the time understood why. Appert thought it was due to the heating and exclusion of air. At that time, it was widely believed that air itself was the cause of spoilage (Fig. 1.2).

There were many other people working on methods of food preservation at the time: including the Englishmen, Donkin, and Saddington (who, in 1807, were awarded five guineas for their method of using heat to preserve fruits without using sugar).

Glass jars were soon largely replaced in commercial canneries with cylindrical tin or wrought-iron canisters (later shortened to 'cans'), following the work of Peter Durand in 1810. The cans were cheaper and quicker to make and much less fragile than glass jars. In England, the firm of Donkin and Hall manufactured large quantities of 'canisters', some of which were taken on expeditions to Baffin Bay in 1814 and on Arctic explorations in 1815.

The reason for lack of spoilage was unknown at the time, since it would be another 50 years before Louis Pasteur (1822–1895), a French chemist and microbiologist proved, by demonstration, that fermentation is caused by the growth of micro-organisms, and not due to spontaneous generation or by exposure to air.

Fig. 1.2 Nicolas Appert, drawing on Commemorative Stamp issued by Monaco in 2010.

1.2 FOOD MICROBIOLOGY AS A SCIENCE

In 1859, Louis Pasteur demonstrated the role of micro-organisms in food spoilage. In his experiment, he heated broth in long swan necked jars to sterilize them. The jars either had filters on them, or very long necks that allowed only air, but not dust and other particles through. Nothing grew in the broths unless the flasks were broken open. He therefore correctly concluded that the living organisms that grew came from outside, as spores on dust, rather than spontaneously generated within the broth or from the air. He showed that the growth of micro-organisms was responsible for spoiling products like beer, wine, and milk.

He invented a process in which milk was heated to kill most bacteria and moulds already present. This process was soon afterwards known as pasteurization (http://www.wikipedia.org). A little known fact about Pasteur is that he, because of his status as a respected scientist, served on the board of Public Hygiene and Sanitation, in France. One of his responsibilities was food laws.

As a result of his discoveries regarding the cause of food spoilage, various methods of preserving foods were researched by many others. Some of these involved adding chemicals to the food. In 1870, Pasteur advocated that the public had 'the right to know' what was being put into the food and stated that all ingredients and additives must be declared on the label (Goldblith 1971a) (Fig. 1.3).

Microbiology was in its infancy in the nineteenth century. Many people contributed to developing it into the science it is today. A few of these scientists who made significant contributions to the understanding of the science of canning are mentioned below (Goldblith 1972).

Joseph Lister (1827–1912) was an English surgeon who made the important development of antiseptic surgical practice for medicine and microbiology (Fig. 1.4). Lister searched for a way to keep bacteria out of wounds and incisions made by surgeons, as death from sepsis was frequent at the time. He read a paper by Louis Pasteur that described the growth of anaerobic bacteria and suggested killing them by heat, filtering, or with chemicals. Carbolic acid (phenol) was known to deodorize sewage, so Lister soaked surgical dressings, and sprayed carbolic acid into the air of

Fig. 1.3 Louis Pasteur, painting by A. Edelfeldt (1885).

Fig. 1.4 Joseph Lister, painting by *unknown*.

operating theatres. The protected wounds did not become infected, and the patients recovered quickly. The success of this technique was so remarkable that aseptic surgical practice was soon established worldwide.

Robert Koch (1843–1910) isolated the *Bacillus* bacteria responsible for tuberculosis and the one for anthrax. He also isolated the bacteria responsible for cholera, *Vibrio cholera*. In his work with anthrax, he noticed that when the bacteria were exposed to unfavourable conditions, they went into a dormant state, forming internal spores that could survive for extended periods in the soil, causing outbreaks of the disease at opportune times. More of Koch's contributions to microbiology were in the areas of culturing bacteria and examining them. He developed solid culture media by adding gelatine and other solidifying agents to liquid media in order to obtain isolated growths of micro-organisms. These isolated growths were called 'colonies' and were found to contain millions of individual micro-organisms packed tightly together. He noted that the colonies were visible to the naked eye, whereas the individual cells were not. Koch also found that, by adding dyes to micro-organisms smeared onto a glass slide, individual cells could be seen more clearly with a microscope.

All canners had losses due to sporadic swelling and spoilage, but the actual causes for it were unknown, and the canners did not know what they should do to overcome this problem.

In about 1860, Isaac Solomon, a canner in Baltimore in the USA, added calcium chloride to the cooking water enabling it to 'boil' at 116 °C instead of 100 °C and thus drastically reducing the cooking times that had been in use.

In 1895, Dr Harry L. Russell published a paper describing swelling spoilage in canned peas. He did experiments where he processed peas at higher temperatures and longer times and showed that the percent spoilage was significantly reduced.

Two important advances in food science and canning were the invention of the steam pressure retort by A.K. Shriver in 1875 and publication of the time and temperature requirements of canned foods by Underwood and Prescott. William Lyman Underwood was a canner who wanted to understand what caused canned food spoilage (Fig. 1.5). Together with Samuel Cate Prescott, they studied many cases of food spoilage over many years. They demonstrated that spoilage could be caused by inoculation and that the proper time and temperature combinations could prevent spoilage. They showed that bacteria were the causative agents of canned food spoilage, the importance of heat penetration in processing canned foods, the importance of proper cooling of canned foods, and they were also the first to recommend incubation

Fig. 1.5 Underwood and Prescott. Courtesy of the MIT Museum.

testing of canned foods. The significance of their work cannot be over-emphasized; canning passed from an individual skill to a scientific discipline.

There are many excellent scientists who have contributed invaluable insight into the field of thermal processing, but most of the foundation work was done by those mentioned above.

The early bacteriological studies on spore death kinetics were done by different researchers at various temperatures. This work was done between 1921 and 1948. As could be expected, the lower the temperature, the slower the rate of kill. Stumbo took this information and calculated the z-value of 18 °F (10 °C) and an F-value of 2.78 minutes for *Clostridium botulinum* spores on a thermal death time curve that passed through 250 °F (121.1 °C) at 2.78 minutes (Tucker 2008). This temperature was appropriate for practical cooking times as well as being safely achievable in the processing vessels of the day. This was the basis of the Fo3, at 250 °F concept which is still applicable today.

The General Method for calculating scheduled process times was originally described by Bigelow et al in 1920, but contributions by Ball in 1928 and O.T Schulz and F. C. W. Olson in 1940 resulted in a much improved General Method. M. Patashnik published his improvements which are the most widely used today in 1953 (Patashnik 1953; Park 1996).

The invention of a standardized reliable thermocouple probe for measurement of real time heat penetration temperatures was an important contribution made by Ecklund (1949).

J.R. Manson, A.A. Teixeira, and K. Purohit were three of Stumbo's graduate students who also contributed significantly to the field of thermal processing. They were engineers, and the first to apply engineering mathematics to simulate the coupling of heat transfer with thermal inactivation kinetics in thermal processing of canned foods. Teixeira used this approach to find optimum retort the time and temperature combinations that would maximize quality retention while delivering specified target lethality. Manson carried Teixeira's work further by improving the mathematical model to simulate convection as well as conduction heat transfer. Working together, Teixeira and Manson demonstrated how such models could be employed in real time online control of batch retorts by automatically extending process time to precisely compensate for unexpected process deviations (Teixeira et al. 1969). The research mentioned above took place largely in the USA.

In the United Kingdom, T.G. Gillespy and his team did valuable work on processing times and temperatures of a large range of foods at Campden Experimental Factory (now Campden BRI) (Tucker 2008). In France, H. Cheftel at Carnaud Research did extensive work on canned foods and published *Principles and methods for establishing thermal processes for canned foods* in French. In South Africa, G.G. Knock did much to increase the understanding and hence reduce the incidence of thermophilic 'flat sour' spoilage in canned peas (Knock 1954). These researchers and others collaborated from all over the world to improve the understanding of the mechanism of canned food spoilage and the requirements necessary to ensure that safe canned food is produced. Improvements in product safety, quality, and energy usage are still continuously being made. G.S. Tucker at Campden BRI further developed numerical techniques for thermal process calculations and a computer program to recalculate process deviations in real time.

1.3 PACKAGING FOR HEAT PRESERVED FOODS

Nicolas Appert's first products were packed in glass. Soon after his discovery was published, Peter Durand, a British merchant patented the idea of preserving food using tin cans. The patent (No 3372) was granted on 25 August 1810 by King George III of England. After receiving the patent, Durand did not make any canned food himself, but in 1812 sold his patent to two other Englishmen, Bryan Donkin, and John Hall, for £1000. Donkin was involved with tinning of iron sheets from 1808 and was keen to expand it to the food industry. Donkin and Hall set up a commercial canning factory and by 1813 were producing their first canned goods for the British army. In 1818, Durand introduced tin cans in the United States by re-patenting his British patent in the United States of America (http://www.wikipedia.org).

The first 'canisters' were made from iron plates that were dipped into molten tin to stop it from corroding. The ends were soldered closed with molten lead. The metal was thick and the cans were heavy and strong. The cans often weighed more than the food that was in them. They had to be cut open with a hammer and chisel.

Appert also started using cans. He made them himself in his cannery. They had the capacity of between 4 and 45 pounds and could be reused. He also added handles to some of them so that they could be used as cooking pots once opened. Although he preferred round cans, he made oval and rectangular ones too, at the request of his customers (Goldblith 1971a).

Improvements to the can came when steel was invented, and this allowed for a much thinner metal which had the same strength to be used. In 1888, the hermetic double seam was invented by Max Ams. This paved the way for automated can lines to be made, whereas before about 6 cans per hour were handmade, the first automated can lines could make about 60 cans per hour. Can making lines today can run as fast as of 1500 cans per minute.

Tin is an expensive metal. In the 1930s, hot dipping of tinplate was replaced by electroplating, where much less tin can be used to perform the same job. Improvements in steel making technology have resulted in even lighter weight cans. Single reduced tinplate of 0.19–0.21 mm thickness and double reduced (DR) tinplate as thin as 0.10–0.15 mm are now used to make cans all over the world.

Although improvements in can design (e.g. by beading the body walls) can compensate to some extent for the loss of strength due to the thinner metal, many of the down-gauging improvements are possible only due to improvements in can handling.

1.3.1 Convenience – the can opener is invented

Only when thinner steel cans came into use could the can opener be invented. Before then, canned food used to come with the written instructions: 'Cut round the top near the outer edge with a chisel and hammer'. The first can openers were primitive claw-shaped or 'lever-type' design. Robert Yeates, a cutlery and surgical instrument maker patented the first can opener in Great Britain in 1855. It was a wooden handled leaver-type cutting blade. Three years later, in the USA, Ezra Warner patented another design (Warner et al. 1858). His looked like a bent bayonet. The large curved blade was driven into a can's rim and then forcibly worked around its edge. This first type of can opener was deemed to be too dangerous for ordinary people to use it, and the store assistants opened each can before it was taken away (http://www.wikipedia.org).

Another can opener was invented by William Lyman of the United States in 1870. It had to be pierced into the centre of the end and had a cutting wheel that rolled around the rim of the can end. It was difficult to operate as it had to be adjusted to the size of the end. A breakthrough came in 1925 when a second, serrated wheel was added to hold the cutting wheel on the rim of the can. The basic principle of this opener was the same as is used on the modern can openers. The first electric can opener was introduced in December 1931.

The easy opening end is the ultimate in convenience as far as can openers go. In the 1960s, a pull tab was patented by Ermal C. Fraze owner of the Dayton Reliable Tool Company in Ohio, USA, for aluminium ends for beverage cans. A lot of work then went into the development of an easy opening tinplate end that could withstand the requirements of being retortable and had a good shelf life. By the1980s, these ends were available commercially.

1.3.2 Other forms of packing for 'canned foods'

Other forms of packaging for 'canned foods' have also become popular and practical; including glass, various plastics, and composites. Although glass was not new – being the packaging that Appert used to develop his method, it was difficult to seal adequately and cans soon became the packaging of choice. Developments in the closures for glass, starting with the Mason Jar in 1858, resulted in glass becoming a popular alternate to cans. Improvements in glass making technology have resulted in bottles that don't break easily under the high heat and pressure conditions experienced during thermal processing. Improvements in lids, caps, and closures that form hermetic seals are easy to open and reclose and have tamper evident features such as buttons that 'pop' on first opening have helped to make glass a viable alternate.

Developments in other packaging types are more recent and restricted to the past 40 years or so. Rigid plastics are useful for ready meals, as they are not breakable and can be heated in the microwave oven. Pouches are flexible and during processing the flat dimensions result in faster heat penetration, therefore shorter cooking times and better quality product. Pouches and some rigid containers come with their own challenges for thermal processing as their irregular shapes make it more difficult to measure the cold spot during processing. Tetra Recart's 'square' format saves space when packing and stacking. All of these options have their advantages and disadvantages. None of the plastics and laminates are as good a gas barrier as glass and metal. Light can also cause deterioration in some products, requiring tinted glass or opaque metal or cardboard casing.

The ideal packaging choice depends on the requirements of product type, processing conditions, required shelf life, and target market.

1.4 DEVELOPMENTS IN CANNERY EQUIPMENT

1.4.1 Seaming

The equipment used in canneries also had to develop significantly. Originally, cans were manually soldered closed, and good artisans could do about 6 an hour. Once the hermetic double seam was invented, closing of cans became much faster. Today, there are seamers that can close over 1000 cans per minute.

1.4.2 Processing

Processing (heating) of the sealed cans started off as being an all-day event. Appert processed most of his products in boiling water. He did experiment with pressure processing, but at that time 'digesters' were quite dangerous and it was not the norm. Around 1863, processors used 'chemical baths', where high concentrations of calcium chloride enabled 'water' to boil at temperatures approaching 121 °C. This allowed for significantly shorter cooking times. By 1870, basic steam retorts were being used to temperatures up to 121 °C, but they were still quite dangerous and hand operated.

In 1950, the first reel and spiral continuous cooker was introduced and was something very innovative. It was the first retort that did not process in batches. It was called the Anderson-Barngrover Cooker and was taken over by FMC and is currently John Bean Technology (JBT). These enabled cans to be processed much more efficiently at higher speeds, thus reducing production costs.

Around the same time, aseptic filling was started. In 1961, flame sterilization (using direct gas flame heating of rolling cans) was introduced in France, as was the continuous hydrostatic pressure sterilizer. Today, with computers being able to control equipment we have many more options and more precise control. There are combinations of steam and air, raining hot water, and rotation. All of these developments have the objective to optimize temperature distribution and product heating to improve product quality. Modern retorts can process at temperatures up to 145 °C, yielding faster throughput.

1.5 FOOD SAFETY

The canning process was originally developed to preserve food safely and for long periods of time. Food safety is often taken for granted with thermally processed foods; however, this understates the efforts made by food companies to ensure their products are safe for consumers to eat. Safety of thermally processed foods is closely monitored using a system called Hazard Analysis and Critical Control Point, or HACCP (Bauman 1974). HACCP is a system that identifies areas of potential contamination within the food process and builds checkpoints, or CCPs, to ensure that the product safety is maintained at all times. Validation of a thermal process and the determination of appropriate CCP levels is a challenging exercise that requires in depth product knowledge and accurate temperature measurement tools.

HACCP originated in the 1950s, when the National Aeronautics and Space Administration (NASA), the Pillsbury Company, and the U.S. Army Laboratories (Natick) collaborated to devise a food safety system to ensure that food for upcoming space expeditions was guaranteed to be safe. Critical Control Points (CCPs) were part of NASA's engineering management requirements (used to test weapon and engineering system reliability), and it was decided to use this concept for this food safety initiative to eliminate the potentially 'critical failure areas' in the food processing procedures. The programme was a success and the food produced proved to be safe.

There were many food safety incidents over the years, but in 1971 two incidents are identified as being tipping points for the food processing industry to realize that they needed a more formal food safety programme: The first was when Pillsbury had a recall on a product called Farina, a cereal used in infant food in which glass pieces were found; and the second was a botulism incident from canned vichyssoise soup made by Bon Vivant. Pillsbury was fortunate to have had direct involvement in the HACCP programme developed for the production of the space mission food and so implemented a HACCP programme to convince their customers that a similar incident would not happen again. The National Canners Association and FDA learnt about the HACCP programme while deciding how to ensure better canned food safety as a result of the soup incident and so asked Pillsbury to give some of their inspectors HACCP training and by 1973 HACCP was used as part of the low acid food regulations that were published in the USA and soon used internationally (Dick and Launius 2007).

Today, HACCP is used throughout the food industry and is generally considered to be part of Good Manufacturing Practice. It is published by the Food And Agriculture Organization (FAO) of the United Nations World Health Organization (WHO) as part of the Recommended International Code of Practice General Principles of Food Hygiene (Codex Alimentarius 1997).

HACCP is important because it prioritizes and controls potential hazards in food production. By controlling major food risks, such as microbiological, chemical, and physical contaminants, the industry can better assure consumers that its products are as safe as good science and technology allows (http://www.safefoodalliance.com).

References

Appert, N. (1810). *The Book for All Households or the Art of Preserving Animal and Vegetable Substances for Many Years*. Glass Container Association of America. Translated by K.G. Bitting (1920).

Bauman, H.E. (1974). The HACCP concept and microbiological hazard categories. *Food Technology* 28 (9): 30–32. 74.

Bitting, A.W. (1937). *Appertizing or The Art of Canning; its History and Development*. San Francisco, CA: The Trade Pressroom.

Codex Alimentarius (1997). Hazard Analysis and Critical Control Point (HACCP) System and Guidelines for its Application. Annex to CAC/RCP 1-1969, Rev. 3.

Dick, S.J. and Launius, R.D. (2007). *Societal Impact of Spaceflight*, NASA History Series. Washington, DC. Chapter 12.

Ecklund, O.F. (1949). Apparatus for the measurement of the rate of heat penetration in canned foods. *Food Technology* 3 (7): 231–233.

Featherstone, S. (2012). A review of developments in and challenges of thermal processing over the past 200 years – a tribute to Nicolas Appert. *Food Research International* 47 (2): 156–160.

Goldblith, S.A. (1971a). Pasteur & truth in labelling "Pro bono publico" – in the best scientific tradition. *Food Technology* 25: 228–229.

Goldblith, S.A. (1971b). A condensed history of the science and technology of thermal processing – part 1. *Food Technology* 25: 1256–1262.

Goldblith, S.A. (1972). A condensed history of the science and technology of thermal processing – part 2. *Food Technology* 26: 64–69.

Knock, G.G. (1952). Survey of soils for spores of Clostridium botulinum (union of South Africa and South West Africa). *Journal of the Science of Food and Agriculture* 2: 86.

Park, D.K. (1996). *Establishing Safe Thermal Processes by Calculation Method Selection*. TechniCAL, Inc. Library.

Patashnik, P. (1953). A simplified procedure for thermal process evaluation. *Food Technology* 7: 1.

Safe Food Alliance Team 2013. The History of HACCP. http://www.safefoodalliance.com.

Teixeira, A.A., Dixon, J.R., Zahradnik, J.W., and Zinsmeister, G.E. (1969). Computer determination of spore survival distribution in thermally-processed conduction heated foods. *Food Technology* 23: 352–354.

Tucker G.S. (2008). History of the Minimum Botulinum Cook for Low-Acid Canned Foods. Confidential R&D Report No 260, Campden BRI.

Warner, E.J. and Waterbury, C.T., Hitchcock, E.E. (1858). Instrument For Opening Cans. US Patent 1 90 63A. Issued 5 January 1858. Patent Number 19,063 – Can Opener. 19,063.

Wikipedia (2020). Peter Durand, Nicolas Appert, French Revolution, Napoleonic Wars, Canning, Can Opener. http://en.wiki.org/wiki/ (accessed 2020).

2 Microbiology of Heat Preserved Foods

Microbiology refers to the study of life and organisms that are too small to be seen with the naked eye. This includes eukaryotes, such as fungi, and prokaryotes, such as bacteria. Microbiology is researched actively, and the field is advancing continually. We have probably only studied and understand a very small percentage of all of the micro-organisms on Earth. Although micro-organisms were first observed over 300 years ago, new micro-organisms are continually being discovered as some can live in extremely adverse conditions and can mutate to adapt to new environments or threats. Extremophiles are micro-organisms that can survive under conditions once thought to have been impossible for life to survive: e.g. thermophiles can thrive at temperatures up to 122 °C; halophilic extremophiles can survive up to 33% salt, as in the Dead Sea; acidophiles can survive in pH as low as 1 and alkaliphiles have optimal growth around pH 10. Fortunately, only a small percentage of micro-organisms are a threat to human health.

2.1 FOOD MICROBIOLOGY

Ever since man has been gathering food, he has had the problem of preserving it. Very soon after food is harvested or slaughtered it starts deteriorating. Food microbiology is concerned with the organisms that cause loss of food quality and safety, as well as those organisms that help to preserve the food and keep other organisms at bay (Frazier 1958; Kay 1992). The micro-organisms of importance in the food industry are fungi and bacteria. Figure 2.1 shows the size of animal cells relative to mould, yeast and bacteria.

Essentials of Thermal Processing, Second Edition. Gary Tucker and Susan Featherstone.

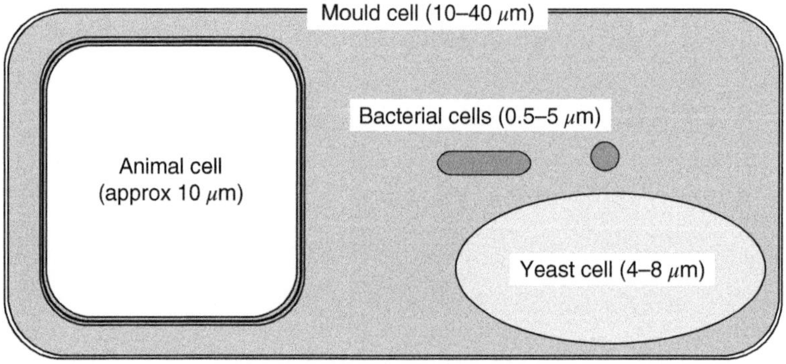

Fig. 2.1 Diagram showing relative size of bacteria, yeast and mould.

2.1.1 Fungi

Fungi are an interesting group of eukaryotic organisms that are neither plant nor animal. Like plants they possess a cell wall and can reproduce both asexually and sexually. But like animals they do not contain chlorophyll and are heterotrophic organisms. The majority of fungal species grow as multicellular filaments, called hyphae, forming a mycelium (mould). Some fungal species also grow as single cells (yeasts). Sexual and asexual reproduction of the fungi is commonly via spores, often produced on specialized structures or in fruiting bodies. The discipline of biology devoted to the study of fungi is known as mycology.

The use of fungi for food preservation is extensive. Yeasts are required for fermentation of beer, wine, and bread, and moulds are used in the production of cheese and soy sauce. Some fungi are also used for antibiotic production. Many fungi produce biologically active compounds, several of which are toxic and are therefore called mycotoxins. Of particular relevance to humans are those mycotoxins that are produced by moulds causing food spoilage, an example being the aflatoxins, which are liver toxins and carcinogenic metabolites.

2.1.1.1 Moulds

Moulds include all species of microscopic fungi that grow in the form of multicellular filaments, the hyphae. The individual mould structures, single hypha or spores, can only be seen under a microscope. The term 'mould' describes the mass of mycelium, made up from many hyphae, that is visible as a colony with the naked eye. They are widely distributed in Nature, in both the soil and in the air. Under suitable conditions (e.g. moisture, temperature, and oxygen), they can grow on almost any food source. They can grow at many extremes, highly acidic or very salty, low temperature, and on carbon sources that are not generally considered to be food (e.g. ceilings, walls, books). One important feature of mould metabolism that is important to note is the ability to consume acids, thus raising the pH of a foodstuff, and potentially allowing micro-organisms to grow that were otherwise inhibited by the low pH. All moulds are aerobic, but some can grow in relatively low oxygen concentrations. Generally, moulds are not very heat-resistant, but some spores can survive heating to about 90 °C for a few minutes. Mature spores (ones that have had several days to mature) are generally more resistant. These present problems for fruit canning where the process severity is limited by the breakdown in fruit texture caused by heat.

Food safety risks from mould contamination are limited to mycotoxins, which are secondary metabolites produced by moulds. The name 'mycotoxin' comes from the Greek words, mykes (mould) and toxicum (poison). There are many known types of mycotoxins, and they can contaminate a wide variety of food and animal feeds. Examples of common mycotoxins include Aflatoxin, Ochratoxin A, Patulin, Fusarin, Fumonisin, deoxynivalenol (DON), and Zearalenone. Mycotoxins are not essential to maintaining the life of the mould in a primary way (i.e. obtaining energy or synthesizing structural components). They are chemical compounds that give the mould a competitive advantage over other moulds or bacterial species in their environment. They are almost all cytotoxic, disrupting various cellular structures such as membranes and interfering with vital cellular processes such as RNA and DNA synthesis.

The consumption of mycotoxin contaminated food is related to several acute and chronic diseases in humans and animals. An example of a serious food-related problem caused directly by mycotoxins occurred in 1960, where 100 000 turkey poults died after eating peanut meal that was contaminated with aflatoxins. Aflatoxins are a group of toxins produced by *Aspergillus* spp. that have a similar structure and form highly oxygenated heterocyclic compounds. There are four major aflatoxins: B1, B2, G1, and G2. Aflatoxins have a potent carcinogenic effect. Aflatoxins occur in crops (most commonly corn, peanuts, and cottonseed) post-harvesting if the moisture levels in the foodstuff is allowed to exceed the critical values for mould growth. They can also be found in milk, cheese, and eggs, if the animals are fed contaminated feed.

Patulin is a mycotoxin that is produced by certain species of *Penicillium*, *Aspergillus,* and *Byssochylamys* moulds. The moulds that produce patulin grow on a variety of foods, including fruit (e.g. apples, grapes, and pears), vegetables, grains (e.g. flour and malt) and cheese. However, given the nature of the food (e.g. pH, protein amino acid composition), the manufacturing processes, or consumption practices for many foods, patulin does not appear to pose a public safety concern, with the exception of apple juice. Patulin is relatively stable in acid solutions, but is susceptible to alkaline hydrolysis and is destroyed by fermentation, which means it is not found in either alcoholic fruit beverages or vinegar produced by fruit juices, but will survive a pasteurization process.

Ochratoxin (sp) A is a mycotoxin produced by *Apergillus ochraceus* and a few other moulds, for example *Penicillium viridicatum* and *Penicillium verrucosum*. It has both antibiotic and toxic (carcinogenic, teratogenic, and nephrotoxic) properties.

2.1.1.2 Yeasts

Yeasts are heterotrophs that use organic compounds as a source of energy and do not require sunlight to grow. The main source of carbon is from hexose sugars such as glucose and fructose, or disaccharides such as sucrose and maltose, although some species can metabolize pentose sugars (e.g. ribose), alcohols, and organic acids. Yeasts are either obligate aerobes or are facultative anaerobes. Unlike bacteria, there are no known yeast species that grow only anaerobically (obligate anaerobes). Yeasts grow best in a neutral or slightly acidic pH environment. They will grow over a temperature range of 10–37 °C, with an optimal temperature range of 30–37 °C, depending on the type of species. Above 37 °C, yeast cells become stressed and above 50 °C most will die. The cells can survive freezing under certain conditions, with viability decreasing over time. Although yeasts are substantially larger than bacteria, they are only visible with the naked eye as colonies and must be viewed through a microscope.

Although they are very useful in the food industry – they are used for bread, beer, and wine manufacture and as a vitamin source – they can spoil some food because they have certain traits that enable them to opportunistically infect and spoil food. These are the ability to grow at low pH (or in high acid conditions below pH 4.0) and the ability to grow at high sugar concentrations (or low water activity down to 0.62 a_w). These characteristics make yeasts ideal candidates for food spoilage in fruit products (e.g. juices, canned fruits, and jams) and other high acid foods such as pickles and yoghurt. Yeasts actively produce carbon dioxide gas in their metabolism, which can cause destructive spoilage (especially in products in sealed containers).

2.1.2 Bacteria

Bacteria are a large group of prokaryotic, unicellular micro-organisms that are very important to the food processor. They are typically only a few micrometres (10^{-6} m) in length (Fig. 2.1). In contrast to higher organisms, bacteria exhibit an extremely wide variety of metabolic types, as highlighted in Table 2.1.

Bacteria are ubiquitous; growing in soil, acidic hot springs, radioactive waste, water, deep in the Earth's crust, as well as on organic matter and the bodies of living plants and animals.

Most bacteria are harmless to humans, but some can secrete enzymes that can cause changes in food or toxins that can contaminate food. Intimate knowledge of food poisoning and food spoilage organisms is critical to the food processor, as this knowledge can assist in controlling microbial outbreaks and contamination.

Microscopic examination is a quick tool used to identify bacteria. Characteristics such as their shape, size, aggregation, structure, and staining reactions (e.g. Gram stain) all are typical to certain groups of bacteria. Colonies of bacteria comprise millions of cells and are visible with the naked eye.

Bacteria are found in a wide range of shapes, such as spheres, rods, and spirals. Certain species of bacteria are motile, propelling themselves using tails called flagellae. Some bacteria produce external slime (or capsules). Some species of bacteria, such as *Bacillus* and *Clostridium*, can form (internal) highly resistant, dormant structures called endospores. Endospores have a central core of cytoplasm containing DNA and ribosomes, which is

Table 2.1 Nutritional types in bacterial metabolism.

Nutritional type	Energy source	Carbon source	Examples
Phototrophs	Sunlight	Organic compounds (photoheterotrophs) or carbon fixation (photoautotrophs)	*Cyanobacteria*, Green sulphur bacteria, *Chloroflexi* or Purple bacteria
Lithotrophs	Inorganic compounds	Organic compounds (lithoheterotrophs) or carbon fixation (lithoautotrophs)	*Thermodesulfobacteria*, *Hydrogenophilaceae* or *Nitrospirae*
Organotrophs	Organic compounds	Organic compounds (chemoheterotrophs) or carbon fixation (chemoautotrophs)	*Bacillus*, *Clostridium* or *Enterobacteriaceae*

protected by an impermeable and rigid coat. They show no detectable metabolism and can survive extreme physical and chemical stresses, such as high levels of UV light, gamma radiation, detergents, disinfectants, heat, pressure, and desiccation. In this dormant state, bacteria may remain viable for thousands of years and can even survive exposure to the vacuum and radiation in space. Endospore-forming bacteria are often implicated in diseases and food poisoning, for example anthrax can be contracted by the inhalation of *Bacillus anthracis* endospores, and botulism can be caused by the consumption of the toxin formed by *Clostridium botulinum* endospore outgrowth in canned food.

2.1.2.1 Growth and reproduction of bacteria

Unlike multicellular organisms, bacteria grow to a fixed size and then reproduce through binary fission (a form of asexual reproduction). Under optimal conditions, bacteria can grow and divide extremely rapidly, and bacterial populations can double as quickly as every 10 minutes. In this sort of cell division, two identical clone daughter cells are produced.

A typical growth cycle of bacteria is shown graphically in Fig. 2.2. When introduced into a food, the bacteria require some time to adapt to their new surroundings, during which very little cell multiplication takes place (called a lag phase). Once adapted, cell division occurs at an exponential rate (called the log phase), during which time cell numbers increase dramatically. When certain factors become limiting (such as the food source and space) and the amount of waste products produced by the bacteria build-up, cell division slows down again (called the stationary phase), and eventually cells stop dividing, resulting in their death. The time taken for the population of cells to double is known as the genera-

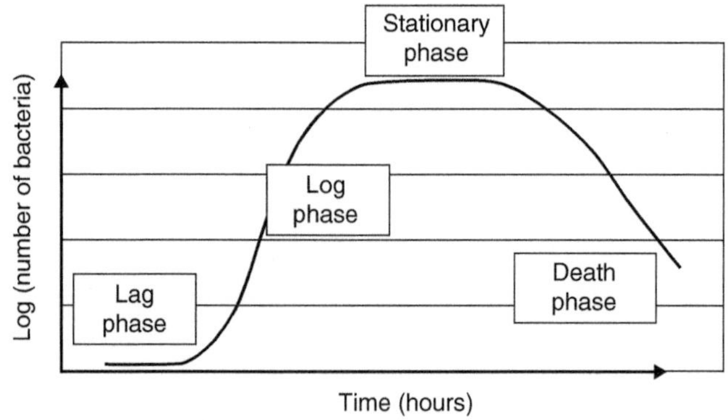

Fig. 2.2 Growth rate versus time for the various phases of microbial growth.

Table 2.2 Theoretical increase in population of bacteria under ideal conditions.

Time	Theoretical number of bacteria	
Start	1	1.0×10^0
30 min	2	2.0×10^0
1 h	4	4.0×10^0
2 h	16	1.6×10^1
5 h	1024	1.0×10^3
6 h	4096	4.1×10^3
8 h	65536	6.5×10^4
12 h	1677216	1.6×10^7
24 h	281474976710656	2.8×10^{15}

tion time. The generation time of a micro-organism will depend on the type and its environment. Typically, under favourable conditions bacteria have a generation time of 20–30 minutes. Thus, every hour, the population can double two or three times over (Table 2.2).

If binary fission growth is not limited, the population can amount to a staggering figure of over 2.8×10^{15} in just 24 hours. Fortunately, while rapid growth of bacteria and other micro-organisms is a reality there are always factors that do limit the growth, so these sorts of numbers do not occur.

Understanding the growth of micro-organisms is important because they can multiply very rapidly. Once accustomed to their environment, in the presence of optimal growth conditions, if the numbers are high to start with then adaptation is quicker and numbers will increase faster and spoilage will occur sooner.

2.2 FACTORS THAT AFFECT THE GROWTH OF MICRO-ORGANISMS

There are a number of factors that affect microbial growth, and it is important that these are understood in order to be able to control the micro-organism numbers (Prescott et al. 1996). In the list below, the first six are called intrinsic factors and are Nature's ways of protecting foods from micro-organisms, and the last three are called extrinsic factors and are properties of the storage environment that foods are kept in:

(i) pH;
(ii) moisture;
(iii) nutrients;
(iv) redox potential;

(v) antimicrobial resistance;
(vi) biological structures;
(vii) relative humidity;
(viii) oxygen content/gases in the environment;
(ix) temperature.

2.2.1 pH

The pH is a measure of the solution's acidity and refers to the hydrogen ion concentration of a solution. It is defined as the negative logarithm of the concentration of H^+ ions, as given in Equation (2.1):

$$pH = -\log_{10}\left[H^+\right] \tag{2.1}$$

where $[H^+]$ is the concentration of H^+ ions in moles per litre.

The pH scale ranges from 0 to 14, with 7 being the pH of pure water and is regarded as neutral. Less than 7 is acidic, and greater than 7 is alkaline. Most micro-organisms grow best at neutral pH, and only a few are able to grow $pH<4.0$. Most foods are acidic (pH 3–7). Chapters 4 and 5 deal with acid and acidified foods in greater detail, as well as the theory of acidity measurement using pH.

Bacteria are more selective about their pH requirements than yeasts and moulds. The fact that pH can limit microbial growth is one of the basic principles of food preservation and has been used for thousands of years. For example, fermentation and pickling have been used to extend the shelf-life of food products by lowering the pH. The fact that no known spore-forming pathogenic bacteria can grow at $pH<4.6$ is the basis for the food sterilization principle for low-acid and acid foods. Figure 2.3 gives some examples of the pH growth ranges of various micro-organisms.

Although each micro-organism will have an optimum pH, most can grow over a fairly wide range. The kind of acid used to change the pH will also affect the ability of micro-organisms to grow (refer to Chapter 6 for details). It is important to note that on either side of the optimum pH the lag phase of the growth curve will be extended while the micro-organism is adapting to the conditions, so a delay in growth does not necessarily mean no growth.

Some micro-organisms can survive in adverse pH conditions, but not exhibit growth. Other micro-organisms can alter the pH of a substrate (due to the metabolites they produce). This can cause other micro-organisms to thrive, which were previously only just surviving.

Many micro-organisms produce acids and lower the pH of a substrate (this is the basis for preserving food by pickling). It is

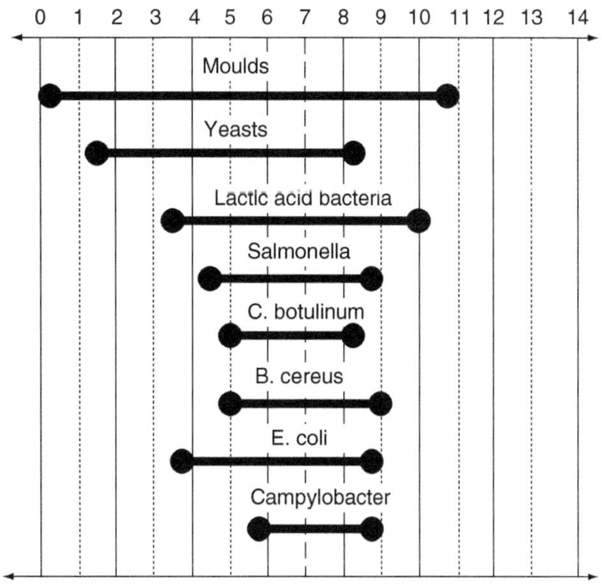

Fig. 2.3 pH growth ranges of micro-organisms.

important to realize that the pH can also be increased by the growth of certain micro-organisms. For example, the pH of a substrate can be raised when *Clostridium acetobutylicum* metabolizes butyric acid and *Enterobacter aerogenes* metabolizes pyruvic acid during growth. This pH increase can allow other organisms that are kept dormant by the lower pH to grow.

Other environmental factors also interact with pH, for example the pH of a substrate increases as the temperature increases (albeit slightly).

2.2.2 Moisture

Micro-organisms cannot grow in a water-free environment, because enzyme activity is absent, and most chemical reactions are greatly slowed down. Fresh vegetables, fruit, meat, and fish naturally have a high moisture content, which averages about 80%. This effect is used in drying, which is one of the oldest methods of food preservation. Drying reduces the availability of the moisture, thereby limiting the numbers and types of micro-organisms that can grow and reduces the rate at which they can do so. A measure of this parameter is called *water activity* and is defined by the ratio of the water vapour pressure in the food substrate to the vapour pressure of pure water at the same temperature and is denoted by the term a_w, as given in Equation (2.2):

$$a_w = p / p_o \qquad (2.2)$$

where p is the vapour pressure of the solution (food), Pa, and p_o is the vapour pressure of pure water, Pa.

Water activity is a measure of the water that is available to the micro-organisms. Pure water has a water activity of 1.0, while most fresh foods have a water activity of about 0.99–0.98. The growth of most micro-organisms is confined to the a_w range above 0.90, but some organisms that are of great importance in food preservation and food spoilage can grow at lower levels. These are termed halophiles, xerophiles, and osmophiles. Halophiles are unable to grow in environments with no salt (NaCl) and often require substantial amounts of salt to grow. Xerophiles are organisms that can grow in relatively dry (low a_w) conditions. Osmophiles can grow in environments where the osmotic pressure is high, for example in high sugar solution (e.g. jams and glacéed fruit).

In general, bacteria require an environment with a higher a_w than yeasts and moulds. Most spoilage bacteria cannot grow at $a_w < 0.90$. The minimum a_w value for a halophilic bacterial growth is 0.75. *C. botulinum* has a minimum growth level of a_w 0.94, whereas the salt tolerant organism *Staphylococcus aureus* can grow at a_w values as low as 0.84.

Fungi (yeasts and moulds) are more resistant to low moisture conditions than bacteria. Most spoilage moulds cannot grow at $a_w < 0.80$, although the minimum a_w for xerophilic mould growth is 0.61. Most spoilage yeasts can grow in a minimum of a_w 0.88, although osmophilic yeasts can grow at a_w as low as 0.61.

2.2.3 Nutrients

To be able to grow and function normally, micro-organisms require water, a source of energy, a source of nitrogen, vitamins, and other growth factors and minerals. Food is made up from water, proteins, fats, carbohydrates, and a host of vitamins and minerals. Some micro-organisms are better suited to food spoilage using specific nutrients than others.

Carbohydrates are organic compounds that contain carbon, oxygen, and hydrogen. They can be simple sugars or complex molecules. They have the general formula $C_n(H_2O)_n$. Food carbohydrates include monosaccharides (e.g. glucose), disaccharides (e.g. lactose, sucrose), and polysaccharides (e.g. dextrins, starches, celluloses, pectins). Monosaccharides and disaccharides are also referred to as sugars. They are readily digested and metabolized by the human body to supply energy, but can also be easily metabolized (fermented) by micro-organisms.

Fats are the second most important source of energy in the micro-organism's diet, after carbohydrates. The yield of energy from fats is more efficient, with fats yielding more than double the amount of energy as an equivalent amount of carbohydrate. They are also an essential part of the diet and are utilized in membrane, cell, tissue, and organ structures. Fats or oils (triglycerides) are a group of naturally occurring organic compounds (esters) comprised of three molecules of fatty acids covalently bonded to one molecule of glycerol. The properties of the fat are determined by the type and length of fatty acids that are bonded to the glycerol molecule.

Fats are designated as saturated or unsaturated, depending on whether the chemical bonds between the carbon atoms of the fatty acid molecule contain all the hydrogen atoms they are capable of holding (saturated), or whether they have capacity for additional hydrogen atoms (unsaturated). Saturated fats generally are solid at room temperature, and unsaturated and polyunsaturated fats are liquids. Unsaturated fats may be converted to saturated fats by adding hydrogen atoms (hydrogenation).

Proteins are the most abundant molecules in cells, making up about 50% of the dry mass. Protein molecules range from soluble globules that can pass through cell membranes and set off metabolic reactions (e.g. enzymes and hormones) to the long, insoluble fibres that make up connective tissue and hair. Proteins are made up from amino acids, of which 20 are used by living organisms. Each amino acid has specific properties, depending on its structure, and when they combine together to form a protein, a unique complex molecule is formed. All proteins have unique shapes that allow them to carry out a particular function in the cell. All amino acids are organic compounds that contain both an amino (NH_2) and a carboxyl (COOH) group. Proteins are very important foods, both nutritionally and as functional ingredients, because proteins in the diet serve primarily to build and maintain cells, but their chemical breakdown also provides energy, yielding almost the same as carbohydrates.

In addition to the main groups of carbohydrates, fats, and proteins, living organism need countless numbers of minerals and trace elements for them to be able to function adequately. Among these are calcium, iodine, iron, magnesium, manganese, phosphorus, selenium, and zinc.

2.2.4 Oxidation–reduction potential

The oxidation–reduction (redox or O/R) potential is the tendency of a substance to acquire electrons and thereby convert to its reduced state. O/R potential is measured in millivolts (mV) or Eh (1 Eh = 1 mV) relative to the standard hydrogen electrode (SHE), which is arbitrarily given a potential of 0.00 V. Different foods have their own intrinsic reduction potential; the more positive the potential, the greater the affinity for electrons and tendency to be reduced. Micro-organisms display varying degrees of sensitivity towards the redox potential.

As micro-organisms grow, they can affect the Eh of their environment. The Eh of the food is also affected by the characteristic O/R of the food, the resistance to change exhibited by the food, the oxygen in the atmosphere and the aeration of the foodstuff. Generally, aerobic micro-organisms prefer positive Eh for growth (oxidized), while anaerobic micro-organisms prefer negative Eh values (reduced). As a general rule, food made from plant origin is more likely to have a redox potential of +300 to +400 Eh and therefore are more likely to spoil by aerobic micro-organisms; while fresh meat usually has a redox potential of −200 Eh and is more likely to spoil by anaerobic micro-organisms. It is important to note that any processing can dramatically change the redox potential e.g. Minced meat often has a positive redox potential.

2.2.5 Antimicrobial resistance

Some foods have inherent antimicrobial activity, that is, they contain naturally occurring substances that restrict microbial growth. Some examples include essential oils (e.g. allicin in garlic, eugenol in cloves, and thymol in sage) and enzymes (e.g. lysozyme in eggs). While these are hugely beneficial for the stability of each product they are found in, they only have a minor role in thermally processed products and will not be considered in detail. Organic acids like acetic and lactic acid are very important for the stability of fermented foods.

2.2.6 Biological structures

The natural covering of some foods (e.g. skins and shells) provides a natural physical barrier to prevent ingress to the moist, nutrient rich, fleshier parts of the plant or animal. Thermal processing usually requires these protective barriers to be removed during preparation, and it is after this point that the foods are at their most vulnerable to microbiological growth and cross-contamination.

2.2.7 Relative humidity

Relative humidity (RH) is a term used to describe the amount of water vapour that exists in a gaseous mixture of air and water, as given in Equation (2.3)

$$RH = \rho(H_2O) / \rho^*(H_2O) \times 100\% \tag{2.3}$$

where RH is the relative humidity of the mixture being considered, $\rho(H_2O)$ is the partial pressure of water vapour in the mixture, Pa, and $\rho^*(H_2O)$ is the saturated vapour pressure of water at the mixture temperature, Pa.

The RH of the storage environment is important for microbial growth, because it can affect the water activity of the food. Foods and microbial hosts can pick up or lose moisture to the environment as a direct result of the relative humidity, as well as allow growth of micro-organisms on the surface of foods if the RH is high enough. Foods with low water activity will pick up moisture from the atmosphere until equilibrium is reached.

The use of RH as a hurdle to microbial growth is considered in greater detail in Chapter 3.

2.2.8 Oxygen content/concentration of gases in the environment

Controlling the availability of free oxygen is one means of controlling microbial activity within a food. Although oxygen is essential to carry out metabolic activities that support all forms of life, some micro-organisms use free atmospheric oxygen, while others metabolize the oxygen (reduced form) that is bound to other compounds such as carbohydrates.

Micro-organisms can be broadly classified into two groups – aerobic and anaerobic. Aerobes grow in the presence of atmospheric oxygen, while anaerobes grow in the absence of atmospheric oxygen. In between these two extremes are the facultative anaerobes, which can adapt and grow in either the absence or presence of atmospheric oxygen and microaerophilic organisms, which grow in the presence of reduced amounts of atmospheric oxygen. Micro-organisms have very specific oxygen requirements and can be classified as:

- obligate aerobes (need oxygen);
- obligate anaerobes (need an absence of oxygen);
- facultative anaerobes (grow in the absence or presence of oxygen).

The gaseous composition of the environment will therefore affect the growth rates of specific micro-organisms. Modified atmosphere packaging describes the practice of modifying the composition of the internal atmosphere of a food package in order to improve the shelf-life. The modification process lowers the amount of oxygen (O_2) in order to slow down the growth of aerobic organisms and the speed of oxidation reactions. The removed oxygen can be replaced with nitrogen (N_2), or carbon dioxide (CO_2), which can lower the pH and/or inhibit the growth of bacteria.

2.2.9 Temperature

Micro-organisms grow over a wide temperature range, with the lowest reported being –34 °C and the highest close to the boiling point of water (100 °C). All micro-organisms do, however, have an optimum temperature range in which they will grow best.

This preference for temperature forms the basis of dividing micro-organisms into groups, for example:

- Psychrophiles have an optimum between 12 and 15 °C, but can grow up to 20 °C.
- Psychrotrophs have an optimum between 20 and 30 °C, but can grow as low as 0 °C.
- Mesophiles have an optimum between 30 and 42 °C, but can grow between 15 and 47 °C.
- Thermophiles grow optimally between 55 and 65 °C, but can grow between 40 and 90 °C.

Just as moulds are able to grow over a wide range of pH values and moisture conditions, they can also tolerate a wider temperature range than bacteria. Yeasts are not usually found growing in the thermophilic temperature range, but prefer psychrophilic and mesophilic temperatures. Some bacteria produce very heat-resistant endospores, and even though most vegetative cells of yeast, mould or bacteria are destroyed at 65 °C, the bacterial spores can persist for many minutes at high temperatures (e.g. 121.1 °C).

The concepts of *D*-value, *z*-value, and the heat resistance of micro-organisms are very important and will be dealt with several times in this book. It can be difficult to understand why you can get a small percentage of cans spoiling due to under-processing; if the batch was under-processed why all of the cans don't spoil. It is important to remember that because micro-organisms are so small, there are usually many, many organisms in a sample; there is a population in the food that is being processed. Sterilizing food is not about killing one micro-organism, it is about reducing the numbers of the population that is there to as close to zero as possible. During a thermal process the rate of destruction of micro-organisms is generally found to be logarithmic; i.e. they are killed at a rate that is proportional to the number of organisms present. The longer the exposure to heat, the more micro-organisms get killed. The process is dependent both on the temperature of exposure and the time required at this temperature to accomplish to desired rate of destruction. Different organisms react differently to different temperatures, so the same time and temperature exposure could result in vastly different rates of population destruction when considering different micro-organisms. A simplistic explanation of why the micro-organisms populations are reduced logarithmically is because although, the maturity and size of the individuals are varied, and therefore, the composition of the individuals are varied and so the resistance to heat is varied, The population is heterogeneous with respect to heat resistance. On exposure to heat the cells are either damaged and either die or recover, with recovery dependent on various factors (e.g. environment, time, stage in life cycle). However, most of the population is similar, so destruction follows a logarithmic pattern, with larger numbers of the most susceptible organisms dying first.

D-Value: The kinetics of thermal destruction of micro-organisms has been extensively studied. Most commonly it is found that the number of micro-organisms decrease exponentially at a constant temperature over time. When these numbers are plotted on a graph with the log of the survivors on the *y*-axis versus time, a straight line can be drawn through the values. The *D*-value or decimal/reduction or decay time is the time taken for the population to be reduced by one log value; i.e. from 100% of the population to 10% of the population (see Fig. 2.4).

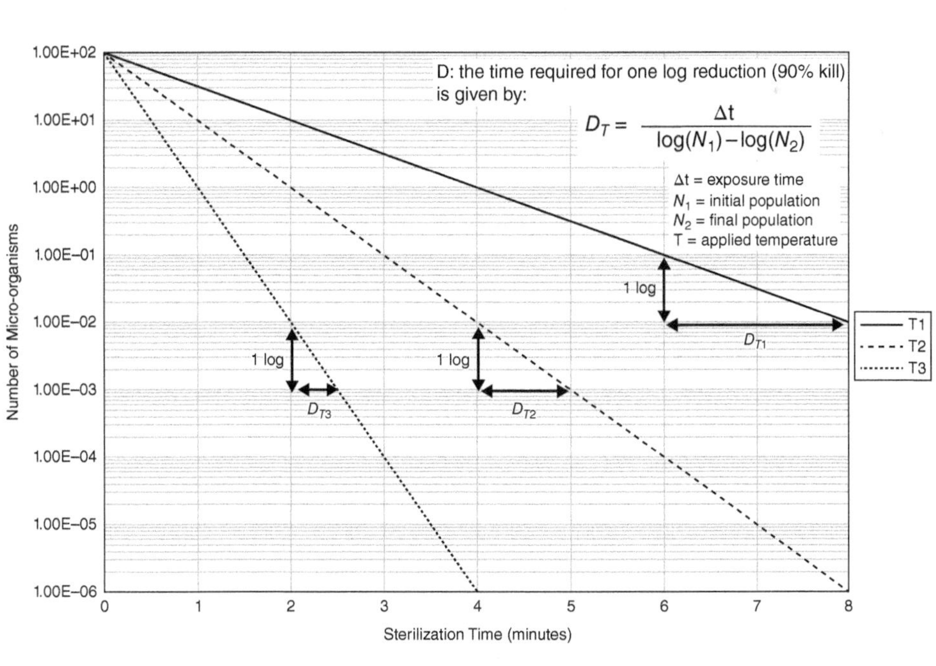

Fig. 2.4 Graph showing survivor curves at three different temperatures.

2.2.9.1 *D*-value

The *D*-value is a measure of the heat resistance of a specific micro-organism, in a particular substrate/food, at a specific temperature. It is calculated by heating a known concentrations of a specific micro-organism at a specific temperature (that would be important for the kind of food under investigation), for specific times and counting the number of survivors. A higher *D*-value indicates that it will take longer to reduce the population by one log and therefore indicates a more heat resistance micro-organism. At a temperature of 121.1 °C the *D*-values found for most spore-forming bacteria are between 0.2 and 2 minutes.

2.2.9.1.1 Probability of a non-sterile unit (PSNU)

Assuming an initial population of 1000 micro-organisms which has a *D*-value = 1 at 121 °C. After 1 minute at 121 °C there is be 100 micro-organisms; after another minute there will be 10 micro-organisms; after a total of 5 minutes there would be 1/100 micro-organisms or the chance of finding 1 micro-organism in 100 samples (see Fig. 2.5).

2.2.9.2 *z*-value

The *z*-value is the increase in temperature required to reduce the *D*-value by one log cycle (or 1/10 of the *D*-value). A *z*-value of 10 means that an increase of 10 °C in the processing temperature results in a 10× decrease in the *D*-value; i.e for every 10 °C increased in processing temperature, it take 1/10th of the amount of time to reduce the population by 90%.

The *D*-value is quoted for a specific temperature, but during actual processing it is not possible to achieve one constant

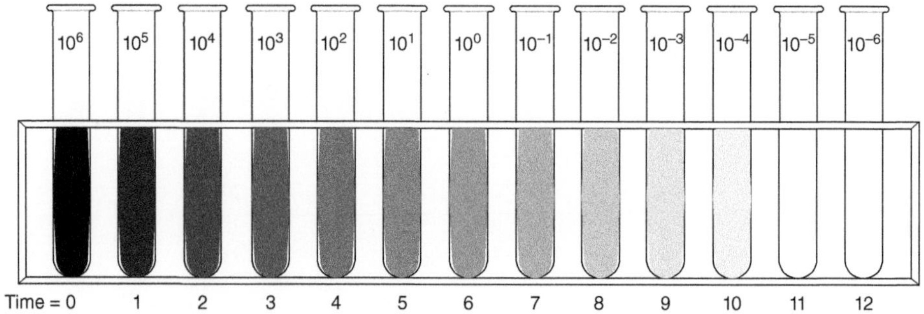

Fig. 2.5 Number of survivors in samples taken after every minute of heating; *D* value = 1.

temperature. To account for the different rate of killing at different processing temperatures (during heating and cooling), the *z*-value is calculated using the relationship between the *D*-value vs the specific temperature it was measured at for that organism. The *z*-value is expressed as C° because it is a rate, note °C, which refers to temperature (Gaze 2006).

The *z* value can be described as being the relative killing power of the heating temperature. In the temperature range 100–130 °C, the *z*-value can range between 6 and 13 C° (typically 10 C°) for the spores of bacteria associated with canned foods and is 6–7 C° for most vegetative cells (Fig. 2.6).

For commercially produced foods, safety is of the utmost importance, but shelf-stability and quality of the products is also very important, because food that is spoilt cannot be sold (Goldblith 1971, 1972). Commercial sterility of thermally processed foods means the condition achieved by the application of heat, which renders the food free from micro-organisms capable of reproducing in the food under ambient (i.e. normal non-refrigerated) conditions of storage and distribution as well as any viable micro-organisms, including spores, of public health significance. Below is a list of many of the micro-organisms that can affect the commercial sterility of thermally processed foods.

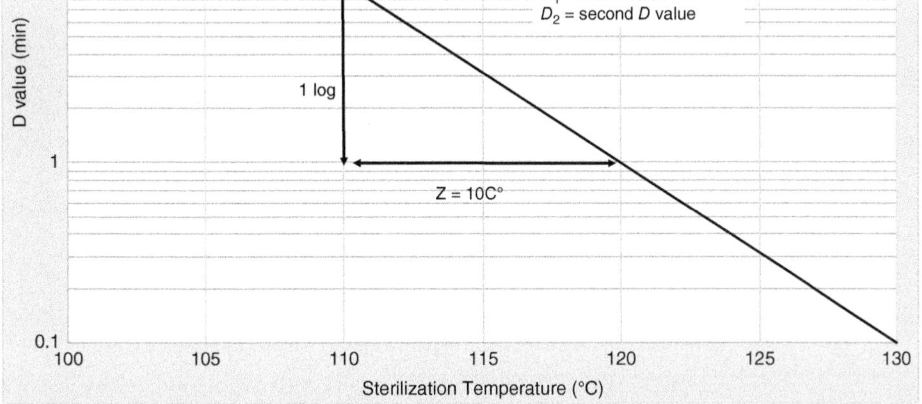

Fig. 2.6 Graph showing *z*-value calculated from *D*-values at various temperatures.

2.3 DESCRIPTION OF SOME MICRO-ORGANISMS OF IMPORTANCE TO THERMAL PROCESSING

All food poisoning micro-organisms are of concern in any prepared foods, but for sterilized processed foods, those that can form heat-resistant spores are of particular interest as they can potentially survive the thermal process (Stumbo 1965). Some thermophilic bacteria can form spores that are extremely heat-resistant (e.g. spores of *Clostridium thermosaccarolyticum* have been recorded with *D*-values as high as 195 minutes at 121 °C). Fungi are less of a concern to sterilized products as even the most heat-resistant fungi cannot survive about 95–100 °C and subsequently are more of a problem for pasteurized foods or as a result of post-processing contamination.

2.3.1 Moulds

Vegetative mould filaments (hyphae) are destroyed by relatively mild heat (65 °C). Moulds of greatest importance in foods multiply by zygospores, conidia or ascospores that are all more heat-resistant than the vegetative organisms. Ascospores in particular can be highly heat-resistant (e.g. *Byssochlamys fulva* produces ascospores that have a *D*-value at 90 °C of between 1 and 12 minutes). Although rarely implicated in food poisoning incidents with thermally processed foods, as mentioned previously, some moulds produce mycotoxins that can cause food poisoning or illness. Most moulds are aerobic. The mould spreads and infects food via airborne spores. Techniques like ultraviolet light treatment and modified atmosphere packaging (to limit the oxygen) are used to control mould growth. Some of the moulds found associated with food spoilage are listed below:

- *Byssochlamys* spp. can cause spoilage in canned fruit as a result of its heat-resistant ascospores and the pectinases that they produce. They produce mycotoxins.
- *Rhizopus stolonifer* is readily killed by heating, but can produce heat-stable pectinases that can persist and cause soft rot that makes even canned fruit or vegetables soft and mushy.
- *Aspergillus* has been implicated in the spoilage of a large number of foods including bacon, bread, peanuts, and fish. They produce mycotoxins.
- *Fusarium* spp. have extensive mycelium with tinges of brown, red, purple and pink. They cause brown rot on citrus fruit and

pineapples and can grow on cereal crops. They produce mycotoxins.

- *Penicillium* moulds typically spoil fruits, causing blue and blue-green colouration. They produce many different mycotoxins.

2.3.2 Yeasts

Yeasts can grow over a wide pH range and also exhibit tolerance to compounds traditionally used in food preservation. For example, some can grow in up to 18% alcohol and others in up to 60% sucrose, 24% NaCl or in a_w solutions as low as 0.65. Yeasts are not very heat-resistant and are generally destroyed at 60–65 °C, but can cause food spoilage as a result of organisms entering leaking containers post-process.

Some examples of yeasts that can cause food spoilage include the following:

- *Rhodotorula* is a diverse organism that can grow in high sugar concentrations and is implicated in the spoilage of jams, jellies, and candies. It can grow well at refrigerator temperatures as well as on the surface of butter.
- *Zygosaccharomyces* spp. can grow in high sugar concentrations and can cause spoilage of jams and jellies. *Zygosaccahromyces rouxii* can grow at pH values as low as 1.8 and water activity as low as 0.62. *Zygosaccahromyces balii* is resistant to some organic acid preservatives. Growth of *Zygosaccahromyces* can result in a lot of carbon dioxide being produced and has been known to cause explosion of glass bottles and cans.
- *Brettanomyces* spp. (*Dekkera*) produces acetic acid from glucose under aerobic conditions and can cause spoilage in beer, wine, and other fermented products. *Brettanomyces bruxellensis* can grow in up to 35% glucose and 45% sucrose.
- *Debaryomyces* spp. can grow in 25% NaCl and at water activities as low as 0.65. It has been found to cause a slimy growth on salted meats, cheeses and in brines. It is also one of the causes of spoilage of yogurt.
- *Saccharomyces bailii* is a common spoilage yeast causing spoilage in mayonnaise, tomato sauce, fruit drinks, and wine. It is resistant to benzoate and sorbate preservatives.

2.3.3 Bacteria

Bacteria are the most import group of food poisoning and food spoilage organisms that must be considered in heat processed foods (Pflug 2003). Bacteria can be divided into groups depending on the temperature at which they grow, as indicated in the temperature classification earlier in this chapter. Tables 2.3 and 2.4 present some heat resistance data for spores of various bacteria that are important in thermal processing. Note that the terms *D*- and *z*-value are described further in Chapter 4.

2.3.3.1 Thermophiles

Clostridium thermosaccharolyticum is one of the most heat-resistant micro-organisms that food processes come into contact with. It is a Gram-positive, rod-shaped anaerobic bacterium. These bacteria are saccharolytic and produce large quantities of CO_2 and H_2 (which causes blowing of the cans). They are not tolerant to high levels of acidity but are important in products of moderate acidity. Where these organisms are implicated in spoilage, they are usually introduced to the product with the raw materials, because conditions for their proliferation in the plant and processing equipment seldom exist (the bacteria being anaerobic).

Geobacillus stearothermophilus (used to be *Bacillus stearothermophilus*) is a rod-shaped, Gram-positive bacterium. It is a thermophile and is widely distributed in soil, hot springs, ocean sediment, and is a cause of spoilage in food products by producing various organic acids. No gas is formed, and it produces 'flat sour' spoilage in canned foods. It will grow within a temperature range of 30–75 °C but optimally at 55 °C. Spores of *G. stearothermophilus* may enter a cannery in soil, on raw foods, and in ingredients such as spices, sugar, soya meal, flour, and starches. Populations may increase at any point where proper environmental conditions exist (i.e. food handling equipment in a canning line that is operated within the thermophilic growth range, such as blanchers). Their presence in some processed containers of commercially sterile low-acid foods may be considered normal and not of particular concern because, although *G. stearothermophilus* can grow at temperatures as low as 30–45 °C, the spores will not develop if the product is stored at temperatures below 43 °C. Therefore, proper cooling after thermal processing and the avoidance of high temperatures during warehouse storage and distribution are essential in controlling its outgrowth.

Desulfotomaculum nigrificans is a strict anaerobic thermophile. These bacteria can cause the so-called 'sulphide stinker' spoilage. They produce hydrogen sulphide (H_2S) in the product, resulting in a bad odour (resembling the odour of rotten eggs). The H_2S dissolves in the contents of the containers, and the containers

Table 2.3 Table showing heat resistance characteristics of some food-related micro-organisms over a wide temperature range (Gaze 2006).

Organism	Heating substrate	Temperature (°C)	*D*-value (min)	*z*-value (C°)	References
Aeromonas hydrophila	Liquid whole egg	48	3.62–9.43	5.0–5.6	Schuman et al. (1997)
Lactobacillus plantarum	Buffer pH 4.0	50	1.81	6.5	Tait et al. (1991)
Aeromonas hydrophila (five strains)	Saline	51	8.08–122.8	5.2–7.7	Palumbo et al. (1987)
Salmonella species	Chicken broth	55	4.87	6.4	Juneja et al. (2001)
Lactobacillus plantarum	Buffer pH 7.0	55	3.64	6.1	Tait et al. (1991)
Escherichia coli O157:H7	Ground pork	55	33.4	4.9	Murphy et al. (2004)
Listeria monocytogenes	Ground pork	55	47.2	5.9	Murphy et al. (2004)
Escherichia coli	Broth	56	4.5	4.9	Tomlins and Ordal (1976)
Saccharymyces cerevisiae	Apple juice pH 2.8	57	9.4	5.0	Shearer et al. (2002)
Saccharymyces cerevisiae	Apple juice pH 3.9	57	32.0	5.0	
Enterobacter sakazakii	Re-hydrated infant formula	58	0.5–9.9	5.6	Edelson-Mammel and Buchanan (2004)
Escherichia coli	Beef gravy	60	1.0–2.0		Juneja and marks (2005)
Stapphyococcus aureus	Pea soup	60	10.4	4.6	Thomas et al. (1966)
Salmonella Senftenberg	Pea soup	60	10.6	5.7	Thomas et al. (1966)
Aspergillus niger	Pilsen beer	60	0.04	3.7	Reveron et al. (2005)
Lactobacillus paracasei	Pilsen beer	60	0.02	6.5	Reveron et al. (2005)
Lactobacillus plantarum	Buffer pH 4.0	60	0.05	6.5	Tait et al. (1991)
Streptococcus thermophilus	Skim milk	60	14.0		Flint et al. (2002)
Pediococcus species	Skim milk	60	12.1		Annous and Kozempel (1998)
Zygosaccharomyces balii (vegetative cells)	Buffer pH 4.5	60	0.4	3.9	Put and de Jong (1982a)
Zygosaccharomyces balii (ascospores)	Buffer pH 4.5	60	14.2	5.0	Put and de Jong (1982a)

(Continued)

Table 2.3 (Continued)

Organism	**Heating substrate**	**Temperature (°C)**	***D*-value (min)**	***z*-value (C°)**	**References**
Zygosaccharomyces balii (ascospores)	Broth	60	8.1	5.0	Put and de Jong (1980)
Saccharomyces cerevisiae (ascospores)	Buffer 4.5	60	22.5	5.5	Put and de Jong (1982b)
Yersinia enterocolitica (three strains)	Physiological saline	60	0.40–0.51	4.0–5.2	Sorqvist (1989)
Enterococcus faecalis	Fish	60	15.7	6.7	Ott et al. (1966)
Pseudomonas fluorescens	Broth	60	3.2	7.5	Tomlins and Ordal (1976)
Salmonella species	Egg yolk and egg white pH 8.2	60	0.3	4.2	Schuman and Sheldon (1997)
Listeria monocytogenes	Egg yolk and egg white pH 8.2	60	1.3	7.0	Schuman and Sheldon (1997)
Listeria monocytogenes	Beef	62	2.9–4.2	6.0	Gaze et al. (1989)
Salmonella species	Chicken broth	62	0.41	6.4	Juneja et al. (2001)
Salmonella senftenberg	Beef	62	2.65	5.9	Thomas et al. (1966)
Staphylococcus aureus	Beef	62	1.36	5.6	Thomas et al. (1966)
Escherichia coli O157:H7	Beef	62.8	0.47	4.7	Line et al. (1991)
Salmonella species	Beef	62.8	0.65	5.6	Goodfellow et al. (1978)
Salmonella enteritidis	Liquid whole egg	62.8	0.06	3.3	Shah et al. (1991)
Rhodotorula rubra (ascospores)	Orange juice, 52° Brix	64	3.5		Medina-de-Salcedo et al. (2002)
Salmonella enteritidis	Egg yolk + 10% salt	64.4	6.4		Palumbo et al. (1995)
Listeria monocytogenes	Egg yolk + 10% salt	64.4	8.3		Palumbo et al. (1995)
Streptococcus faecium	Nutrient broth	65	4.5	9.4	Ghazala et al. (1995)
Lactobacillus fructivorans	Tomato ketchup	65	1.2		Bjorkroth and Korkeala (1997)
Lactobacillus plantarum	Buffer 7.0	70	0.02	6.1	Tait et al. (1991)

Lactobacillus plantarum	Tomato juice	70	11.0	12.5	Tomlins and Ordal (1976)
Listeria monocytogenes	Carrot	70	0.27	6.7	Gaze et al. (1989)
Listeria monocytogenes	Ground pork	70	0.085	5.9	Murphy et al. (2004)
Escherichia coli O157:H7	Ground pork	70	0.048	4.9	Murphy et al. (2004)
Hepatitis A virus	Cream	71	0.52		Bidawid et al. (2000)
Salmonella species	Ground chicken	71.1	0.007 (3 s for 7D)	5.7	Mazzotc (2000)
Clostridium Pasteueianum	Papaya pulp pH 3.8	80	2.7	8.8	Magalhaes et al. (1997)
Clostridium botulinum non-proteolytic E	Water	80	3.3	9.4	Robert and Ingram (1965)
Clostridium botulinum non-proteolytic Type E	Carrot	80	4.33	9.8	Gaze and Brown (1990)
Clostridium botulinum non-proteolytic Type E	Cod	80	15.1	8.3	Gaze and Brown (1990)
Clostridium botulinum non-proteolytic Type B	Cod	80	18.3	8.6	Gaze and Brown (1990)
Clostridium botulinum non-proteolytic Type B	Carrot	80	4.24	9.8	Gaze and Brown (1990)
Clostridium botulinum non-proteolytic Type B	Buffer pH4.5	82.2	32.3	9.7	Scott and bernard (1982)
Eurotium chevalieri (ascospores)	Apricot juice	83	3.77	8.2	Yildiz and Coksoyler (2002)
Bacillus coagulans	Buffer pH 6.8	85	31.7	12.1	Dincer and Unluturk (2002)
Bacillus coagulans	Tomato juice pH 4.2	85	31.7	12.1	Dincer and Unluturk (2002)
Bacillus cereus (emetic and diarrhoeal strains [6])	Buffer pH 7.0	85	32.1–106 (mean 45.4)	6.9–13.9 (mean 8.9)	Johnson et al. (1982)

(Continued)

Table 2.3 (Continued)

Organism	Heating substrate	Temperature (°C)	*D*-value (min)	*z*-value (C°)	References
Bacillus cereus	Buffer pH 7.0	85	220.0	9.8	Goepfert et al. (1972)
Clostridium butyricum	Phosphate buffer pH 6.5	85	23.0	8.3	Cerf et al. (1967)
Microbacterium species	UHT milk	85	1.1	—	Maggi et al. (1997)
Bacillus megaterium	Soy sauce	87	6.68	—	Yamamoto et al. (1980)
Bacillus coagulans	Buffer 6.8	90	13.7	12.1	Dincer and Unluturk (2002)
Bacillus coagulans	Tomato juice pH 4.2	90	10.7	5.5	Dincer and Unluturk (2002)
Bacillus cereus	Buffer 7.0	90	71.0	9.8	Goepfert et al. (1972)
Clostridium botulinum Type E	Cod	90	0.79	8.3	Gaze and Brown (1990)
Clostridium botulinum Type E	Carrot	90	0.48	9.8	Gaze and Brown (1990)
Clostridium botulinum Type B non-proteolytic	Cod	90	1.1	8.6	Gaze and Brown (1990)
Clostridium botulinum Type B non-proteolytic	Carrot	90	0.36	9.8	Gaze and Brown (1990)
Clostridium butyricum	Peach	90	1.12	11.5	Gaze et al. (1998)
Clostridium tyrobutyricum	Buffer pH 7.0	90	18.0	—	Cerf et al. (1967)
Talaromyces flavus (ascospores)	Fruit juice	90	6.2	6.4	King (1997)
Neosartorya fischeri (ascospores)	Tomato juice	90	4.4–6.6	—	Kotzekidou (1997)
Byssochlamys nivea (ascospores)	Tomato juice	90	1.5	—	Kotzekidou (1997)
Byssochlamys fulva (ascospores)	Tomato juice	90	8.1	—	Kotzekidou (1997)
Byssochlamys fulva (ascospores)	Grape drink	93	5.0	7.8	Hatcher et al. (1979)
Clostridium pasteurianum	Buffer pH 4.5	95	3.95	8	Casolari and Giannone (1966b)
Clostridium butyricum	Peach	95	0.39	11.5	Gaze et al. (1998)
Bacillus coagulans	Buffer pH 6.8	95	4.8	12.1	Dincer and Unluturk (2002)

Bacillus coagulans	Tomato juice pH 4.2	95	4.8	12.1	Dincer and Unluturk (2002)
Bacillus cereus	Buffer pH 7.0	95	13.0	9.8	Goepfert et al. (1972)
Alicyclobacillus Species	Orange juice	95	2.1–2.8	6.4–7.1	Murakami et al. (1998)
Alicyclobacillus acidoterrestris	Buffer pH 3.8	95	2.2–2.8	6.4–7.1	Murakami et al. (1998)
Bacillus coagulans	Buffer pH 4.0	98.9	9.5	—	Kosker et al. (1951)
Bacillus cereus	Buffer pH 7.0	100	8.0	10.5	Molin and Snygg (1967)
Bacillus cereus	Buffer pH 7.0	100	8.0	9.8	Goepfert et al. (1972)
Bacillus coagulans	Red pepper pH 4.5	100	5.5	—	Ferria and Hernandez (1972)
Bacillus megaterium	Water	100	1.6	—	Gibriel and Abd-El (1973)
Bacillus megaterium	Tomato juice	100	1.0	—	Gibriel and Abd-El (1973)
Bacillus megaterium	Mango juice	100	0.8	—	Gibriel and Abd-El (1973)
Bacillus megaterium	Orange juice	100	0.8	—	Gibriel and Abd-El (1973)
Bacillus licheniformis	Buffer pH 4.0	100	1.05	10.2	Tait et al. (1991)
Bacillus licheniformis	Buffer pH 7.0	100	4.29	8.1	Tait et al. (1991)
Bacillus licheniformis	Water	100	13.5	6	Tait et al. (1991)
Bacillus polymyxa	Buffer pH 7.0	100	18.0	—	Walker (1964)
Clostridium butyricum	Peach	100	0.15	11.5	Gaze et al. (1998)
Clostridium butyricum	Red	100	4.5	—	Ferria and Hernandez (1972)
Bacillus coagulans	Buffer	106	6.3–7.3	8.2–9.6	Put and Wybinga (1963)
Bacillus coagulans	Buffer	108	4.2–4.6	8.2–9.6	Put and Wybinga (1963)
Bacillus licheniformis	Buffer pH 4.0	110	0.12	10.2	Tait et al. (1991)

(Continued)

Table 2.3 (Continued)

Organism	Heating substrate	Temperature (°C)	*D*-value (min)	*z*-value (C°)	References
Bacillus licheniformis	Buffer pH 7.0	110	0.27	8.1	Tait et al. (1991)
Bacillus coagulans	Buffer	110	2.2–2.5	8.2–9.6	Put and Wybinga (1963)
Bacillus coagulans	Buffer pH 4.5	110	0.64–1.4	—	Lynch and Potter (1988)
Bacillus coagulans	Buffer pH 4.2	110	0.56–10.6	—	Lynch and Potter (1988)
Bacillus coagulans	Buffer	112	1.2–1.8	8.2–9.6	Put and Wybinga (1963)
Alicyclobacillus acidocaldarius	Orange juice	120	0.1	7	Palop et al. (2000)
Bacillus subtilis	Buffer pH 6.8	121	0.57	9.8	Put and Aalbersberg (1967)
Thermophilic clostridia (isolated from tomato juice)	Tomato juice pH 4.3	121	4.5–6.3	—	Garulli et al. (2005)
Thermophilic clostridia (isolated from tomato juice)	Tomato juice pH 4.5	121	3.3–8.7	—	Garulli et al. (2005)

Table 2.4 Typical heat resistance of spores of some bacteria.

Species	Temperature (°C)	*z*-value (C°)	Typical *D*-value (minutes)
Mesophiles			
Bacillus cereus	100		5.5
	121		2.37
B. coagulans	121		3.0
B. licheniformis	100		13.0
B. subtilis	121		0.3–0.7
B. coagulans	100		9.5
	121		1.6–3.4
Clostridium butyricum	85		12–23
	100		0.1–0.5
C. sporogenes	121		0.2–1.5
C. perfringens	100		0.3–17.6
C. botulinum[a]	121	10.0	0.21
C. botulinum Type A (proteolytic)	121	9.0	0.13
C. botulinum Type B (proteolytic)	121	11.0	0.15
C. botulinum Type B (non-proteolytic)	82	8.3–16.5	1.5–32.3
C. botulinum Type E (non-proteolytic)	77		0.7–1.95
C. botulinum Type F (proteolytic)	121	9.3–12.1	0.14–0.22
C. botulinum Type F (non-proteolytic)	77		1.6–9.5
C. botulinum Type G (proteolytic)	116	20.9–27.3	0.25–0.29
Thermophiles			
Geobacillus sterarothermophilus	121	7.7	6–16.0
Desulphotomaculum nigrificans	121	9.5	55.0
C. thermosaccharolyticum	121	6.9–11.0	68.0–195.0
General micro-organisms			
Vegetative bacteria, yeasts, and moulds	65		0.5–1.00

[a] Most commonly used for thermal process calculations.

often remain 'flat'. In cans, the H_2S interacts with the iron in the containers, resulting in the contents discolouring to black. This bacterium used to be classified as a *Clostridia* and is still referred to as *Clostridium nigrificans* in some textbooks.

Alicyclobacillus acidoterrestris is a thermoacidophilic, non-pathogenic, spore-forming bacterium. *A. acidoterrestris* spores are resistant to low pH pasteurization, and they can germinate and grow causing spoilage, characterized by a medicinal or disinfectant smell attributed to a chemical guaiacol (*o*-dihydroxybenzene) that the organism produces. It is an unusual organism in that it is almost the only thermophilic organism that can tolerate high acid conditions. It is found in soils of fruit orchards and enters processing environments as a contaminant of raw materials. It is most commonly found associated with spoilage of pasteurized fruit juices.

2.3.3.2 Mesophiles – spore-forming bacteria

Food poisoning incidents that result in fatalities are incredibly rare. However, one bacterium that can cause death is *C. botulinum*. The botulinum toxin is one of the most potent neurotoxins known, to the extent that it is used in biological warfare. Food contaminated with only a few nanograms (10^{-9} g) can be toxic. For this reason, while fairly uncommon, the risk is taken very seriously by all producers of products packed in an anaerobic environment and stored for an extended period of time (e.g. canned and vacuum-packed food). All low-acid food processes are designed with organisms such *C. botulinum* in mind (Rees and Bettison 1991).

Clostridium botulinum are Gram-positive, obligate anaerobe, spore-forming bacteria. They are commonly found in soils throughout the world, in any region. Since they are found in the soil, they contaminate vegetables cultivated in or on the soil and will colonize the gastro-intestinal tract of fish, birds, and mammals. Non-proteolytic *C. botulinum* can grow at chill temperatures as opposed to the proteolytic strains which cannot. *C. botulinum* are classified as a single species of bacteria but includes at least three genetically distinguishable groups of organisms that have been recognized to be toxic for humans (*C. botulinum*, *Clostridium Baratii,* and *Clostridium butyricum*). These organisms share the ability to produce the neurotoxins. The toxin types are classified as A, B, C, D, E, F, and G. Human botulism has been described with the strains of *C. botulinum* that produce toxin type A, B, and E. *C. botulinum* bacteria produce spores that are heat-resistant and cannot be killed simply by boiling. Canned food sterilization processes are specifically designed so as to eliminate the possibility of any spores surviving. The toxin is, however, more heat sensitive and heating at 80 °C for 30 minutes or 100 °C for 10 minutes will destroy the active toxin.

Factors that limit the growth of *C. botulinum* include the following:

- *Low pH*: *C. botulinum* will not produce toxin in acid or acidified foods (i.e. below pH 4.6).
- *Low water activity*: A minimum a_w of 0.94 is needed to support bacterial growth and toxin production. This water activity

corresponds to a 10% salt (NaCl) solution, which is why salting is sometimes used as a method of preservation.

- *Temperature*: Most strains grow optimally at 40 °C, but some of the non-proteolytic strains can grow at temperatures as low as 3 °C.
- *Food preservatives*: Many preservatives (nitrite, sorbic acid, phenolic antioxidants, polyphosphates, etc.) inhibit *C. botulinum* growth.
- *Competing micro-organisms*.

There are a number of other mesophilic bacteria of importance in thermal processing. The main ones are described below and will feature in other places in this book (FDA 2017).

Clostridium butyricum is a Gram-positive, spore-forming, anaerobe. These bacteria spoil moderately acid food such as canned peaches and other fruit with pH values above 4.0, as well as canned vegetables (e.g. peas, potatoes) and other products of low acidity. Butyric acid and gas are produced during their metabolism.

Clostridium perfringens are Gram-positive, spore-forming, anaerobic rod-shaped bacteria. They are widely distributed in Nature and also found associated with the intestines of animals and humans. The spores are not particularly heat-resistant (when compared with spores of *C. botulinum*) and food poisoning from this organism is usually associated with cooked foods that have been inadequately cooled and held for several hours before consumption. This organism produces a toxin that causes intense abdominal cramps and diarrhoea.

Bacillus cereus are Gram-positive, facultative aerobic, spore-forming rod-shaped bacteria. The spores of this organism are also not particularly heat-resistant. Presence of this organism in large numbers can result in two types of food poisoning illnesses: A protein toxin that causes diarrhoeal symptoms, similar to that caused by *C. perfringens*, and a heat-stable peptide that causes vomiting symptoms, similar to that caused by *S. aureus*. The diarrhoeal type of illness has been diagnosed after consumption of a variety of foods (e.g. milk, meat, fish, and vegetables), while the vomiting type is associated with consumption of rice products and other starchy foods such as potatoes and pasta. Puddings, soups, casseroles, pastries, and salads have all been implicated with food poisoning by this bacterium.

Bacillus coagulans are thermoduric and acid-tolerant bacteria that have been isolated from canned tomato products, where they cause a reduction in the pH of spoilt products by acid production without CO_2 formation. The ends of the spoilt containers therefore remain flat, and the bacteria can also be termed 'flat sour' bacteria. Optimal growth is between 37 and 45 °C and at pH levels between 5.0 and 7.0. *Bacillus coagulans* is not as heat-resistant as *G. stearothermophilus* but can grow at lower pH values (e.g. down to pH 4.2).

2.3.3.3 Mesophiles – non-spore-forming pathogenic and spoilage bacteria

Salmonella spp. are Gram-negative, rod-shaped, motile bacteria that do not form spores. They are widespread in occurrence, being found for example in fresh and salt water, soil and animal faeces. Various foods have been found to be associated with Salmonellosis, the illness caused by the *Salmonella* spp., including raw meat, poultry and seafood, raw eggs and foods made from raw eggs, dried gelatine, cocoa, chocolate, peanut butter, yeast, and coconut. Salmonellosis is caused by infection with the organism and extremely low doses (levels as low as 15 cells) can cause the disease.

Staphylococcus aureus are Gram-positive, small round bacteria (cocci) that can produce a heat-stable toxin. Less than 1 µg of toxin can cause illness. *Staphylococcus aureus* are found everywhere, but the most common source of contamination of food is via humans, i.e. food handlers. *Staphylococci* are found in the nasal passages, throats, on the skin and hair of more than half of healthy people. Intoxication is cause by contaminated food being held either not hot enough (i.e. <60° C) or not cold enough (i.e. >8 °C), which allows the organism to grow and produce its toxin. The symptoms (nausea, vomiting, cramping) of Staphylococcal food poisoning are very fast (a few hours) and usually acute.

Escherichia coli are Gram-negative bacteria that are found in the intestines of all mammals, including humans. *Escherichia coli* produce toxins that can cause four different classes of illnesses that are of concern to the food industry. The first class is enterovirulent (EEC) that causes gastroenteritis; this class includes *E. coli* 0157:H7 (EHEC), which is enterohaemorrhagic. The second type is referred to as enterotoxigenic (ETEC), the third is enteropathogenic (EPEC), and the fourth is enteroinvasive (EIEC). Outbreaks of food poisoning due to *E. coli* are usually associated with faecal contamination of water by food handlers or due to unprocessed or under-processed (and under-cooked) foods. The recent interest in organic food production systems has enabled *E. coli* to grow and contaminate fruit and vegetable raw materials, resulting in outbreaks in pasteurized apple juice.

Lactic acid bacteria are a large diverse group comprising of bacteria from *Lactobacillus*, *Lactococcus*, *Leuconostoc*, *Pediococcus*, *Streptococcus*, etc. These bacteria are commonly involved with fermentation reactions that are desirable, but can also cause spoilage by acid and gas formation in some products. Post-process recontamination is the most likely cause, if a spoilage incident is attributed to lactic acid bacteria. This is because these bacteria are not at all heat-resistant.

2.3.3.4 Psychrophiles

There are very few psychrophiles of concern in thermal processing; however, one of the few that can cause problems with cook-chill foods is *Pseudomonas* spp. This is a large group of Gram-negative rod-shaped bacteria that are responsible for spoilage of many refrigerated proteinaceous products (e.g. meat, fish, eggs). Some strains produce blue-green pigments. They are implicated in many food spoilage conditions (e.g. green rot, black rot, pink rot, and red rot). They have very varied nutritional requirements and can even cause spoilage of bottled water.

2.4 RISK OF LEAKER SPOILAGE FROM DAMAGED OR COMPROMISED PACKAGING

Although the incidence of leaker spoilage in canned food is extremely low, it is the most common type of spoilage that is found in canned food (Featherstone 2020), as there is such stringent control of the thermal processing. Food authorities (FDA, USDA, Health Canada, Australian Institute of Food Safety) cite the reason for discarding of dented or damaged cans as the risk of *C. botulinum* contaminating the food (very small risk, but the risk is present).

Usually in leaker-type spoilage incidents you will find heat sensitive organisms and often, mixed cultures (Shorten 1993). The contaminating bacteria can be any food spoilage or food poisoning bacteria (e.g. *E. coli*, *Salmonella*, *B. cereus*, *C. botulinum*, *S. aureus*).

Not all bacteria will cause gas production or obvious food spoilage, so it will not necessarily be possible to see or smell that the food has been compromised.

The most common source of contamination is the cooling water that is used at the end of processing. Depending on the water treatment and the environment, the kinds of micro-organism that

A study done by the USDA (Anon 1983), showed that regulatory bodies have seen some significant incidents over the years. i.e.:

- 1957 Staphylococcal food poisoning from canned peas in England.
- 1963 Botulism from canned tuna in USA (due to compromised can).
- 1964 Typhoid outbreak in canned corned beef in Aberdeen (due to a puncture in the cans).
- 1978 Botulism from Alaskan canned salmon exported to the UK (due to leakage through cable burn on double seam).
- 1982 Botulinum from canned Salmon (due to tear in can body).

In 2002 in South Africa there was a botulism incident with two deaths from Pilchards in Tomato Sauce from a can that was found to have leaked due to external corrosion.

are present can vary and include endospores of *Clostridia* and *Bacillus* species which are known to be able to become resistant to sanitizers including chlorine.

There is also the case of good thermal process, good cooling water, etc., but that the double seam is compromised as a result of imperfection in the can flange. Contamination could occur during storage, labelling, or anywhere further along the distribution channel due to poor hygiene in handling or the environment.

With regards to contamination from *C. botulinum* due to leaker spoilage, the risk would depend on the environment in which the compromised pack is processed and stored. Most pathogenic bacteria will be sourced from soil, sediment or other organic matter. It is possible to become contaminated from the air (dust, etc.) but far less likely.

We know that physical abuse once the product has left the processor and is in the retail distribution system does happen. Assuming that the thermal process is good: a full botulinum cook was successfully given; cooling water hygiene was good, and biocides were well controlled; post-process line hygiene was good and physical can abuse was minimal; and seams and seals are within specified measurements. The risk from routine handling during shelf packing or loading into trolleys or unloading at home would be extremely low. It is very difficult to quantify this numerically, as it is not possible to reliably gather this sort of detail for incidents where food poisoning has occurred due to post-processing leaker contamination once spoilage has occurs and the incident is investigated.

2.5 A GUIDELINE FOR IDENTIFYING SPOILAGE IN CANNED FOODS

When resolving the cause of spoilage of shelf-stable canned (cans, pouches, jars, etc.) foods isolation of the spoilage micro-organisms is essential. Before this is done gather as much data about the spoilt samples as possible e.g. number of affected packs, production details and records, retort process information, duration before spoilage was observed, place where spoilage as observed. Information about the spoilt samples, specifically are also important e.g. fill mass, production codes involved, physical condition, degree of swelling, etc.

Information about the spoilt product must be recorded e.g. pH of product, physical attributes of the product compared with a control, e.g. discolouration, thinning or clotting of sauce, and off odours.

Isolation of the spoilage micro-organisms is done by inoculating specific microbial media with a sample of the spoilt product and incubating them at specific temperatures (between 30 and 55 °C, depending on the test). Samples may need to be sub-cultured onto solid media to separate out mixtures of micro-organisms.

Direct microscopy should be done to describe the morphology of the micro-organisms as well as Gram stains, spore stains, catalase, and oxidase reactions, and assessment of motility.

In Table 2.5, there is a guideline to the most probable cause of spoilage given the results of the microbiological test results (D. Shorten, CarnaudMetalBox Technology plc, personal communication). Extreme care must be exercised in drawing conclusions as to the cause of spoilage. All possible relevant information must be gathered and assessed together with the microbiological culture results.

Table 2.5 Most probable cause of spoilage of canned foods.

Type of cultures found	Most probable cause
Gram-negative bacteria	Generally, indicators of leaker spoilage/post-processing reinfection of packs. Causes include seam damage, wet can handling, failure of cooling water disinfection, defective double seams or seals, can or package structural defect.
Mixed bacteria: Gram-negative bacteria and Gram-positive cocci	Wet can handling, in combination with structural defect/strain of cans or seam damage.

(Continued)

Table 2.5 (Continued)

Type of cultures found	Most probable cause
Gram-positive bacteria: non-heat-resistant cocci alone	Wet can handling, in combination with structural defect/strain of cans or seam damage.
Wide spectrum mixed flora: both heat-resistant and non-heat-resistant	Gross under-processing (virtually no process at all); or a can that missed the retort altogether.
No isolate recovered, but seen in microbiological examination of the spoilt product.	Either culture died out during the time between spoilage and examination (no conclusion can be drawn); or pre-process spoilage on seamer delay, preparation abuse, delay between seaming/retorting.
Mixture of thermophilic spore-forming bacteria: usually only aerobic spore formers of obligate and sometimes of facultative forms.	Insufficient cooling of cans in retort (stack-burn); or elevation of temperatures on can storage.
Thermophilic spore-forming bacteria, obligate thermophiles, predominantly aerobic types	Marginal under-processing; or high loading of highly heat-resistant spores in raw materials; or temperature-abuse in can storage after processing.
Thermophilic spore-forming bacteria, obligate thermophiles, possible mixture of aerobes and anaerobes	Marginal under-processing (worse than above); or high loading of highly heat-resistant spores in raw materials; or temperature-abuse in can storage after processing.
Thermophilic spore-forming bacteria, obligate and facultative thermophiles, mixture of aerobes and anaerobes	Under-processing (worse than above), or excessive loading of heat-resistant spores in raw materials; or temperature-abuse in can storage after processing.
Mixture of facultative mesophilic and thermophilic spore-forming bacteria, possible mixture of aerobes and anaerobes	Most likely as significant under-processing (worse than above); or possibly, excessive loading of heat-resistant spores in raw materials.
Mixture of facultative mesophilic and obligate mesophilic spore-forming bacteria, possible mixture of aerobes and anaerobes	Most likely gross under-process (worse than above); or possibly excessive loading of heat-resistant spores in raw material.

References

Anon (1983). Container Integrity: Regulatory Focus Centre in 1983. 1983 Annual Report. NFPA Research.

FDA (2017). *Bad Bug Book*. https://www.fda.gov/food/foodborne-pathogens/bad-bug-book-second-edition (accessed 5 December 2020).

Featherstone, S. (2020). A study of causes of spoilage found in canned foods. FST Magazine South African Food Science and Technology 9 (2): 18–19.

Frazier, W.C. (1958). *Food Microbiology*. New York: McGraw-Hill Book Company, Inc.

Gaze, J.E. (2006). *Pasteurisation: A Food Industry Practical Guide*, 2e. Campden BRI. Guideline 51.

Goldblith, S.A. (1971). The science and technology of thermal processing, part 1. Food Technology 25: 1256–1262.

Goldblith, S.A. (1972). The science and technology of thermal processing, part 2. Food Technology 26: 64–69.

Kay, J.M. (1992). *Modern Food Microbiology*, 4e. New York: Chapman & Hall.

Pflug, I.J. (2003). *Selected Papers on the Microbiology and Engineering of Sterilization Processes*, 6e. St. Paul, MN: Environmental Sterilization Laboratory, University of Minnesota.

Prescott, L.M., Harley, J.P., and Klein, D.A. (1996). *Microbiology*, 3e. Dubuque: Brown Publishers.

Rees, J.A.G. and Bettison, J. (1991). *Processing and Packaging of Heat Preserved Foods*. London: Blackie and Son Ltd.

Shorten, D. (1993). *Bacillus* and *Clostridia* in relation to the spoilage of canned food. *Society for Applied Bacteriology Summer Conference*, University of Nottingham, UK, 16 July 1993.

Stumbo, C.R. (1965). *Thermo-Bacteriology in Food Processing*. New York: Academic Press Inc.

3 Hurdles to Microbial Growth

Food preservation is the process of treating food to stop or slow down spoilage (loss of quality or nutritional value) caused or accelerated by micro-organisms and or intrinsic enzymes. A number of hurdles to microbial growth are described, including water activity, pH, storage temperature, and preservatives such as salt and sugar. These can be used alone or in combination.

Heat treatment is an extremely effective means of food preservation, but a full sterilisation of foods is not always appropriate, as the nutritional value, texture, and/or colour of the food could be unacceptably changed. In these instances, a mild heat treatment, together with at least one other method of food preservation is used. When a combination of two or more food preservation factors is used, generally each factor can be used at a slightly lower level. Using more than one food preservation factor or hurdle is termed 'hurdle technology'. Hurdle technology uses the intelligent combination of different preservation factors or techniques to achieve multi-target, mild but reliable preservation effects (Leistner 2000). It is usually more effective than single targeting and allows hurdles of lower intensity, with the benefits of improved product quality (Lee 2004).

According to Leistner (1999), the number of already applied and potential hurdles for the preservation of foods amounts to over 100. As our understanding of the mode of action of food preservation improves, factors such as microbial metabolic exhaustion, stress reactions, and homeostasis are taken into account when considering food preservation. The traditional methods of heating, chilling, drying, curing, acidification, oxygen removal, fermenting, and adding chemical preservatives are frequently used (Table 3.1), but newer methods such as pulsed technologies and 'natural' preservatives are also being employed.

Essentials of Thermal Processing, Second Edition. Gary Tucker and Susan Featherstone.

Table 3.1 Selected food preservation techniques and their effect on spoilage micro-organisms.

Method	Effect on microbial growth or survival
Refrigeration	Low temperature to retard growth
Freezing	Low temperature and reduction in water activity to prevent microbial growth, slowing of oxidation reactions
Drying	Reduction in water activity sufficient to delay or prevent microbial growth
Vacuum and oxygen-free modified atmosphere packaging Modified atmosphere packaging (enriched CO_2) Addition of weak acids; e.g. sodium lactate	Low oxygen tension inhibits strict aerobes and delays growth of facultative anaerobes Specific inhibition of some micro-organisms Reduction of the intracellular pH of micro-organisms
Sugar preservation	High osmotic pressure causes inhibition of growth or death
Ethanol preservation	Ethanol is toxic to micro-organisms and causes inhibition of growth or death
Addition of preservatives (e.g. nitrite or sulphite)	Inhibition of specific groups of micro-organisms by interfering with cell membrane functioning
Pasteurisation	Delivery of heat sufficient to inactivate target micro-organisms to the desired extent
Irradiation	Delivery of ionising radiation to disrupt cellular RNA
Pulsed electric field processing	Short bursts of electricity for microbial inactivation

To use all of these food preservation methods, a good understanding of the biochemistry of the mode of action is required.

The most commonly used hurdles that are combined with thermal processing are discussed below. Some discussion of these has already taken place in Chapter 1 but the relevance of these factors to mild pasteurisation is stressed here:

1. control of the load of micro-organisms;
2. use of restrictive pH;
3. use of anaerobic conditions;
4. use of low temperatures;

5. drying/low water activity;
6. chemical preservation;
7. irradiation.

3.1 CONTROL OF THE MICRO-ORGANISM LOADING

Shells, skins, membranes, and fat are the natural defences that prevent microbial contamination of meat, fish, fruit, and vegetables before harvesting. Once these barriers are removed, there are a number of methods that can be used to physically remove micro-organisms from foods, which includes:

- Washing of raw foods, which is an important part of most ingredient preparation, but may result in increasing the microbial load if the water used for washing is not clean. Washing can also increase the food's moisture content and thus increase the spoilage potential.
- Trimming includes the removal of spoiled portions. It is an important step in ensuring that the load of an ingredient batch is kept as low as possible.
- Filtration is limited to clear liquids, but is used successfully with some fruit juices, beer, wine, soft drinks, and water.
- Centrifugation or sedimentation is used in some milk and juice production.

Reducing the initial microbial load is critical because it determines the extent of heat process time and temperature that must be applied to achieve the condition of commercial sterilisation. Although it is important to ensure that a food process starts with a low load, it is also important to understand why each step of a process is being done, because this helps to understand the importance of that step. This is where the HACCP plan is used to ensure levels of microbial contamination are low. However, it is important to note that even if micro-organisms are removed from food, various preformed enzymes and/or toxins might remain and present problems.

3.2 USE OF RESTRICTIVE pH LEVELS

Preventing the growth of micro-organisms by lowering the pH is a very old food preservation technique (Chapters 5 and 6). Pickling refers to preserving foods using inorganic acids (and often salt), originally due to fermentation, but more recently also by addition of acids to the food product. Commercially pickled products are

pH 0	Battery acid (30%) H_2SO_4
pH 1	Gastric acid (1.1)
pH 2	Cola (2.5)
pH 3	Apricots (3.2)
pH 4	Tomatoes (3.8–4.4)
pH 5	Rain (5.5)
pH 6	Egg yolk (5.7)
pH 7	Human blood (7.2)
pH 8	Egg whites (8.5)
pH 9	Baking soda (8.5)
pH 10	Milk of magnesia (10.6)
pH 11	Household ammonia (11.2)
pH 12	Household bleach (NaClO) (12.6)
pH 13	Household lye (NaOH) (13.5)
pH 14	Concentrated sodium hydroxide (NaOH)

Fig. 3.1 Materials of varying pH from battery acid at pH 0 through to strong alkalis at pH 14. For a colour version of this figure, please see the colour plate section.

almost always now given a heat treatment of sorts. Figure 3.1 shows examples of materials of varying pH from battery acid at around pH 0 through to strong alkalis at pH 14.

Most micro-organisms grow best at neutral pH (6.5–7.5) and only a few are able to grow at pH<4.0. Bacteria are more fastidious about their pH requirements than are yeasts and moulds. The fact that pH can limit microbial growth is a basic principle of food preservation and has been used for thousands of years. Fermentation and pickling extend the shelf life of food products by lowering the pH. The fact that no known spore-forming pathogenic bacteria can grow at pH<4.6 is the basis for the food sterilisation principle for low acid and acid foods.

Foods can be classified according to the pH/acidity of the product, with the following classification proving useful when assessing spoilage incidents:

- Low-acid foods have pH values exceeding pH 4.6 and above.
- Medium acid foods range in pH from 4.2 to 4.5.
- Acid foods have pH between 3.8 and 4.2.
- High acid foods have pH values below 3.8.

Under normal conditions of storage, bacterial spores that survive commercially adequate processing are incapable of development in acid foods, and this fact has led to the adoption of the term 'commercial sterility'. This indicates that the microbiological condition is not necessarily completely sterile, but is a safe, marketable food. *Commercially sterile food may be defined as food, which has been so processed, that, under ambient storage conditions, will neither spoil nor endanger the health of the consumer.*

Acid foods need to be heat or otherwise processed to destroy non heat–resistant yeast and bacterial vegetative bacterial cells and mould spores, which are generally not inhibited by low pH alone (Chapter 5).

All micro-organisms have a pH range in which they are most comfortable. This is usually around neutral pH, but some micro-organisms have a larger range that they can tolerate. Adverse pH affects the respiration of micro-organisms; two examples of how they are affected are the functioning of their enzymes and the transport of its nutrients into the cell.

The intracellular pH of a micro-organism is virtually constant and so, if a micro-organism is placed in an environment that is too acid or too alkaline, it will have to expend a lot of energy to maintain its own internal pH balance. If the pH is too extreme, it can result in the death of the micro-organism. Bacterial cells have an overall negative charge and so nonionised compounds can enter the cell, while ionised ones cannot. At neutral and alkaline pH, organic acids do not enter, but at acidic pH these organic acids are nonionised and can therefore enter the negatively charged cell. This is why the effect of an organic acid is enhanced at low pH, even though they are weak acids.

3.3 ANAEROBIC ENVIRONMENT OR MODIFIED ATMOSPHERE ENVIRONMENT

Modified atmosphere describes the practice of modifying the composition of the internal atmosphere of a package in order to improve the shelf life. The process often aims to lower the amount of oxygen (O_2), moving it from 21 to 0% in order to slow down the growth of aerobic organisms and the rate of oxidation. The removed oxygen can be replaced with nitrogen (N_2), which is functionally inert, or carbon dioxide (CO_2), which can lower the pH or inhibit the growth of bacteria. The lack of oxygen will inhibit aerobic micro-organisms only and can provide an atmosphere that is preferred by anaerobic organisms. However, the anaerobes generally have naturally slower generation times than the aerobes and so their numbers will increase slowly if the modified atmosphere is combined with chilled storage. A disadvantage is that anaerobic conditions will give anaerobic spoilage organisms a selective advantage, which might not be advisable for certain foods.

For thermally processed foods, the use of modified atmospheres tends to be for foods that oxidise during the lengthy storage at ambient temperature. It is not usual for modified atmospheres to enhance the microbiological shelf life because the heat process is designed to achieve this aim.

3.4 LOW TEMPERATURES

Low temperatures are used to retard chemical reactions such as the actions of food enzymes, and to slow down or stop the growth and activity of micro-organisms in food. The lower the temperature, the slower will be the chemical reaction, the enzyme actions, and the microbial growth. A temperature below 0 °C will prevent the growth of micro-organisms, although in practice 3 °C is often considered to be a temperature sufficiently low to control (stop) micro-organism growth. Low temperatures are inhibitory to growth, rather than lethal to the micro-organisms. Table 3.2 presents minimum growth temperature for selected micro-organisms.

Chilling is often used together with a mild heat treatment (e.g. two minutes at 70 °C is a guideline to eliminate pathogens such as *Salmonella*, *Escherichia coli*, and *Listeria*). A further hurdle of 'lowering the load' of micro-organisms in the food is usually also used, as described in the section above, by working in a clean environment and applying the appropriate packaging to protect the food or ingredients.

Freezing stops multiplication of microbial cells. The water activity (a_w) of the product will drop due to crystallisation of the available water. Although there might be an initial decrease in cell numbers due to ice crystal formation, which may rupture the cells,

Table 3.2 Minimum growth temperatures for some pathogenic bacteria (taken from Campden BRI 1996).

Organism	Minimum growth temperature (°C)
Bacillus cereus	4.0
Clostridium botulinum (psychotropic)	3.3
E. coli 0157	7.0
Listeria monocytogenes	−0.4
Salmonella sp.	4.0
Staphylococcus aureus	6.7
Vibrio parahaemolyticus	5.0
Yersinia enterocolitica	−1.0

many micro-organisms may survive freezing and continue proliferation and growth once the product is thawed. The spoilage potential of the product will actually increase on thawing, as the texture changes and drip water forms.

Freezing does not destroy microbial toxins and viruses. Food that shows signs of microbial spoilage is not necessarily poisonous, and food that contains microbial toxins does not always appear spoilt. When low temperature is used for food preservation, either solely or in conjunction with mild heat processing, it is important to take cognisance of the 'use by' date.

See Chapter 7 for more discussion on Heat Preserved Chilled Foods.

3.5 DEHYDRATION OR LOW WATER ACTIVITY

Micro-organisms cannot grow in a water-free environment, because enzyme activity is absent and most chemical reactions are greatly slowed down. This is the principle behind dehydration or low water activity foods.

Fresh vegetables, fruit, meat, and fish have naturally high moisture content, which averages about 80%. Drying reduces the amount of moisture, thereby limiting the numbers and types of micro-organisms that can grow, and reduces the rate at which they can do so. There are different methods of drying that can reduce the moisture content slightly or virtually completely, depending on the desired effect, for example:

- sun drying;
- drying by mechanical means; e.g. hot air drying, spray drying, roller drying;
- freeze drying; sublimation of water from a frozen food by means of a vacuum.

Some foods (e.g. jams, marmalades) have relatively high moisture content, but the moisture is chemically bound (to sugar in the case of jam) and not available to the micro-organisms. A measure of the availability of water is called *water activity* and is defined by the ratio of the water vapour pressure in the food substrate to the vapour pressure of pure water at the same temperature and is denoted by a_w. Pure water has a water activity of 1.0, while most fresh foods have a water activity of about 0.99.

In general, bacteria require a higher a_w than yeasts and moulds. Most spoilage bacteria cannot grow at a_w <0.91, with *Clostridium botulinum* having a minimum growth level of 0.94. However, *Staphylococcus aureus* has been found to grow at a_w as low as 0.84. The lowest reported a_w value for a bacterial growth was 0.75. Most spoilage moulds cannot grow at $a_w < 0.80$, with the lowest reported a_w for any mould growth at 0.65 and for yeasts at 0.61.

It is important to note that a_w is not the same as the moisture content of foods. While some high moisture foods will have a higher a_w than dry foods, this is not always the case. Some foods may have similar moisture contents but very different water activities, e.g. jams have high moisture contents; however, the moisture is bound to the sugar in the jam and unavailable for microbial growth, giving jams a low a_w. Not all sugars are the same at lowering the water activity e.g. fructose is a more effective humectant than sucrose or glucose.

High salt and high sugar mixtures preserve food because they either kill the micro-organisms directly due to the difference in osmotic pressure, or they bind the free water and lower the water activity, so that there is no water left for the micro-organisms to use for growth. They effectively dry the product out. With salt or sugar, we can theoretically preserve food for decades without it being consumed by bacteria. The value of this method to reliably preserve food cannot be understated. Before refrigeration (~100 years ago) sugar preserves were one of the only ways to keep fruit throughout the winters. Similarly, salt was, and still is, used to preserve fish, meat, and vegetables for many months.

In thermal processing, it is common to use low water activity as a hurdle technology that is applied in combination with a thermal

Table 3.3 Table showing the water activity of different food products.

Food	**Approximate water activity (a_w) range**
Fresh fruit and vegetables	>0.98
Meat and fish	>0.98
Cheese	±0.97
Salami	±0.82
Fruit juice concentrate	0.80–0.93
Dried fruit	±0.76
Honey	0.60–0.75
Pasta, biscuits	±0.50
Breakfast cereal	0.20–0.30
Instant coffee	±0.10
Milk powder	±0.10

process. Modern jams contain sufficient acidity and sugar to avoid the need for a full botulinum process; however, a mild pasteurisation process is required to provide a shelf-stable product. Other examples of low water activity in combination with a mild process include pesto, sun-dried tomatoes and fermented meats such as salami.

3.6 CHEMICAL PRESERVATION

Some chemicals have specific antimicrobial activity and can be used to control the growth of micro-organisms in food. Naturally occurring antimicrobial chemicals are present in herbs and spices and have been used traditionally for thousands of years. More recently, as our understanding of food science has improved, a range of specific active chemicals is used. The use of chemical preservatives is strictly controlled by legislation in the appropriate country, as is the quantity that may be used.

Chemical preservatives may inhibit micro-organisms by interfering with the cell membranes (influencing cell permeability), enzyme activity, and/or genetic mechanisms. Most chemical preservatives have the highest activity at low pH.

Properties required of an ideal chemical preservative include:

- wide range of antimicrobial activity;
- nontoxic;
- economical;
- must not affect the sensory qualities of a product;
- kill rather than inhibit.

Examples of chemical preservatives are many and include the following:

- benzoic acid, propionic acid, sorbic acid, and their salts;
- sulphur dioxide, methyl and ethyl parabens;
- nitrates and nitrites;
- antioxidants, e.g. BHA, BHT, TBHQ;
- antibiotics, e.g. Pimaricin (Delvocid), Nisin (Nisaplin); however, no antibiotics of medical significance may be used in foods.

The mode of action of preservatives can be very specific and varies from compound to compound, and so the desired effect must be carefully considered when using preservatives.

3.6.1 Organic acids

Organic acids have been used for many years in food preservation. They can be naturally present in foods due to fermentation or added as part of the recipe. Some organic acids behave primarily as fungicides or fungistats, while others tend to be more effective at inhibiting bacterial growth.

The mode of action of organic acids is related to the pH reduction of the substrate, acidification of internal components of cell membranes by ionization of the undissociated acid molecule, or disruption of substrate transport by alteration of cell membrane permeability. The undissociated portion of the acid molecule is primarily responsible for antimicrobial activity; therefore, effectiveness depends upon the dissociation constants (pK_a) of the acid.

Organic acids are usually more effective at low pH. The most commonly used organic acids in food preservation include citric, succinic, malic, tartaric, benzoic, lactic, and propionic acids.

Acetic acid (vinegar) is known to cause disruption of the cell membrane. It also causes denaturation of proteins inside the cell. Acetic acid is highly effective and widely used as an acidulant and preservative. Only some bacteria (*Acetobacter* spp. and some lactic acid bacteria) and some yeast and mould are resistant to acetic acid. The growth of most food poisoning bacteria is inhibited by 0.1% acetic acid. Many pickles, mustards, and dressings use acetic acid in their formulation, with a pasteurisation treatment of two minutes at 70 °C for products below pH 3.7.

Sorbic acid causes membrane damage of many micro-organisms and disrupts enzyme activity inside the cells. It is widely used to inhibit the growth of yeast, mould, and some aerobic Gram-positive bacteria. However, it is relatively infective against lactic acid bacteria. Sorbic acid is more effectual at low pH and has virtually no effect above pH 6. It is widely used in pH-controlled drinks, such as fruit squashes, with a mild process of two minutes at 70 °C.

Potassium sorbate is a white crystalline powder that has greater solubility in water than sorbic acid, which may be used accordingly in making concentrates for dipping, spraying, or metering fruit and vegetable products. It has antimycotic actions similar to sorbic acid, but slightly more potassium sorbate must be used than sorbic acid to ensure the same protection.

Benzoic acid has a similar mode of action to sorbic acid, but is used mainly to control the growth of yeast and mould (at levels of 0.05–0.10%). Often both benzoic and sorbic acids are used together so their effect is increased.

Sodium benzoate is used as a food preservative in products that are acid in nature as an antimycotic agent (most yeasts and moulds are inhibited by 0.05–0.1%). Sodium benzoate is more effective in food systems where the pH is as low as 4.0 or below.

Propionic acid is similar in action to the other organic acids, but is used primarily to inhibit mould growth (0.05–0.10%). It is also effective against many bacteria. It is most commonly used in baking to prevent ropiness caused by *Bacillus subtilis* survival through the baking process.

Citric acid is present in citrus fruits. It has been demonstrated that citric acid is more effective than acetic and lactic acids for inhibiting growth of thermophilic bacteria. Also, combinations of citric and ascorbic acids inhibit growth and toxin production *of C. botulinum* in vacuum-packed foods.

Malic acid is widely found in fruits and vegetables. It inhibits the growth of yeasts and some bacteria due to a decrease in pH.

Tartaric acid is present in fruits such as grapes and pineapples. The antimicrobial activity of this acid is attributed to pH reduction.

Lactic acid is not naturally present in foods; it is formed during fermentation of foods such as sauerkraut, pickles, olives, and some meats and cheeses by lactic acid bacteria. It has been reported that lactic acid inhibits the growth of spore-forming bacteria at pH 5.0 but does not affect the growth of yeast and moulds.

3.6.1.1 *CIMCSEE code*

The Comite´ des Industries des Mayonnaises et Sauces Condimentaires de la Communaute´ Economique Europeenne (CIMSCEE) is an industry guideline for the commercial production of microbiologically safe and stable emulsified and nonemulsified dressings and sauces containing acetic acid (CIMCSEE 1992). It is a model based on the combined preservation effect of pH, acetic acid, sugar, and salt and is designed to control the growth of yeasts, lactobacilli, and moulds. Products are generally Cold fill emulsified or nonemulsified products. The code considers only the aqueous phase, assuming that no microbiological spoilage will occur in the oil phase. The code consists of two equations A safety equations (Equation 3.1) to be considered relating to potential safety of the product derived from inactivation rates of Salmonella and a stability/spoilage equation (Equation 3.3) to determine the potential for spoilage by acetic acid tolerant yeasts, moulds, and lactic acid bacteria (Adams and Moss 2006)

The safety equation is:

$$\sum_{s} = 15.75(1-\alpha)\times(\%\text{total acetic acid}) + 3.08(\%\text{salt}) + (\%\text{hexose}) + 0.5(\%\text{disaccharide}) + 40(4.0-\text{pH}) \quad (3.1)$$

The factor $(1-\alpha)$ is the proportion of disassociated acetic acid across the range pH range 2.0–4.5. It is dependent on a number of factors but principally on the pH of the aqueous phase of the product; using the pK_a (4.76 at 25 °C) for acetic acid and the pH of the aqueous phase, $(1-\alpha)$ can be calculated using the Henderson–Hasselbalch equation (Equation 3.2), where pK_a is the negative logarithm of the ionization constant (K) of an acid, this is the pH of the solution in which half of the acid molecules are ionised. (Note that the pK_a value is not an absolute one; it will vary with the ionic strength of a solution so salt addition for example will change the value.)

$$\mathrm{pH} = \mathrm{p}K_\mathrm{a} + \log_{10}\left[\alpha/(1-\alpha)\right] \tag{3.2}$$

Using this formula, the $(1-\alpha)$ values at pH 3 and pH 5 can be calculated as 0.983 and 0.365 respectively. This means that at pH 3, 98.3% of the acetic acid is undissociated compared with only 36.5 % at pH 5.

Safety from microbiological pathogens has been defined as one which is so formulated that when an inoculum of viable cells of *E. coli* is added to the product this is reduced by 3 log cycles in less than 72 hours. Products exhibiting this level of antibacterial activity have a CIMSCEE safety value, Σ_s, of greater than 63.

The stability/spoilage equation is:

$$\Sigma = 15.75(1-\alpha)\times(\%\text{total acetic acid}) + 3.08(\%\text{salt}) + (\%\text{hexose}) + 0.5(\%\text{disaccharide}) \tag{3.3}$$

The % sugar, acetic acid, and salt are used in terms of weight of aqueous solution (ie water + salt + sugar + acetic acid).

Any sauce with a $\Sigma > 63$ based on this equation is regarded as microbiologically stable, without refrigeration, even after opening.

Below is a worked example (Jones 1994) that shows the for a horseradish sauce, with pH 3.5.

Sauce ingredients	**% in recipe**	**% in water phase**
Acetic acid (80%)	2.6	3.2
Hexose (glucose, fructose)	0.0	0.0
Disaccharide (sucrose)	4.6	5.7
Salt	1.6	2.0
Water	71.4	89.0
Total water phase	80.2	

The value used here for $(1-\alpha)$ is 0.950. Dissociation constant (K_a) of acetic acid is 1.7×10^{-5}). Substituting all of these values into Equations (3.1) and (3.3) gives the following:

$$\sum_{s}=(15.75\times0.950)\times(3.2)+(3.08\times2.0)+(0.0)+(0.5\times5.7)$$
$$+40(4.0-3.5)=76.9$$

$$\Sigma=(15.75\times0.950)\times(3.2)+(3.08\times2.0)+(0.0)$$
$$+(0.5\times5.7)=56.9$$

According to the code, therefore, this sauce is intrinsically safe as Σ_S = 76.9 is greater than 63; but microbiologically unstable as Σ = 57.4 is less than 63. Because of this, a challenge test based on the protocol contained in the CIMSCEE code needs to be carried out to confirm the microbial stability or otherwise of the sauce.

Free CIMSCEE calculators are available on the Internet.

3.6.2 Sulphites and nitrites

Sulphites in the form of sulphur dioxide, sulphite, bisulphites, or metabisulphites are used extensively to control the growth of undesirable micro-organisms in fruit, fruit juice, wine, meat, fish, and pickles. At low pH (<4.0), it inhibits yeast and mould and at higher pH values it inhibits spore-forming bacteria. As the pH gets lower, more of the compound is in the sulphur dioxide form (from bisulphite form). The sulphur dioxide is believed to pass through the cell membrane by diffusion and then disrupt normal cell metabolism.

Nitrites enter the cell via the cell membrane and disrupt normal metabolic pathways and cellular respiration via formation of a variety of complex compounds. Nitrates are mainly used to inhibit bacterial growth. The effectiveness is related to pH, with lower pH being more effective. Nitrites are used widely with meat products so that a lower process can be applied than the full botulinum $F_0 3$, usually of the order of $F_0 1$.

3.6.3 Antibiotics

Nisin is a peptide antibiotic produced by some strains of *Lactococcus lactis*. Its inhibitory effect is restricted to Gram-positive bacteria where it causes death of the target organism by cell lysis. Nisin can be used to prevent the spoilage (by *Clostridium* and *Bacillus* spp.) of cheese, milk products, some canned products, cured meats, and alcoholic beverages.

3.6.4 Antioxidants

Phenolic antioxidants are mainly used to prevent chemical oxidation of foods, but they do have some microbial effect, especially inhibiting bacterial growth (at levels of 100–400 ppm). They have also been shown to reduce toxin production in some bacteria. High protein concentrations reduce the effectiveness of these antioxidants, as does low temperature.

3.7 IRRADIATION

Three different areas of the electromagnetic spectrum are used to preserve food via quite different mechanisms. These are outlined below:

Microwaves, unlike other forms of radiation, microwaves act indirectly on micro-organisms through the generation of heat. When a food containing water is placed in a microwave field, the dipolar water molecules align themselves with the field. As the field reverses its polarity $2–5 \times 10^9$ times each second, depending on the frequency used, the water molecules are continually oscillating. This kinetic energy is transmitted to neighbouring molecules leading to a rapid rise in temperature throughout the product. One problem with microwaves is that they have limited penetrating power, which can lead to cold spots.

UV radiation has wavelengths below 450 nm and contains sufficient energy to excite electrons in molecules making them more reactive. Chemical reactions thus induced in micro-organisms can cause the failure of critical metabolic processes leading to injury or death. The greatest lethality is shown by wavelengths around 260 nm, which correspond to a strong absorption by nucleic acid bases. These wavelengths induce the formation of a covalently linked dimer between adjacent thymine bases in DNA which disrupts transcription and DNA replication in affected cells.

UV light can be used for air disinfection, process water, food surfaces, food containers, and certain clear juices. Unfortunately, UV light does not penetrate very deep (which limits its commercial application) and is easily filtered out by glass. Personnel must be protected from UV light, since the wavelengths used can cause burning of skin and eye damage.

Gamma-irradiation: Gamma rays are produced by the decay of radioactive isotopes. The most commonly used isotopes are Cobalt 60 and Caesium 137. They emit high energy gamma rays which have great penetrating power. Gamma rays are a kind of ionising radiation. It affects micro-organisms directly by interacting with key molecules within the microbial cell, or indirectly through the inhibitory effects of free radicals produced by the radiolysis of water. The main site of damage in the cell is the chromosome. It causes single- and double-strand breaks in the DNA molecule.

Food irradiation has been extensively researched. It can be precisely controlled, penetration is deep, instantaneous, and uniform. It can be used to controls food-borne pathogens and eliminates insects, parasites, and micro-organisms to extend shelf life. In addition, the product retains the fresh character and at low levels, irradiation produces no significant or detectable sensory changes in most products. Other uses of food irradiation include inhibition of sprouting and delaying of ripening consumers experience no harmful effects with the consumption of irradiated foods. Irradiation does not render the food radioactive. Toxins, enzymes, and viruses, however, are not destroyed.

$$\text{Unit of measurement} = \text{kilo Gray}\left(\text{kGy}\right)$$
$$= \text{kilo Joule energy absorbed per kg of material.}$$

3.8 COMBINATION TREATMENTS

There is no universal preservative method for all foods. Each type of food, recipe, and required shelf life and method of storage must be considered and, in many cases, they are best used in combination (also referred to as hurdle technology) (Fig. 3.2).

In many foods, micro-organisms experience a whole battery of suboptimal factors which collectively determine the food's characteristics as a medium for microbial growth. This situation as the hurdle effect, where each inhibitory factor can be visualised as a hurdle contributing to a food's overall stability and safety. The scientific description of this multifactorial technique for preserving foods may be relatively recent, but the concept has been applied in many traditional products such as cheese, cured meats, smoked fish, fruit preserves, etc., all of which rely on a number of different contributing factors for their stability and safety. For detailed use of this concept, see Chapters 5 and 6.

Fig. 3.2 Hurdles involved in making an acid product e.g. canned fruit.

References

Adams, M.R. and Moss, M.O. (2006). *Food Microbiology*, 2e. Royal Society of Chemistry.

Campden BRI (1996). A code of practice for the manufacture of vacuum and modified atmosphere packaged chilled foods with particular regard to the risks of botulism. *Campden BRI Guideline No. 11.* Campden BRI, Chipping Campden UK.

Comité des Industries des Mayonnaises et Sauces Condimentaires de la Communauté Économique Européenne (1992). Code for the production of microbiologically safe and stable emulsified and non- emulsified sauces containing acetic acid. Comité des Industries des Mayonnaises et Sauces Condimentaires de la Communauté Écono- mique Européenne, Brussels.

Lee, S.-Y. (2004). Microbial safety of pickled fruit and vegetables and hurdle technology. *International Journal of Food Safety* 4: 21–32.

Leistner, L. (1999). Combined methods of food preservation. In: *Handbook of Food Preservation* (ed. M. Shafliur Rahman). New York: Marcel Dekker.

Leistner, L. (2000). Basic aspects of food preservation by hurdle technology. *International Journal of Food Microbiology* 55: 181–186.

Jones, A.A. and Man, C.M.D. (1994). *Shelf Life Evaluation of Foods.* Dordrecht: Springer Science+Business Media.

4 Low-acid Canned Foods

Low-acid foods are a group of foods that do not contain any preservation hurdles to micro-organism growth and rely on:

1. the heat process to control micro-organism numbers in the food; and
2. a hermetically sealed package to prevent recontamination.

This means that if the heat process is delivered incorrectly, with the result that containers of food receive sublethal processes, there are no hurdles to the growth of any surviving micro-organisms in the food. It is therefore likely that some of the surviving micro-organisms will germinate from their spores and grow in the food. This will result in either food safety or food spoilage incident, depending on which group of micro-organisms dominates.

Of equal importance to the heat process is the need for the package to prevent recontamination of the food during its long ambient storage. Low-acid foods typically have shelf lives of 18 months or even longer. The declared shelf life will depend on the eating quality of the food at the end of the shelf life. Growth of toxin-producing micro-organisms will not be an issue and so it is chemical degradation of the food over time that gives rise to a limited shelf life. Much of the food degradation is caused by oxygen in the headspace and trapped within the product, and for this reason, most canned foods are closed under vacuum.

The term 'low-acid canned food' suggests the acid concentration in the food is at a low level so that it is not a hurdle to the growth of micro-organisms. However, the low-acid category could also include foods in which other preservation hurdles are present, for example low water activity (e.g. added sugar or salt) or added preservatives. Both of these hurdles are used in thermal processing

Essentials of Thermal Processing, Second Edition. Gary Tucker and Susan Featherstone.

to enable a lower severity of process to be delivered in low-acid foods. For example, the US FDA regulations state that commercial sterility can be achieved by control of water activity (a_w) and heat (FDA: Title 21 Code of Federal Regulations (CFR) Part 113 – Thermally Processed low-acid foods packaged in hermetically sealed containers). The heat is necessary at a_w levels>0.85 to destroy vegetative pathogens and spoilage organisms. Chapter 3 describes hurdle technologies in more detail. Despite this apparent confusion, low-acid foods are generally known as those in which there are no hurdles to microbial growth.

4.1 PRODUCTION OF A THERMALLY PROCESSED FOOD

Without simplifying too much, the manufacture of a thermally processed food can be broken down into two basic operations:

1. The food is heated to reduce the numbers of surviving micro-organisms to an acceptably small statistical probability. This includes both pathogenic and spoilage organisms that are capable of growth under the intended storage conditions (DoH 1994); and
2. The food is sealed within a hermetic package to prevent contamination of the food.

Preservation methods such as traditional canning achieve this by sealing the food in its package before the application of heat to the packaged food product. On the other hand, continuous processing operations heat the food within a heat exchange system before dispensing it into the package. Both methods reduce the numbers of micro-organisms in the food to commercially accepted levels and the packages prevent recontamination over the shelf life. Choice of an in-pack or in-line heat process depends on many factors. Primarily the choice is to apply the most suitable heat process to a food product so that its quality is maximised. It is also dependent on the type of packaging and whether an in-pack process is suitable for that packaging. For example, most fruit juices are pasteurised in heat exchangers and filled into cartons or plastic bottles. Neither of these packaging formats is easy to retort process in the finished package and so a continuous process is an obvious option.

The heat process must target the correct type of micro-organism in order that the product has the correct shelf life. This is because there are millions of types of micro-organisms that can grow within food products and the selection of which ones to target is critical. Fortunately, only a few can cause damage to our health. Of primary concern from a public health perspective are those that produce toxins such as *Clostridium botulinum, Listeria monocytogenes, Salmonella, Escherichia coli O175, Staphylococcus aureus, Bacillus cereus*, and *Campylobacter*. Food poisoning organisms such as *Listeria, Salmonella, Escherichia coli O175*, and *Campylobacter* are not very resistant to heat and therefore are not considered when designing processes for low-acid foods. These organisms are relevant to certain groups of pasteurised foods in which the levels of heat applied are substantially less than for a full sterilisation process. Chapters 5–7 considers the types of products, processes, and shelf-life requirements that make these organisms the targets for the thermal process.

The requirements of a full sterilisation process for low-acid foods are discussed in the next section.

4.2 $F_0$3 STERILISATION PROCESSES

A full thermal sterilisation process is required if no preservation hurdle to microbial growth exists in the food product. For low-acid foods, the most heat-resistant pathogen that might survive the thermal process is *Clostridium botulinum* (Esty and Meyer 1922). This bacterium can form heat-resistant spores under adverse conditions, which will germinate in the absence of oxygen and produce a highly potent toxin that causes a lethal condition known as botulism. This can cause death within seven days and has been implicated in several notorious incidents. The most recent incident in the UK was in 1989, when *C. botulinum* spores survived in canned hazelnut puree, which was used as flavouring for yoghurt. A reformulation of the hazelnut puree to a recipe with lower sugar, but without an increased process, resulted in several blown cans and some containing botulinum toxin.

UK practice is for a commercial sterilisation process to reduce the probability of a single *C. botulinum* spore surviving in a pack of low acid product to one in a UK billion (i.e. 1 in 10^{12}). This is called a 'botulinum cook', and the required heat process to achieve this condition is at least three minutes equivalent at 121.1 °C, referred to as $F_0$3 (DoH 1994). These terms will be used frequently in this book.

Clostridium botulinum spores will germinate in anaerobic conditions where there are available moisture and nutrients, and the acidity levels are low (pH>4.5). Food products with pH>4.5 are referred to as low-acid foods, whereas products with pH values of 4.5 and below are referred to as acid, high acid, or acidified foods. Chapters 5 and 6 describe the groups of foods that rely on acidity as a hurdle to micro-organism growth. In the USA, the critical limit between acid and low-acid foods is taken at pH 4.6 (http://vm.cfsan.fda.gov/~comm/lacf-toc.html). This critical pH limit is an important determinant as to whether heat-preserved foods receive a pasteurisation or sterilisation treatment.

It is also important to ensure that heat resistant spoilage organisms in acid foods do not cause the pH to shift above 4.5 and thereby allow the potential outgrowth of *C. botulinum* spores. This is known as metabiosis (Montville 1982). It has been observed with post-process contamination of tomatoes by moulds that elevated the pH and allowed subsequent toxin production by *C. botulinum* (Huhtanen et al. 1976; Odlaug and Pflug 1978).

Although the mathematics of the minimum botulinum cook are given in various guideline documents (DoH 1994), it is interesting to search out the background to where these recommendations first arose. The $F_0 3$ process actually relates to thermal death kinetic work carried out by researchers in the USA between 1921 and 1950. The original death kinetics work often quoted is that of Esty and Meyer (1922), who investigated the death kinetics of *Bacillus botulinus* (then name for *C. botulinum*). The most prominent of their data was that for the maximum heat resistance, referred to as the terminal death time, for *B. botulinus* grown under optimum conditions. They started with an initial population of approximately 60 billion spores and thermally processed until a negative result was obtained, i.e. all the spores were destroyed and no outgrowth resulted. This number of spores represents the maximum number they could fit into a 1 ml container; in effect, there was nothing present but spores. Note that 60 billion refers to US billion (10^9), which is 60×10^9 spores. Results of maximum heat resistance to moist heat for the most resistant strain of *B. botulinus* were reported as follows:

4	minutes at 120 °C	(248 °F)
10	minutes at 115 °C	(239 °F)
33	minutes at 110 °C	(230 °F)
100	minutes at 105 °C	(221 °F)
330	minutes at 100 °C	(212 °F)

In the original paper, the above data were plotted, as shown in Fig. 4.1, but the data can also be converted to a logarithmic thermal death axis, giving the more conventional approach shown in Fig. 4.2.

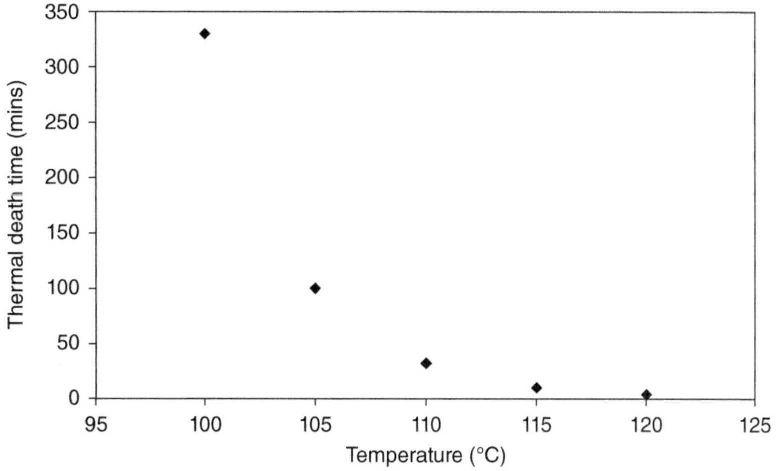

Fig. 4.1 Maximum heat resistance data for *C. botulinum*, taken from Esty and Meyer (1922). Sixty billion spores in a phosphate buffer solution, pH 7.0. *Source*: Modified from Esty and Meyer (1922).

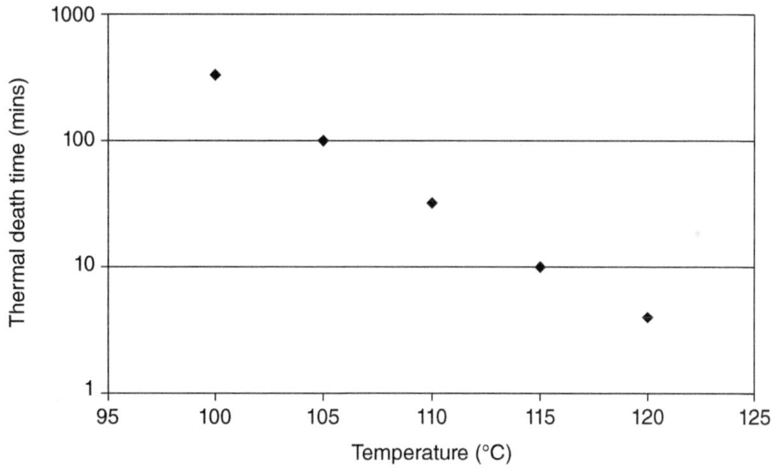

Fig. 4.2 Maximum heat resistance data for *C. botulinum*, taken from Esty and Meyer (1922). Sixty billion spores in a phosphate buffer solution, pH 7.0. Data plotted with thermal death time on a logarithmic axis. *Source*: Modified from Esty and Meyer (1922).

Several years later, Stumbo (1949) took information from the Esty and Meyer (1922) work for the thermal death time data and summarised it using the more familiar concepts of time at a given temperature. He found that an F value of 2.78 minutes was required at 250 °F to process the spores to the point at which none survived. In effect, this was the time to destroy 6×10^{10} spores at

250 °F (121.1 °C) or to achieve almost 11-log reductions of *C. botulinum* spores. Thus, a safe assumption is to process to at least one more log reduction than is required, and so the *F* value of 2.78 minutes was increased to 3 minutes, which is the likely basis of the $F_0$3 value used widely today. We now use 0.21 minutes for the time required to achieve 1-log reduction in *C. botulinum* spores at 121.1 °C, and so the 12-log reduction requires $12 \times 0.21 = 2.52$ minutes, which is rounded up to 3 minutes.

4.3 COMMERCIAL STERILISATION

Any low-acid food that achieves an $F_0$3 process is referred to as commercially sterile and will have a long ambient shelf life. The term 'commercially sterile' is commonly used for low-acid foods; however, both pasteurised and sterilised foods are considered commercially sterile because of the caveat regarding the micro-organisms deemed 'capable of growing in the food'. One of the definitions of commercial sterility is given by the UK Department of Health (DoH 1994):

Commercial sterility is the condition achieved by the application of heat, which renders food free from viable micro-organisms, including those of known public health significance, capable of growing in the food at temperatures at which the food is likely to be held during distribution and storage.

An $F_0$3 process will be safe with respect to *C. botulinum* spores, but it is important to remember that the product may contain a small number of surviving spores of more heat-resistant spoilage micro-organisms. This is because there are many more types of spoilage micro-organisms than those that cause concern for public health, and therefore, it would be expected that a greater range of heat resistance is found. Table 4.1 presents published data on a selection of relevant spoilage organisms, including some data for *C. botulinum* spores.

Data in Table 4.1 are intended to illustrate two main points. The first is that *C. botulinum* spores show a range of heat resistance values and the value chosen for the minimum botulinum process is towards the higher end, but not at the higher limit. The second is that there are other heat resistant organisms likely to be found in low-acid foods, and although they are not pathogenic organisms, they can still spoil the food.

Thus, the high *D*-values in Table 4.1 show that it is not possible to kill all micro-organisms present in a food product without giving the food a very long process and rendering it commercially

Table 4.1 Heat resistance data for spoilage micro-organisms relevant to low-acid foods.

Micro-organism	Temperature range studied (°C)	$D_{121.1}$ (s)	z-value (C°)
C. botulinum			
Phosphate buffer	104–127	8.0	9.0
Water	104–127	3.1	8.5
Pureed peas	104–127	5.3	8.3
Meat and vegetables	100–113	6.6	9.8
Seafood	100–113	3.0	7.4
Poultry	100–113	3.0	7.4
Rock lobster	105–115.5	18.0	10.8
Bacillus stearothermophilus			
Phosphate buffer	100–140	149	14.3
	100–140	170	12.3
	100–140	226	11.7
Bacillus subtilis			
Phosphate buffer	127–144	28.8	9.4
Clostridium sporogenes			
Phosphate buffer	100–120	15.0	9.1
Strained pea	115.5–143.3	60.0	9.8

Source: Data from Holdsworth (1997).

unpalatable. In practice, the food manufacturer makes a decision on the level of commercial risk with the applied thermal process, taking into account the likely spoilage micro-organisms present in the raw materials. For example, a simple canned food such as canned carrots will contain few ingredients (carrots, water) and so the chance of contamination from heat resistant spoilage organisms is low. This is an example of a relatively pure product, and it is acceptable to use a process close to the minimum botulinum cook value of $F_0 3$, perhaps $F_0 4$ to 6 at the end of the heating phase. On the other hand, a formulated soup made from vegetables, starches, and flavourings is likely to contain spores of heat-resistant organisms present with the starches and flavourings. Operating close to $F_0 3$ will result in the occasional spoilt batch of product and so it is typical for formulated products and recipe dishes to work towards higher F_0 values such as 8–10 at the end of the heating phase.

The sterilisation process of $F_0 3$ has been in operation for almost 100 years and with no known cases of *C. botulinum* poisoning from correctly applied processes. It is not intended to take into account spoilage organisms because of the wide variety that exists and the specific nature of the raw materials on which they are found. While the $F_0 3$ concept is not perfect, it has served the industry well. The next section looks at the theory behind how bacterial spores are killed by heat and the models used to calculate their destruction.

4.4 MICRO-ORGANISM DEATH KINETICS

Death of bacteria by moist heat is assumed to be almost logarithmic (Ball and Olsen 1957; Stumbo 1965), i.e. it follows first-order reaction kinetics in which the rate of decomposition is directly proportional to the concentration (Equation 4.1). There are many debates amongst the academic community as to whether first-order kinetics is appropriate. However, for practical purposes, the thermal processing sector continues to use first-order kinetics for two main reasons:

1. There is an overwhelming body of evidence to suggest that correctly applied thermal processes have not resulted in food poisoning incidents arising from *C. botulinum* spore survival. Note that all *C. botulinum* incidents have been traced back to post-process recontamination or the wrong process being applied. Hence, first-order kinetics has served the industry well for almost 100 years and will do so for many years to come.
2. Applying a higher-order reaction rate to time and temperature data is considerably more complicated than for first-order kinetics. It could require a different approach to process validation work.

First-order reaction kinetics are therefore used for calculating integrated *F*-values. Equation (4.1) describes the rate of change in concentration (or numbers *N*) of micro-organisms with time (*t*), where (*k*) is the proportionality constant:

$$-\frac{\mathrm{d}N}{\mathrm{d}t} = kN \tag{4.1}$$

or

$$-\frac{\mathrm{d}N}{N} = k.\mathrm{d}t \tag{4.2}$$

Integrating Equation (4.2) between the limits of N_0 at time zero (the initial number of micro-organisms per unit mass or volume) and N after a time of heating, t (the final number of micro-organisms per unit mass or volume) results in Equation (4.3):

$$k = \frac{\ln(N_0 - N)}{t} \qquad (4.3)$$

This is usually expressed using base ten logarithms ($\log_{10}$), which are referred to in the remaining text without the subscript (log). Hence, Equation (4.3) becomes:

$$k = \frac{2.303 \log(N_0 / N)}{t} \qquad (4.4)$$

The conventional microbiological approach to quantifying thermal processing uses the decimal reduction time (D_T), which is defined as the time required to destroy 90% of the organisms by heating at a single reference temperature (T). This is calculated by the time required to traverse one log cycle on a micro-organism survivor curve, as shown in Fig. 4.2.

Substituting terminology from microbiological death kinetics into the general equation for the straight line Fig. 4.3, Equation 4.5 is obtained.

$$\log N_0 - \log N = \frac{t}{D_T} \qquad (4.5)$$

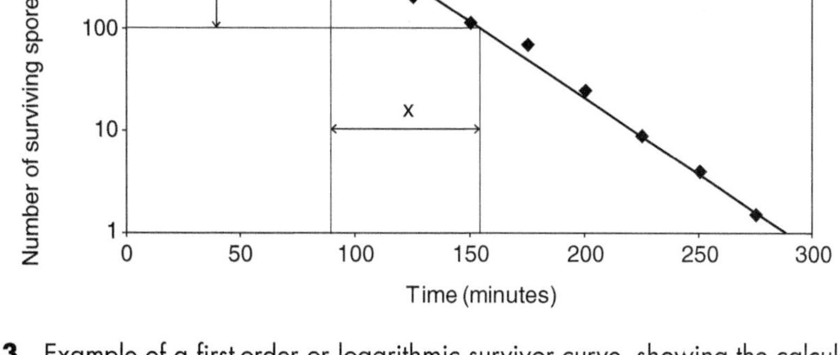

Fig. 4.3 Example of a first-order or logarithmic survivor curve, showing the calculation of decimal reduction time (D), the time required to decrease the concentration of organisms by a factor of ten.

or

$$t = D_T \cdot \log\left(\frac{N_0}{N}\right) \tag{4.6}$$

By comparing Equations (4.4) and (4.6), the decimal reduction time and proportionality factor can be equated. Decimal reduction time is the more convenient term used in thermal processing:

$$D_T = \frac{2.303}{k} \tag{4.7}$$

Equation 4.6 presents the heating time (t) needed at a constant reference temperature in order to reduce the number of micro-organisms from their initial population (N_0) to a final population (N). This heating time is also referred to as a sterilisation or F-value and represents the target number of minutes at temperature (T) to achieve the desired log reduction in micro-organisms (Equation (4.8)).

$$F = D_T \cdot \log\left(\frac{N_0}{N}\right) \tag{4.8}$$

Thus, for a sterilisation process where 12-log reductions are required, the target F-value for an organism with a D-value of 0.3 minutes at 121.1 °C is 3.6 minutes. As given earlier, the conventional approach in the UK (DoH 1994) uses a D-value of 0.21 minutes at 121.1 °C for *C. botulinum* spores, which equates to a minimum F-value of 2.52 minutes. For convenience, this is rounded up to three minutes.

4.5 LOG REDUCTIONS

It was stated earlier in this chapter that 12-log reductions should be achieved for the complete destruction of mesophilic strains of *C. botulinum* spores. Table 4.2 provides process times at 121.1 °C required to achieve the 12-log reduction in *C. botulinum* spores in addition to the same information for certain spoilage organisms of greater heat resistance.

The organisms in Table 4.2 are by no means the most heat resistant that can be found in low-acid foods. However, they are used to illustrate that a 12-log reduction in a spoilage organism

Table 4.2 Process times at 121.1 °C to achieve 12-log reductions in spores of *C. botulinum* and selected spoilage organisms of greater heat resistance.

Micro-organism	Medium	$D_{121.1}$ (s)	$D_{121.1}$ (min)	Time for 12-log reduction (min)
C. botulinum	Pureed peas	5.3	0.09	1.1
	Meat/vegetables	6.6	0.11	1.3
	Seafood	3.0	0.05	0.6
	Poultry	3.0	0.05	0.6
	Rock lobster	18.0	0.30	3.6
B. subtilis	Phosphate buffer	28.8	0.48	5.8
B. stearothermophilus	Phosphate buffer	149	2.48	29.8
C. sporogenes	Strained pea	60.0	1.00	12.0

such as *C. sporogenes* will require an excessive process time that will be detrimental to the food quality. Put another way, an $F_0 3$ process will not be sufficient to achieve 12-log reductions in any of the spoilage organisms in Table 4.2.

Commercial practice is to use a lower number of log reductions for spoilage organisms, because the 12-log scenario deals with the extreme condition in which a food is contaminated at its maximum physical level. This is not an unreasonable approach for the lethal *C. botulinum* organism, because food products contaminated with botulism toxin result in human deaths and usually bankruptcy for the manufacturer. There may nowadays be a prison sentence for those deemed responsible. Spoilage outbreaks differ markedly in that, whilst undesirable, the impact is significantly less. They result in commercial losses from recalled products and damage to the brand image; however, manufacturers can survive these situations.

The reality with spoilage incidents is that the actual numbers of likely organisms present in the food immediately before the process are taken into account. Final numbers of organisms must be sufficiently low that there are minimal chances of survivors (see Worked Examples 4.1 and 4.2).

Calculations on actual reductions in spoilage organisms, as shown in the above worked examples, suggest that the required number of log reductions can be less than 12. In practice, it is common to use log reductions between 4 and 5, typically 5. For the heat resistance values given in Table 4.2, *C. sporogenes* and *B. stearothermophilus* require process times (equivalent at 121.1 °C) of 5 and 12.5 minutes, respectively, to achieve the 5-log reduction. If the process times cannot be achieved without detriment to the product quality or cause production losses, it is

Worked example 4.1 *Clostridium sporogenes*

Measured $D_{121.1}$ value of 1.0 minutes.

Initial numbers of spores (N_0) are measured at 850/g of food.

To ensure removal of spores, it is necessary to process until <1 spore remains viable at the end of the process (N).

Equation 4.6 can be rewritten using the above data to calculate the process time (t) to achieve the above condition:

$$t = 1.0 \log\left(\frac{850}{1}\right) = 2.93$$

Hence, the process time required to eliminate these spores is a minimum of 2.93 minutes. Note that this does not allow margins for variability.

Worked example 4.2 *Bacillus stearothermophilus*

Measured $D_{121.1}$ value of 2.48 minutes.

Initial numbers of spores (N_0) are measured at 1100/g of food.

To ensure removal of spores, it is necessary to process until <1 spore remains viable at the end of the process (N).

Equation 4.6 can be rewritten using the above data to calculate the process time (t) to achieve the above condition:

$$t = 2.48 \log\left(\frac{1200}{1}\right) = 7.64$$

Hence, the process time required to eliminate these spores is a minimum of 7.64 minutes. Note that this does not allow margins for variability.

necessary to investigate how the initial number of organisms can be reduced.

Chapter 11 also discusses log reductions in micro-organisms, but from a process validation perspective when using a log reduction method. It introduces the idea of defining a starting number of micro-organisms per unit mass of food and processing until the risk of survivors is vanishingly low.

There is an interesting difference in the numbers of log reductions required for pasteurised and sterilised foods. A pasteurisation process usually operates to 5 or 6 log reductions of the target organism (Campden BRI 1992, 2006), and this differs from sterilised foods where the intention is to achieve at least 12-log reductions in *C. botulinum* spores. The lower log reduction for pasteurisation has evolved because of the reduced risks associated with the target microbial species when compared with the lethal botulinum toxin. Many of the early pasteurisation processes were spoilage processes, and so the mathematics were similar to those presented above for the spoilage risks to sterilised foods. For example, with fruits with high levels of natural acidity, spoilage from yeast and mould is relevant, which is a much-reduced risk than with a food safety organism such as *C. botulinum*. This justified the decision to reduce the population of yeast and mould by only 6-log reductions. With pasteurised foods, there is also the protection offered by the presence of additional preservation hurdles such as acidity, low water activity, and low-temperature storage.

References

Ball, C.O. and Olsen, F.C.W. (1957). *Sterilization in Food Technology*. New York: McGraw-Hill Book Co.

Campden BRI (1992) Food pasteurisation treatments. *Technical Manual No. 27*. Campden BRI, Chipping Campden UK.

Campden BRI (2006) Pasteurisation: a food industry practical guide, 2nd edn. *Guideline No. 51*. Campden BRI, Chipping Campden UK.

DoH (1994). *Guidelines for the Safe Production of Heat Preserved Foods*. London: The Stationery Office. ISBN: 0 11 321801X. (out of print).

Esty, J.R. and Meyer, K.F. (1922). The heat resistance of the spores of *B. Botulinus* and allied anaerobes. XI. *Journal of Infectious Diseases* 31: 650–663.

Holdsworth, S.D. (1997). *Thermal Processing of Packaged Foods*. London: Blackie Academic & Professional.

Huhtanen, C.N., Naghski, J., Custer, C.S., and Russell, R.W. (1976). Growth and toxin production by *Clostridium botulinum* in moldy tomato juice. *Applied and Environmental Microbiology* 32: 711–715.

Montville, T.J. (1982). Metabiotic effect of *Bacillus licheniformis* on *Clostridium botulinum*: Implications for home-canned tomatoes. *Applied and Environmental Microbiology* 44 (2): 334–338.

Odlaug, T.E. and Pflug, I.J. (1978). *Clostridium botulinum* and acid foods. *Journal of Food Protection* 41: 566–573.

Stumbo, C.R. (1949). Further considerations relating to the evaluation of thermal processes for foods. *Food Technology* 3: 126.

Stumbo, C.R. (1965). *Thermobacteriology in Food Processing*. New York: Academic Press.

5 Acid Foods and Other Pasteurised Products

This chapter discusses processing foods with naturally low pH (<4.6) and is therefore pasteurised. The principles of pasteurisation are explained, and examples of pasteurising some acid products are given. There is a section of fruit processing, as this is an important product group in this food category. There are also examples of pasteurised products that are not acid products, but are pasteurised due to the low water activity.

5.1 BACKGROUND

The objective of thermal processing is to eliminate all microorganisms that can grow in a product and cause spoilage or food poisoning. Acid foods mean foods that have a natural pH of 4.6 or below[1] Acidified foods are low-acid foods to which acid(s), or acid food(s) are added and which have a water activity (a_w) greater than 0.85 and a finished equilibrium pH of 4.6 or below after heat processing. *Clostridium botulinum* is the most heat resistant spore-forming bacteria known to produce a lethal toxin, but it cannot germinate or grow below pH 4.6. *Bacillus cereus* is another heat resistant, toxin-producing spore-former, and it is also sensitive to pH. Many of the principles for processing acid foods and acidified foods are the same.

Some products can withstand a fairly severe heat process, while others will lose some of their texture, colour, and flavour. Some products with lower pH (more acid), such as fruits, are more sensitive to

1 Codex Alimentarius and the USA, Food and Drug Administration define acid foods as 'a food that has a natural pH of 4.6 or below, but in the United Kingdom, it is defined as less than and equal to 4.5'.

Essentials of Thermal Processing, Second Edition. Gary Tucker and Susan Featherstone.

thermal processing and, for these products, other hurdles to microbial growth are often used in conjunction with a milder heat process.

Other examples of organisms of concern with regard to food poisoning are *Listeria monocytogenes*, *Salmonella*, *Escherichia coli O175*, *Staphylococcus aureus,* and *Campylobacter* and are all easily destroyed by mild heat treatments. These organisms display a range of tolerances to pH levels in foods.

There are a host of spoilage organisms (i.e. capable of producing gasses, taints, or enzymes that degrade the texture of the food) that are sensitive to pH and heat to differing degrees. The thermal process chosen must therefore take into account the actual pH of the product, as well as the kinds of micro-organisms that are most likely to contaminate that product; e.g.

- pH of <4.6 will inhibit the growth of most spore-forming bacteria, and specifically *C. botulinum.*
- A mild heat treatment of 70 °C for a few minutes will kill off all vegetative organisms, but not the bacterial spores.
- Many micro-organisms are inhibited by low water activity (a_w).

It must be noted that at a pH of 4.2–4.6 there is a risk of *C. botulinum* outgrowth if other spoilage organisms are able to grow and elevate the pH from their metabolic by-products. This risk is more of a concern in acidified foods, and some processors or authorities prefer to acidify to below 4.2 or treat products with >4.2 as a low-acid food.

Studies have been done with tomato products and it has been found that these products are safe from *C. botulinum* where the pH does not exceed 4.7.

Acid foods typically have a shelf life of 1–3 years. The end of the shelf life is often determined by:

- *Texture*: the natural acidity can chemically cause the cell walls to weaken and for there to be a gradual loss of texture.
- *Colour*: often products in this group, such as fruits, contain pigments that will darken if exposed to light when packed in glass or clear plastic.
- *Tin pickup*: products in this group are often packed in internally plain (unlacquered) cans to enhance their bright fruity colour (see discussion below).

5.1.1 Naturally acid foods

This category includes foods, such as fermented foods, that have a natural pH of 4.6 or below. Natural pH means the pH before

Table 5.1 pH values of some fruits[a].

Fruit	pH	Fruit	pH
Pome fruits		*Stone fruits*	
Apple	3.3–3.9	Peaches	3.6–4.1
Pear	3.6–4.2	Apricots	3.4–3.8
Citrus fruits		Cherries	3.2–4.5
Grapefruit	3.0–3.7	Plum	2.8–3.0
Orange	3.0–3.5	*Tropical fruitsa*	
Lemon	2.2–2.4	Pineapple	3.2–4.0
Soft fruits		Mango	5.8–6.0
Strawberry	3.3–3.9	Litchi	4.7–5.0
Blackberry	3.9–4.5	Guava	3.8–4.4
Raspberry	3.2–3.6	Granadilla (passion fruit)	3.3–3.6
Blueberry	3.1–3.4	Papaya	5.2–6.0
Gooseberry	2.8–3.1	Banana	4.5–5.2
Cranberry	2.3–2.5	*Jams*	3.3–3.7
Tomatoes	4.0–4.6		

[a] Some of the tropical fruits have relatively high pH and are actually classed as acidified products.

processing. However, if during the preprocessing stages, the pH rises above 4.6 (through washing, lye peeling, etc.) and an acid or acid-containing food needs to be added to reduce the pH back to 4.6 or below, that product is considered an acidified food (see Chapter 6). The most frequently found products in the naturally occurring acid and high acid group include most canned fruit and fruit juices, tomatoes, and jams. Table 5.1 presents pH information of many of the fruits used in canning.

5.2 PASTEURISATION

Pasteurisation is a relatively mild heat treatment that is applied to acid foods, whereas sterilisation processes are aimed at destroying all pathogenic micro-organisms and their spores. Pasteurisation processes are designed to eliminate vegetative spoilage micro-organisms that are not very heat-resistant, and the low product pH inhibits the growth of any surviving micro-organisms and prevents spores germinating. Pasteurisation also inactivates enzymes, naturally present, that could cause chemical deterioration of the product on storage.

Acid and acidified foods are heat treated in one of two ways:

- Processing in water, steam, or a combination of the two (see Chapter 8).
- Hot fill and hold (see Section 5.8).

Regardless of the kind of process, pasteurisation may be defined as any heat treatment that has a lethality value (F_0) <3, that reduces the number of pathogenic and spoilage organisms.

The mathematical theory of pasteurisation is the same as for sterilisation, but the target micro-organisms that populate acid foods are less heat resistant and so effectively the reference temperatures (T_{ref}) for the micro-organism death kinetics are lower. There is a wide range of decimal reduction times (D_T-values) and z-values quoted for the relevant micro-organisms. Some pasteurisation processes are described by Pasteurisation Units (PU), similar to the F_0 value for a low-acid product. PU are also referred to as P-values, but the meaning is the same. The PU was originally designed for the pasteurisation of beer, where it has a reference temperature of 60 °C and a z-value of 7 °C.

When calculating pasteurisation units, it is important to give the reference temperature, as there is no international value as there is for low-acid canned foods (i.e. $F_0 3$). There are some general guidelines that can applied when deciding the level of thermal process to apply. However, for many canned acid foods, the processes are designed so that a minimum temperature is achieved at the coldest part of the product.

Pasteurisation value (P) must be defined by quoting the z-value and reference temperature (T_{ref}) that were used for the calculations.

$$\text{i.e.} \quad P^{Z}_{T_{ref}}$$

Controlling the product pH is critical so that the types of micro-organisms that can grow in the food are also controlled. Table 5.2 presents the minimum growth pH for some of the micro-organisms of concern for acid foods.

5.2.1 Considerations when designing a safe pasteurisation process

There are several considerations to take into account when designing a safe thermal process for an acid food. In order to be assured the pasteurisation treatment given to a product is effective the following are examples of what should be considered.

- Microbiological quality/load of the raw ingredients. The higher the load, the more severe the heat treatment will need to be. This relates to the worst-case number of the target micro-organisms; to be able to design the process to reduce the load appropriately.

Table 5.2 Limiting pH values for growth of micro-organisms of concern in the acid foods group.

Organism	Minimum growth pH	Other factors	Medium/substrate
Bacillus licheniformis	4.2		Tomato puree
Clostridium butyricum	4.0		Canned mung beans
Clostridium pasteurianum	3.8	a_w 0.98–0.99	Mango preparation
Clostridium pasteurianum	4.1	a_w 0.975	Mango preparation
Clostridium barati	3.7		Canned mung beans
Bacillus coagulans	3.85		Tomato juice
Bacillus polymyxa	3.8–4.0		Sucrose nutrient broth
Bacillus macerans	3.8–4.0		Sucrose nutrient broth
Lactobacillus brevis	3.5		Citrus fruit
Aspergillus spp.	2.0		Laboratory media
Fusarium spp.	2.0		Laboratory media
Penicillium spp.	2.0		Laboratory media

Source: Campden BRI (1992). © 1992, Campden BRI.

- The types of micro-organisms that are likely to be associated with the product. The most heat resistant pathogens and spoilage organisms need to be considered for the process targets, i.e. the target micro-organisms for the process to reduce numbers to an acceptable risk level.
- It is important to know the death kinetics of those micro-organisms in the product that the process is being designed for, as they can change depending on the type of product.
- Understand the pH, water activity, salt content, and presence of any other compound or condition that will add to the hurdles for microbial growth.

There are other considerations that should also be taken into account when developing a safe and shelf-stable product e.g.

- Storage temperature and conditions of the product.
- Will the product be consumed in one opening, or will it be stored for some time (refrigerated?) with small amounts being used at a time?

5.2.2 Calculation of pasteurisation values

The aim of pasteurisation is to reduce the number of pathogenic and spoilage organism by a specific amount; dependent on the kind of product and type of organisms likely to contaminate it and the likely load; typically, 5 or 6 log reductions. The reduction in the load of the target micro-organisms, together with the hurdle of low pH and/or water activity, will be sufficient to ensure that a satisfactory shelf life is attained.

The death kinetics of spores and vegetative cells (*D*- and *z*-values) represent the heat resistance of particular organisms and are affected by the environment (product type) that they are in. Thus, the lethal effect of heat exposure can be enhanced, reduced, or synergised by the presence of sugars, acids, fats, and even the concentration of the specific product in question. Tables 5.3–5.5 present a selection of industrially relevant data for pasteurisation processes. Much of this was taken from Campden BRI guidelines (Campden BRI 1992, 2006).

The data on the heat resistance are usually taken from work done and published by researchers, as it usually is impractical to

Table 5.3 Heat resistance data for Bacilli and Clostridia applicable to pasteurised products.

Organism	Heating substrate	Heating temperature (°C)	*D*-value (minutes)	*z*-value (°C)
B. cereus	Buffer (pH 7.0)	100	8.0	10.5
B. coagulans	Buffer (pH 4.0)	98.9	9.5	–
B. coagulans	Buffer (pH 4.5)	110	0.64–1.46	–
B. coagulans	Buffer (pH 4.2)	110	0.56–10.6	–
B. coagulans	Red pepper (pH 4.5)	100	5.5	–
B. coagulans	Buffer	106	6.3–7.3	8.2–9.6
B. licheniformis	Buffer (pH 4.0)	100	1.05	10.2
		105	0.19	
		110	0.12	
B. licheniformis	Buffer (pH 7.0)	100	4.29	8.1
		105	0.98	
		110	0.27	
B. polymyxa	Buffer (pH 7.0)	100	18 (approx)	–
B. subtilis	Buffer (pH 6.8)	121	0.57	9.8
C. botulinum non-proteolytic type B	Buffer (pH 7.0)	82.2	32.3	69.7
C. botulinum non-proteolytic type E	Water	80	3.3	9.4
C. butyricum	Buffer (pH 7.0)	85	23	–
C. butyricum	Peach	90	1.12	11.5
C. pasteurianum	Buffer (pH 4.5)	95	3.95	–
C. tyrobutyricum	Buffer (pH 7.0)	90	18	–

Sources: Campden BRI (1992) and Gaze (2006).

Table 5.4 Heat resistance data for other bacteria applicable to pasteurised products.

Organism	Heating substrate	Heating temperature (°C)	*D*-value (minutes)	*z*-value (°C)
Enterococcus (Strep) faecalis	Fish	60	15.7	6.7
E. coli	Broth	56	4.5	4.9
Lactobacillus plantarum	Tomato juice	70	11	12.5
L. plantarum	Buffer (pH 4.0)	50	1.81	6.45
L. plantarum	Buffer (pH 7.0)	55	3.64	6.1
		60	0.20	
		65	0.05	
		70	0.02	
Listeria monocytogenes	Carrot	70	0.27	6.7
Pseudomonas fluorescens	Broth	60	3.2	7.5
Salmonella senftenberg	Pea soup	60	10.6	5.7
S. aureus	Pea soup	60	10.4	4.6

Sources: Campden BRI (1992) and Gaze (2006).

Table 5.5 Heat resistance data for other yeasts and moulds applicable to pasteurised products.

Organism	Heating substrate	Heating temperature (°C)	*D*-value (minutes)	*z*-value (°C)
Byssochlamys fulva	Grape drink	93	5.0	7.8
Saccharomyces cerevisiae (ascospores)	Buffer (pH 4.5)	60	22.5	5.5
Zygosaccharomyces bailii (vegetative cells)	Buffer (4.5)	60	0.4	3.9
Z. bailii (ascospores)	Buffer (4.5)	60	14.2	3.9
Z. bailii (ascospores)	Broth	60	8.1	5.0

Sources: Campden BRI (1992) and Gaze (2006).

determine the z-value required to calculate a process by doing the microbiology. Using the values from the literature, the time required to process a product to achieve a specific reduction in microbial load can be calculated. The method uses an equation first published by Stumbo in 1973 (Equation 5.1):

$$F = D_{\mathrm{T}} \times \log 10^{n} \quad (5.1)$$

where F is the required lethality of the process (minutes), D_T is the D-value at the reference temperature, T_{ref} (minutes) and n is the number of log reductions required (usually 6).

Once the required lethality has been calculated, the equivalent process at other temperatures can be calculated, using Equation (5.2):

$$L = \log 10^{-1}\left(\left(T - T_{ref}\right)/z\right) \quad (5.2)$$

where L is the lethal rate (equivalent to a P-value of one minute at T_{ref}), T is the temperature under consideration (°C), T_{ref} is the reference temperature (°C), and z is the z-value of the organism under consideration (°C).

Equation (5.3) is used to calculate the required number of minutes to process at the chosen temperature, to achieve the required lethality (L):

$$\text{Process Time} = F / L \quad (5.3)$$

5.3 INHIBITORY FACTORS TO MICRO-ORGANISM GROWTH

The efficacy of heat in destroying organisms is influenced by various factors. It is for this reason that mild heat alone cannot be used to produce safe stable pasteurised foods. The most common factors used in conjunction with heat are pH, undissociated organic acids, water activity, and low-temperature storage. Even if an appropriate heat treatment is applied at the desired level, if these other factors are not in control, there is the potential for micro-organisms to grow and spoil the food.

The following statements are usually true:

- Most spore-forming bacteria are inhibited by a combination of pH <4.6 and $a_w <0.90$.
- Acid tolerant spore-forming bacteria will survive and grow at pH greater than 3.8. This includes many of the butyric anaerobes, a group of bacteria that under anaerobic conditions can ferment sugars to butyric acid and includes *Bacillus macerans* and *Bacillus polymyxa*.
- Some xerophilic, osmophilic, spore-forming yeasts, and moulds can grow below a_w 0.85.

- Antimycotic agents (preservatives) have been successfully used in foods to prevent the outgrowth of yeasts and moulds.
- *Clostridium botulinum* does not usually germinate and grow in foods with a pH of <4.6 or a water activity (a_w) of <0.94, although some experimental conditions have shown growth and toxin production by *C. botulinum* at pH<4.6 (Raatjies and Smelt 1979).
- Chilling is only a short-term barrier to microbial growth.

It should be noted that the thermal destruction of some yeasts and moulds is more complex than simple logarithmic destruction. Ascospores can become activated at temperatures as high as 85 °C. Unusual heat resistance was reported in pasteurised beverages and sauces that contained ascospores of *Talaromyces trachyspermus* and *Neosartorya fischeri* (Campden BRI 1992).

In a product where the pH is less than 3.5, there is very low risk of germination and outgrowth of bacterial spores, but certain yeasts and moulds can thrive. The heat process for food pH<3.5 therefore only needs to be designed to destroy those yeasts and moulds. In products with pH>3.5, other organisms are capable of surviving, and so the heat process applied must be sufficient to ensure that their numbers are destroyed. As a generalisation, within the acid foods group, products can be further divided into pH categories so that a minimal but effective heat treatment can be applied (Campden BRI 2006).

5.4 PASTEURISATION VALUE (P-VALUE) GUIDELINES

5.4.1 High acid: pH<3.5

This group includes cherries, plums, apricots, berries, citrus, and sometimes pineapples (Table 5.1). The spoilage organisms include yeasts, moulds, and lactic acid bacteria. Due to the high acidity of these fruits, they generally are very susceptible to softening during heat processing, and processors usually try to process them as close to the minimum requirement as possible. As this group is very susceptible to heat damage, it is advisable to process in water as the rate of heat transfer is lower than through steam, and so is gentler on the product (see Examples 5.1 and 5.2).

Products with a pH value <3.5 only require a process equivalent to 70 °C for five minute (*z*-value 7.5). This will give a 6-log reduction of most lactobacillus species, acidophilic yeast, and mould, but will not be effective against certain heat-resistant acid-tolerant

Example 5.1 Apricot softening

Undesirable softening of apricots is sometimes found in canned apricots. This can be caused by enzymatic softening due to contamination by one of two moulds; *Byssochlamys* produces heat-stable ascospores that can germinate after processing, or *Rhizopus* that produces a heat-stable enzyme, that even though the mould itself may be destroyed during processing, is sufficiently heat stable to cause softening of the fruit. The softening can also be caused by acid in the fruit causing the cell walls to disintegrate during and after processing.

Example 5.2 Processing of grapefruit

Grapefruit is a highly acidic product (pH 3.0–3.3) that is very heat sensitive. Segments must be prepared using either chemical or manual processes. No agitation during cooking is recommended. Low process temperatures are preferred, and a minimum CCT of 80 °C is targeted and hygiene must be excellent. After processing, the best quality products are achieved if there is efficient cooling (Fig. 5.1).

Fig. 5.1 Grapefruit cannery.

mould species like *Byssochlamys fulva*. This mould is difficult to control through the heat process because of its high D_T values. Thorough washing of fruit and good line hygiene is very important to ensure the microbial load is low.

Depending on the product, process recommendations for these highly acid products sometimes target a can centre temperature (CCT) in the 70 °C, with a holding time of only 2–3 minutes, but practically, a CCT of 80–82 °C is usually targeted for most products in this group.

Ascospores of *Byssochlamys fulva* or *Byssochlamys nivea* can survive this process and good hygiene in production is essential to ensure that they are not present in the canned product, because if they grow, they can produce enzymes that soften the fruit.

5.4.2 Acid: pH 3.5–4.0

This group of fruits includes peaches, apples, and pears (Table 5.1). General process recommendations are a *P*-value of five minutes at 85 °C; *z*-value 8.3. This is equivalent to the coldest spot in the can reaching 85 °C for five minutes or 95 °C for 19 seconds (see Examples 5.3 and 5.4).

The process is intended to target spoilage organisms and includes yeasts, moulds, lactic acid bacteria, and some acid-tolerant bacteria, including the group of butyric anaerobes. For all acid and high acid fruit products, good hygiene during product peeling and sorting is critical as many of the organisms that can spoil canned fruit are sufficiently heat-resistant that they cannot be processed away without damaging the texture and colour (Fig. 5.2).

5.4.3 Acid: pH 4.0–4.2

A process equivalent to 95 °C for five minutes (*z* value 8.3 °C) is deemed appropriate.

Example 5.3 Canning of peaches

The pH of peaches varies from 3.6 to 4.1. Canned peaches are processed in continuous cooker coolers with axial agitation. This results in good heat distribution and uniform cooking. A CCT of 91 °C minimum is often used, even though most guidelines state that 85 °C is sufficient.

Example 5.4 Canning of pears

Pears are a difficult fruit to process because their ripening is unpredictable and so they must be held in cold storage for about 10 days to minimise the texture variability. Also, the pH can be as high as 4.5 and this must be monitored and the fruit acidified when necessary. It is recommended to acidify to pH 3.8–4.0, but it must be pH < 4.2. As with other fruits, hygiene is important to minimise spoilage. Processing to a minimum CCT of 96 °C is recommended. Green fruit is significantly harder and takes longer to process. There is an underprocessing risk if fruit of differing textures are processed together – as well as the other microorganisms that can cause spoilage in this group. Spoilage by *C. butyricum* can be found.

Fig. 5.2 Peaches being conveyed into a cannery.

The primary organisms of concern in this category are butyric anaerobes.

5.4.4 Medium Acid: pH 4.2–4.6

For products with a pH in the range 4.2–4.6, a process between 95 and 100 °C for 10 minutes (z value 10 °C) is suggested.

Fig. 5.3 Tomatoes being conveyed into the cannery.

This group includes commonly canned products such as tomatoes and some pears. The primary organisms of concern in this category are butyric anaerobes. The heat resistance of microorganisms is greater at higher pH levels, which requires processing to cold spot temperature in the range 100–104 °C. If the fruit texture is unable to withstand this high process, the product must be acidified so that a less severe heat process can be applied (Fig. 5.3).

Many of the fruits are very heat sensitive, and so the applied processes are often close to the minimum *P*-value requirements. These are often established by trial and error, because the low pH ensures that food poisoning is not a risk with these products. Hence, the processes are based on spoilage and there are more variations in the guidelines than are found with fully foods that must achieve $F_0 3$.

5.5 GUIDELINES AND GENERAL RECOMMENDATIONS

It has been discussed earlier that products of different pH can support the growth of different spoilage organisms. This gives rise to a wide variation in recommended P-values based on measured pH

Table 5.6 Suggested *P*-values (T_{ref} 93.3 °C, z 8.9 °C) for fruit and vegetables.

Product	pH	P-value (minutes)
Lemon juice	2.5	0.1
Plums	2.8	0.2
Gooseberries	3.0	0.5
Pickled vegetables	3.0	0.5
Greengages	3.2	0.8
Rhubarb	3.2	0.2–0.4
Mandarins	3.2–3.4	1.0–2.0
Grapefruit juice	3.2	0.2–0.4
Apricots	3.2–4.0	1.0–8.0
Apples	3.3	0.2–0.6
Blackberries	3.3	0.2–0.6
Orange juice	3.5–3.8	0.2–0.6
Pineapples	3.5	0.6–0.8
Strawberries	2.3–4.0	0.8
Jams	3.5	0.8
Sour cherries	3.5	0.2–0.4
Sauerkraut	3.5–3.9	0.5
Pickled gherkins	3.5–3.8	0.5–1.0
Bilberries	3.7	0.5
Sweet cherries	3.8	0.6–2.5
Guavas	3.8	0.8
Nectarines	4.0	1.5–8.0
Peaches	4.0	1.5–8.0
Pears	4.0	1.3–10.0
Sweet and sour gherkins	3.6–4.1	0.5–1.0
Tomatoes	4.2–4.5	2.0–10.0
Tomato paste	4.2–4.5	1.0–5.0

Source: Eisner (1988).

value. The recommendations in Table 5.6 were taken from Eisner (1988). This is one of the few information sources that provide processes for products over a wide range of pH ($2.5 < pH < 4.6$).

Another source of information is the National Food Processors Association (USA), which give the following recommendations for processing acid products that may contain butyric anaerobes (e.g. tomato).

For products between pH 4.0 and 4.3, the process should be equivalent to five minutes at 200 °F (93.3 °C):

$$\text{P-value} > \text{five minutes}\left(T_{\text{ref}}\ 200°\text{F}, z\,15°\text{F}\right)\text{or}\left(T_{\text{ref}}\ 93.3°\text{C}, z\,8.3°\text{C}\right)$$

For product between pH 4.3 and 4.6, the process should be equivalent to 10 minutes at 200 °F (93.3 °C)

$$\text{P-value} > 10\,\text{minutes}\left(T_{\text{ref}}\ 200°\text{F}, z\,15°\text{F}\right)\text{or}\left(T_{\text{ref}}\ 93.3°\text{C}, z\,8.3°\text{C}\right)$$

5.5.1 Guidelines to critical factors in thermal processing of acid foods

Although the types of micro-organisms that can spoil acid and high acid foods are limited, and the risk of food poisoning from consumption of food from this group is low, commercial production of these products demands commercial sterility for economic reasons (see Example 5.5). The commercial sterility of acid and high acid products is dependent on many complementary factors that pertain to the product and to the processing. Critical factors (as defined by the FDA) are 'any property, characteristic, condition, aspect or other parameters, which when varied may affect the scheduled process and the attainment of commercial sterility.' Product and process-related critical factors are many and include the following:

- microbial load/bio-burden;
- pH of the product;
- water activity;
- consistency;
- product composition/formulation, including sizes of solid
- ingredients;
- in-going weight of product(s), including liquor where appropriate;
- the presence (if any), concentration, and types of preservatives;
- drained weight;
- headspace;
- the numbers and kinds of spoilage micro-organisms;
- container size and geometry;
- container vacuum;
- processing method;
- type and characteristics of heat processing system;
- processing temperature;
- minimum initial temperature;
- processing time.

Small deviations from the product or process specification, which may seem negligible or insignificant, may seriously affect the adequacy of the process for that product, therefore any changes or deviations should be evaluated as to the effect on the adequacy of the process. This subject is considered in more detail in Chapter 13.

Example 5.5 Thermo-tolerant aciduric spore-forming bacteria

Alicyclobacillus spp. are thermophilic, acid-tolerant, spore-forming, aerobic bacteria that are implicated in the spoilage of fruit juices and can cause spoilage in canned fruit and tomatoes. The bacteria are capable of growing at pH as low as 2.5. Two different species that have been found in food spoilage are *Alicyclobacillus acidoterrestris* and *Alicyclobacillus acidocaldarius*. These bacteria are not a health risk, but cause spoilage by producing halogenated phenols (e.g. guaiacol) that cause taints at low concentrations. A typical description of a spoilage complaint is often as bitter, 'TCP', 'disinfectant', 'chemical'. Contamination from *Alicyclobacillus* is generally from the soil. The organisms cannot be successfully 'processed away', as the temperature-time combinations would damage the heat-sensitive acid products. They should be eliminated through good hygiene practices.

5.6 THERMAL PROCESSING OF FRUIT

Canned fruit is usually processed in hot water, raining water, or steam or steam systems. The low pH of the fruit allows for a milder thermal process; a pasteurisation process. Some fruits are more sensitive to heat than others and therefore the selection and control of processing temperature is very important in determining the quality of the final product. As most of the thermal processes discussed in this chapter are below 100 °C, for very heat sensitive products, it is best to choose hot water as the processing medium, because the heat transfer is gentler than from steam.

The heat resistance of most micro-organisms is closely linked to pH; the lower the pH, the more sensitive the micro-organisms are to heat destruction. Although the pH of products discussed in this group can go as high as about pH 5, it is often more effective to acidify the product to pH $<$ 4.2 or even $<$ 4.0. Citric acid is most commonly used for this purpose, as it imparts only a slight fruity flavour and is relatively inexpensive (Fig. 5.4).

Fig. 5.4 Pineapple packed in internally plain cans.

5.6.1 Packaging selection

The most common type of packaging for high acid fruits is tinplate cans. Fruit can also be packed in plastic (HDPE, PP, or multilayer) tubs or in glass. However, one of the factors that can limit the shelf life of fruits packed in plastic is the barrier properties of the plastics, and specifically its ability to limit oxygen ingress.

A factor that limits the shelf life of fruits packed in metal cans is the colour of the fruit. Internally, plain cans are used to pack 'white' fruit (e.g. pears, peaches, apricots, pineapples, and citrus) as the product has a much superior colour when compared with the same product packed in internally lacquered cans (Fig. 5.1). In lacquered cans, white fruit tends to darken, and the end of the shelf life is usually determined by an unacceptably dark colour after 18–24 months. This darkening is caused by oxidation reactions.

5.6.2 Oxidation reactions inside an internally plain can of acid fruit

When these acidic fruits are packed in plain tinplate cans, a reducing environment is created when tin, from the tinplate, is dissolved in the acids from the fruit. The hydrogen ions (H^+) from the acid, instead of reacting with other compounds of the fruit and causing discolouration, react with the electrons from the tinplate and hydrogen is formed. This reducing environment restricts the oxidation-type reactions that cause colour and flavour changes. This system works very well to maintain the bright colour of 'white' fruits (such as those mentioned above), but can cause the pigments of some fruits, such as strawberries or cherries, to lose their colour.

A disadvantage of using plain cans is that the product picks up tin over time. The tin pickup rate is difficult to accurately predict, as there are many factors that can influence it. In general, in the first month after canning, 10–20 ppm of tin is dissolved by the product. Thereafter, tin is dissolved at a slowly decreasing rate of 2–4 ppm per month. The maximum permissible amount of tin allowed in canned fruit is 200 ppm in most countries, but can vary from 150 to 250 ppm, depending on the legislation of the particular country.

Factors that can affect the rate of tin pickup include the amount of oxygen in the can, the temperature of storage, the length of storage, and the presence of detinning accelerators. When using internally plain cans, good manufacturing practices include:

- Packing the can with a high vacuum (>30 kPa);
- Packing the can with a low-to-moderate headspace (4–8 mm);
- Controlling delays during packing. For example, delays after syruping and before seaming can result in detinning at the headspace interface as a result of the oxygen from the atmosphere reacting with the tin and the product;
- Storage at a low temperature (<20 °C), because a high temperature accelerates the corrosion reactions;
- Controlling product quality so that compounds such as sulphur dioxide, nitrates, and copper are avoided, because they can all accelerate corrosion.

Fruits that benefit from being packed in internally plain cans are pears, peaches, apricots, pineapples, and citrus.

5.6.3 Pigments that discolour in internally plain cans

The pigment anthocyanin, which gives many berry fruits their blue/red colour, is sensitive to the reducing environment created inside the plain can and becomes colourless in this environment. Anthocyanin is also a detinning accelerator and will speed up the rate of detinning, and thus shortening the shelf life. For these reasons, berry fruit and fruits with red colour are not packed in internally plain cans.

5.7 THERMAL PROCESSING OF PRODUCTS WITH LOW WATER ACTIVITY

There are many products that rely on reduced water activity (a_w) to inhibit microbial growth. Table 5.7 gives examples of the types of products that can be thermally processed at different a_w levels. Two of the food categories (jams and cakes) are outlined in the sections that follow the table.

5.7.1 Jam and high sugar preserves

Jam is made from fresh, frozen, concentrated, or previously canned fruit that is cooked with sugar (and added pectin, if required), until enough water has evaporated, and the pectin forms a gel on cooling.

In jam production, the water from the fresh fruit is evaporated off either under vacuum or atmospheric pressure. The pH of most jam is low and, in the region, 3.0–3.4. The total soluble solids (TSS) are 65–68°Brix with a pectin content of 0.5–1.0%. The pectin can be naturally present in the fruit or added if the type of fruit does not have enough. The low pH has a dual effect of limiting the spoilage micro-organisms and allowing the pectin to set.

The final jam mixture is filled at >90 °C (if prepared under vacuum where lower temperatures are used the product must be heated before filling). No further processing is required, but a hot fill and hold is necessary to ensure the container and headspace are sterilized (see Section 5.8).

Critical control points for jams are TSS (°Brix), pH, and fill temperature.

Table 5.7 Examples of different water activity food groups.

Water activity level (a_w)	Product categories
<0.70	Dehydrated products e.g. powdered milk, rice, pasta. There is no microbial growth at this level, but spores and some vegetative cells will remain dormant and grow if moisture is added.
0.70–0.85	Some yeasts and moulds can grow at these low moisture levels. The kinds of foods include jams and jellies, cakes, some cured meats.
0.85–0.90	More micro-organisms can grow at this water activity, including some bacteria as well as most yeast and mould types, but growth is still limited by the lower available moisture. A typical spoilage bacterium is *Bacillus licheniformis.*

5.7.2 Canned cakes and sponge puddings

Many cake and bread recipes contain very little or no acid and thus have the potential for supporting the growth of food poisoning bacteria. The hurdles to prevent microbial growth are:

- The low level of water in the product as measured by the low water activity (a_w). The recipe of the product will determine the amount of water available for micro-organisms to grow.
- The anaerobic environment inside the package, although it is important to note that limited oxygen in the headspace does not prevent the growth of harmful bacteria. Low oxygen is more important to reduce oxidation that can change the colour of the product.
- The formulation of the product including active compounds like humectants, including sugars, glycerol, and salt, which assist in a_w control.
- The microbiological load on the product must be low.

There three main ranges of water activity within this reduced water activity group of products.

a. Products that have water activity <0.80 cannot support the growth of any spoilage organisms once they have received an adequate thermal process. This sort of product should be heated to a pack centre temperature of 70 °C after seaming on the end.
b. Products formulated to have water activity in range 0.80–0.85 do not need to have a significant thermal process (e.g. fruitcakes, cakes, and sweetened pastry goods). The kinds of micro-organisms that can grow at this water activity range are not pathogenic and include osmophilic yeasts and xerophilic moulds. A process of 70 °C for 5–10 minutes after the end is seamed on is considered sufficient. If there is reason to suspect that the initial microbiological load includes osmotolerant yeasts and moulds, then a pack centre temperature of 90 °C should be targeted.
c. Products that have water activity in the range 0.85–0.90 can support growth of a limited number of micro-organisms (e.g. sponge puddings). A bacterium that can grow in this water activity range is *Bacillus licheniformis*. A pressure process equivalent to Fo 1.0 is considered sufficient to ensure that spoilage does not occur (Gaze 2006).

One of the important aspects to consider when packing these sorts of products is to reduce the potential for any microbiological contamination. Cans and ends are sterilized and any exposure to the atmosphere and aerial contamination is kept to a minimum.

There are cakes packaged in barrier film that are relatively new onto the market. One example of these is made by the UK company Everfresh Natural Foods. Everfresh operates a retort pasteurisation process after baking and repacking of bread and cake. Both are made from sprouted cereal grains, which has the advantage of converting some of the cereal starch into sugars for water activity control. The post-bake retorting process extends the shelf life at ambient conditions beyond 12 months, which makes the product unique within the bakery categories. Worked example 5.6 presents information about the thermal processes given to the bread and cake products.

5.7.3 Bacteria of concern

Clostridium botulinum requires an $F_0 3$ process to reduce its numbers to acceptable levels. However, water activity <0.94 is a controlling factor that prevents out-growth of surviving *C. botulinum* spores over the ambient shelf life. There are other Clostridia that can cause food poisoning, for example, *Clostridium perfringens*, but none are as lethal or heat stable as *C. botulinum*. Other

Example 5.6 Everfresh retorted bread and cake products

Everfresh bread uses a sprouting process that results in higher sugar content than with conventional bread. This lowers the a_w to around 0.92–0.94. Everfresh cakes also contain sprouted grains but have a wider range of a_w values, often from 0.75 to 0.90, depending on the sugar content of the ingredients. Both products are first baked in a conventional oven where they reach 94–96 °C at the centre. They are removed from the baking tins when cooled and repacked into barrier film for retorting in a steam/air overpressure retort. This process gives a shelf life of 18 months. The thought process behind selecting the retorting conditions is highlighted below with details of the micro-organism of concern summarised in Table 5.8.

Table 5.8 Micro-organisms of concern, risks, and their control in heat-treated bakery products.

Micro-organism	Minimum a_w	Risks
Clostridium botulinum	0.94	Spores are very heat stable and will survive the baking and retort pasteurisation processes. Toxin needs boiling for a few minutes to inactivate.
Clostridium perfringens	0.97	As for *C. botulinum* but less heat stable
Bacillus cereus	0.93	As for *C. botulinum* but less heat stable
Listeria, E. coli, Salmonella	0.92–0.95	Recontamination and transfer from food handlers during repacking. An issue if damage allows organisms to be located deep inside the product.
Staphylococcus aureus	0.84	Recontamination and transfer from food handlers during repacking. It can grow and produce toxin if left for periods >2 h. Toxin is heat stable.
Aspergillus spp. (mould)	0.80	Airborne spores or surface contact during repacking operation prior to retorting. Easy to inactivate by retorting.
Penicillium spp. (mould)	0.79–0.82	As above
Zygosaccharomyces spp. (yeast)	0.62	As above
Saccharomyces cerevisiae (yeast)	0.90	As above

food poisoning bacteria such as *Bacillus cereus*, which causes severe stomach cramps and projectile vomiting, can produce spores with a moderate tolerance to heat. As with *C. perfringens*, *B. cereus* spores as less tolerant to heat than those of *C. botulinum*.

A second relevant group of food poisoning micro-organisms was the 'aerobic pathogens' (*Listeria, E. coli, Salmonella*). These have similar tolerance to heat and can cause food poisoning from very low numbers of cells, sometimes as low as 10–20 cells. The main route in which contamination could occur was from surfaces in contact with the product, for example during the detinning and repacking operations. Cells could be transferred from food handlers but also are found in the environment on surfaces and tools. A thermal process equivalent to 70 °C for two minutes (T_{ref} 70 °C, z 7.5 °C) will achieve a 6-log reduction in cell numbers.

The third bacteria of concern were *S. aureus*, which is known to be transferred from food handlers. The cells are not heat tolerant but the toxin is, to the extent that boiling was not sufficient to destroy the toxin. Control of *S. aureus* was through hand washing and testing of staff, by preventing it from being transferred to the bread and cake. *S. aureus* has a limiting a_w for growth of around 0.84 so was a risk for both bread and cake. A two-hour maximum period between packing and retorting was set to control growth and toxin production.

5.7.4 Yeast and mould

Yeast and mould both form spores that are readily distributed by air. Contamination was mainly from air but could also be from surfaces where yeast and mould were growing or where spores had settled. The detinning and repacking operations had potential for yeast and mould to recontaminate the product surfaces. A thermal process equivalent to 70 °C for five minutes (T_{ref} 70 °C, z 7.5 °C) will achieve a 6-log reduction in yeast and mould spores at the surface.

5.7.5 Process recommendations – cake

A surface pasteurisation was recommended. Contamination from airborne and contact micro-organisms was only likely on the surfaces. Cakes had a lower water activity (0.75–0.90) than bread (0.92–0.94), which offered greater protection from microbial growth. The recommended process was to achieve a surface process equivalent to 70 °C for five minutes (T_{ref} 70 °C, z 7.5 °C). The process times were established by positioning probes 5 mm below the cake surface and between two packs taped together which was the worst case. A maximum of two hours was allowed between repacking and retorting.

5.7.6 Process recommendations – bread

A core pasteurisation was recommended because of the higher water activity than with cakes. The recommended process was to achieve a centre thermal process equivalent to 70 °C for five minutes (T_{ref} 70 °C, z 7.5 °C). A maximum of two hours was allowed between repacking and retorting.

5.8 THERMAL PROCESSING OF CURED MEATS

Products like traditional hams and corned beef are cured products and therefore are preserved both by the thermal process that they received as well as the added salts. These salts, typically sodium chloride, potassium chloride, nitrates, nitrites, and other ingredients, including polyphosphates and ascorbate have antimicrobial properties and work together to ensure that the product is shelf stable. This represents a hurdle approach in which water activity and preservatives are used to control micro-organism growth.

The levels of these compounds are critical and must be controlled and monitored. Sodium chloride lowers the water activity and is the most important ingredient in cured products. Reformulated low salt versions of the product may require a full sterilization process to be shelf-stable.

Anaerobic spores are susceptible to the action of nitrites and so the levels of nitrites are very important when thermal processing to a target F_0 value of <3. Typically, a cured meat requires 3–4% sodium chloride in the aqueous phase of the product and 150 ppm nitrite (0.015%), with an even distribution of the salts essential for food safety. The guideline heat process of F_0 1.0 – 1.5 (Betts and Gaze 1992) requires a competent thermal processing specialist to be consulted when any hurdle process recommendation is required.

Critical control points for heat-treated cured products include:

- Control of pH; nitrite levels; ascorbate; water activity/moisture levels.
- Control of the ingredients and their concentration in brines.
- Control of injection, tumbling and soaking times and temperatures to ensure a consistent and even distribution of the salts.
- Control of thermal processing time and temperature.
- Ensuring that no post-processing contamination can occur.

References

Betts, G.D. and Gaze, J.E. (1992). *Guidelines for food food pasteurisation treatments. Technical Manual No. 27.* Chipping Campden UK: Campden BRI.

Campden BRI (1992). *Food Pasteurisation Treatments. Technical Manual No. 27.* Chipping Campden UK: Campden BRI.

Campden BRI (2006). *Pasteurisation: A Food Industry Practical Guide. Guideline No. 51*, 2ee. Chipping Campden UK: Campden BRI.

Eisner, M. (1988). *Introduction into the Technique and Technology of Rotary Sterilization.* Milwaukee WI: Private Author's Edition.

Gaze, J.E. (2006). Pasteurisation: a food industry practical guide (second edition). *Campden BRI Guideline No. 51.*

Raatjies, G.J.M. and Smelt, J.P. (1979). *Clostridium botulinum* can grow and form toxin at pH value lower than 4.6. Nature (London) 281: 398–399.

6 Acidified Foods

This chapter covers all aspects of acidified foods, including the acidification of foodstuff; some information about pH – what it is and how it is measured; and processing requirements of acidified foods. There is also a description of hot fill and hold for products that are not processed in-pack. The chapter concludes with a section on the critical factors relevant to acidified products.

6.1 BACKGROUND

Acidified foods are low-acid foods to which acid components are added to reduce the equilibrium pH to 4.6[1] or below after heat processing. These foods have a water activity greater than 0.85 and are packaged in hermetically sealed containers. Examples of these products include, but are not limited to:

- Fresh pack pickles (Figs. 6.1 and 6.2) and acidified peppers.
- Fruit that has natural pH higher than 4.6, which are acidified to pH 4.6 or below e.g. some pears and tropical fruits.
- Fermented olives that have been subjected to processes like lye treatment which increases the pH above 4.6; so that acid is required to reduce the pH to below 4.6.
- Fermented products that require the addition of some acid to ensure the pH is less than 4.6 or lower.
- Sauces/salsa (e.g. tomato-based) that contain a significant amount of low-acid ingredients and/or the resultant pH differs significantly from that of the predominantly acid ingredient.

1 In the United Kingdome (UK) a value of 4.5 is used not 4.6.

Essentials of Thermal Processing, Second Edition. Gary Tucker and Susan Featherstone.

Fig. 6.1 Pickled cucumbers ready to be processed.

Fig. 6.2 Beetroot in vinegar brine with temperature measuring probe at geometric can centre.

Legislation states (Codex and FDA) that acidified foods shall be manufactured, processed, and packaged so that a finished equilibrium pH value of 4.6 or lower is achieved within the time designated in the scheduled process and maintained in all finished foods. Acidified foods are thermally processed to an extent that is sufficient to destroy the vegetative cells of micro-organisms of public health significance and those that can cause spoilage.

Clostridium botulinum is an anaerobic micro-organism, which is almost universally present. Under certain conditions, *C. botulinum* can grow in foods and produce a powerful toxin that affects the nervous system. *C. botulinum* will only grow where there is an absence of oxygen, a 'favourable' pH and temperature, and sufficient water and nutrients necessary for its growth. Low-acid canned foods provide this favourable environment. When a product is acidified to a pH of 4.6 or less, *C. botulinum* cannot grow. The pH is critical and therefore a critical part of the manufacture of acidified products is control of the finished equilibrium pH.

In acidified foods, between pH of 4.2–4.6, there is a risk of *C. botulinum* outgrowth if other spoilage organisms are able to grow and elevate the pH from their metabolic by-products. Some processors and authorities therefore prefer to acidify to below 4.2 or treat products with >4.2 as a low-acid food.

The Codex Alimentarius *Code of Hygiene Practice* states that:

> It should be emphasised that the acidification and heat processing of low-acid canned foods are very critical operations involving public health risks and appreciable losses of product if inadequately processed.

The United States Food and Drug Administration (FDA) has defined an 'acidified food' as:

> . . .a low-acid food to which acid/s or acid food/s has been added to the product to produce a finished equilibrium pH of 4.6 or below and a water activity greater than 0.85. Foods such as fresh packed pickles, peppers and marinated vegetables are acidified foods. Foods such as jams, jellies, carbonated beverages; acid foods such as condiments, sauces and dressings that contain small amounts of low-acid foods with a resultant pH that does not differ significantly from the predominant acid or acid food ingredient; naturally acid foods like tomatoes are not considered to be acidified foods. Fermented foods (some of cucumber pickles, most green olives and sauerkraut) are not considered to be acidified foods because pH reduction is not accomplished by the addition of acids or acid foods.

Table 6.1 Table showing recommended criteria for judging whether a pH shift is significant.

If the equilibrium pH of the predominant acid or acid foods is:	Then one should consider a shift in pH to be significant when
>4.2	Any shift in pH
4.2	The shift in pH is >0.2
≥3.8 and <4.2	The shift in pH is >0.3
<3.8	The shift in pH is >0.4

Three important concepts in the acidified foods group are (Barron and Fraser 2013):

- Small amount of low-acid ingredients: the FDA recommends no more than 10% by mass.
- Predominantly acid food: The acid ingredients must have a pH of <4. And must be at least 90% of the total mass.
- pH does not differ significantly: The FDA recommends the following criteria to judge whether the pH shift is significant see Table 6.1.

6.2 ACIDITY MEASUREMENT USING pH

6.2.1 The history of pH

The term 'pH' is used to designate the intensity or degree of acidity. It was developed by a Danish scientist, Soren Peter Lauritz Sørensen in 1909. Until Sørensen developed the pH scale, there was no widely accepted way of expressing hydrogen ion concentrations. Sørensen suggested that the power could be represented by a pH scale in which 7 is neutral, and 1 and 14 are the extremes of acidity and alkalinity, respectively. The letters pH are an abbreviation for 'pondus hydrogenii' (translated as potential hydrogen), meaning hydrogen power, as acidity is caused by a predominance of hydrogen ions (H^+).

6.2.2 The chemistry of pH

pH is defined as the negative logarithm of the hydronium ion concentration or $-\log[H_3O^+]$. Pure water auto-ionises to produce equal concentrations of hydronium and hydroxide ions, as in Equation (6.1) (see Example 6.1):

$$2H_2O \leftrightarrow H_3O^+ + OH^- \tag{6.1}$$

This equilibrium obeys the law of mass action in the form:

$$K_W = [H_3O^+][OH^-] = 1.0 \times 10^{-14} \text{ (at } 25°) \tag{6.2}$$

K_w is an equilibrium constant for the chemical reaction, $mol^2\ dm^{-6}$.

This 'modern' form of the equation for water autoionisation recognises that protons do not exist in solution but instead are bound to an electron lone pair in water, as in Equation (6.3):

$$H^+ + H_2O \leftrightarrow H_3O^+ \tag{6.3}$$

The obsolete hydrogen ion term has been replaced by 'hydronium ion', but it is common practice to continue to use pH (and not pH_3O^+). Since the hydronium ion concentration and the hydroxide ion concentration are equal in pure water, it follows that:

$$[H_3O^+] = [OH^-] = 1.0 \times 10^{-7} \tag{6.4}$$

The value of pH, the logarithm of the reciprocal of the hydronium ion concentration in solution, is determined by measuring the difference in potential between two electrodes immersed in a sample solution. A suitable system consists of a potentiometer, a glass electrode, and a reference electrode. A precise pH determination can be made by making an electromotive force (emf) measurement of a standard buffer solution whose pH is

Example 6.1 Calculating the pH of a solution

Pure water is neutral and has a pH of 7.0. This indicates that the hydronium ion concentration is 0.0000001 g per litre, and so the pH=−log $(1.0\times 10^{-7}) = 7.0$.

In dilute acid solution, the hydronium ion concentration is higher, for example in a 10^{-6}M solution, $[H_3O^+] = 10^{-6}$ mol/L so that the pH = 6.0. That is, one step lower, or higher on the pH scale, represents 10 times higher (or lower) hydronium ion concentration.

Table 6.2 Concentration of $[H_3O^+]$ and $[OH^-]$ in various liquid foods at 25 °C.

Food	$[H_3O^+]$ (mol/L)	$[OH^-]$ (mol/L)	pH
Cola	2.24×10^{-3}	4.66×10^{-12}	2.65
Grape juice	5.62×10^{-4}	1.78×10^{-11}	3.25
Seven Up	3.55×10^{-4}	2.82×10^{-11}	3.45
Schiltz beer	7.95×10^{-5}	1.26×10^{-10}	4.10
Pure water	1.00×10^{-7}	1.00×10^{-7}	7.00
Tap water	4.78×10^{-9}	2.09×10^{-6}	8.32
Milk of Magnesia	7.94×10^{-11}	1.26×10^{-4}	10.10

Source: Nielsen (2003). Springer Nature.

known, and then comparing that measurement to an emf measurement of a sample of the solution to be tested. Table 6.2 presents examples of liquid foods of varying pH, showing that most are acidic.

Acidification of foods requires that acids are added to foods to change the pH of that food. Acids can be divided into two groups: strong acids and weak acids. Weak acids are more common in Nature and are most important in acidification of foods because they are effective antimicrobial agents. Weak acids are mostly organic and include:

- benzoic acid;
- sorbic acid;
- propionic acid;
- lactic acid;
- citric acid.

Strong acids tend be inorganic and less common in Nature. Some examples of strong acids are

- hydrochloric acid;
- sulphuric acid;
- phosphoric acid.

Weak acids (HA) ionise in water and dissociate incompletely into positive and negative compounds; the hydronium ion and the conjugate base (A^-), as given in Equation (6.5). For simplicity, the hydronium ion is represented by H^+.

$$HA \leftrightarrow H^+ + A^- \tag{6.5}$$

The degree of dissociation into the positive and negative ions is a characteristic of different acids. When the amount of HA, H^+ and A^- do not change, they are said to be in equilibrium. This is calculated using the dissociation constant (K_a), which is written as the quotient of the equilibrium concentrations, as in Equation (6.6):

$$K_a = \frac{[H^+][A^-]}{[HA]} \tag{6.6}$$

A logarithmic measure of the acid dissociation constant is more commonly used in practice, pK_a, which is equal to –log10 K_a, and is also be referred to as the acid dissociation constant, as in Equation (6.7):

$$pK_a = -\log_{10} K_a \tag{6.7}$$

A high value of pK_a indicates a low level of dissociation of the acid into its constituent ions. A weak acid has a pK_a value in the approximate range –2 to 12 in water. Acids with a pK_a value of less than about –2 are said to be strong acids and are almost completely dissociated in aqueous solution, to the extent that the concentration of the undissociated acid becomes undetectable. Table 6.3 presents examples of pK_a values for some weak acids used in food preservation.

The pH of the aqueous solution will affect the concentrations of the undissociated and the dissociated acid, as in Equation (6.8):

$$pH = pK_a + \log[A^-]/[HA] \tag{6.8}$$

Table 6.3 Table of pK_a for some weak acids.

Weak acid	pK_a
Acetic acid	4.75
Ascorbic acid	4.17
Benzoic acid	4.19
Citric acid	3.12
Lactic acid	3.86
Malic acid	3.40
Propionic acid	4.87
Sorbic acid	4.75

At pH values $<\mathrm{p}K_a$, the equilibrium shifts to [HA] and, at pH values $>\mathrm{p}K_a$, the equilibrium shifts to $[H^+]$ and $[A^-]$.

The undissociated form of the acid is more effective as an antimicrobial agent, hence weak acids used for food preservation are more effective at lower pH values.

6.2.3 Measurement of pH

If the finished equilibrium pH is 4.0 or below, the acidity of the final product may be determined by any suitable method. If the finished equilibrium pH of the food is above 4.0, the measurement of the finished equilibrium pH shall be by a potentiometric method (see Examples 6.2 and 6.3).

6.2.3.1 Potentiometric method

The primary instrument for use in pH determination is the pH meter or potentiometer. The typical pH meter is equipped with a glass membrane electrode and a reference electrode or a single probe combination electrode. The most commonly used reference electrode is the calomel electrode, which incorporates a salt bridge filled with saturated potassium chloride solution. Calomel electrodes

Example 6.2 Measuring the pH of salad dressing

Separate out the oil and water phases (by centrifugation). Measure the pH of the aqueous phase only. It is not possible to measure the pH of oil. Measuring samples high in oil will affect the pH electrode and thus the pH measurement.

Example 6.3 Measuring the pH of acidified bean salad

Separate out the solids and the liquid portions. Measure the pH of the liquid portion directly. Collect the solids (filter out on a screen). Wash with distilled water. Homogenise. Measure the pH, adding a little distilled water is required. The pH after processing should be ≤4.6.

should be kept filled with saturated potassium chloride solution or other solution specified by the manufacturer, because they may become damaged if they are allowed to dry out. For best results, electrodes should be soaked in buffer solution, distilled or deionised water. Electrodes should be rinsed with water before immersing in the standard buffers and rinsed with water or the solution to be measured next between sample determinations. To obtain accurate results, a uniform temperature should be maintained for the electrodes, the standard buffer solutions and the samples. Tests should be made at a temperature between 20 and 30 °C, the optimum being 25 °C. The accuracy of most pH meters is stated to be approximately 0.1 pH unit, and reproducibility is usually ±0.05 pH units or less.

6.2.3.2 *Colorimetric measurement*

This method may be used instead of the potentiometric method, if the pH is 4.0 or lower. The colorimetric method for pH involves the use of indicator dyes in solutions that gradually change colour over limited pH ranges. An indicator that has the greatest colour change at approximately the pH of the sample being tested is selected. The pH is determined by the colour of the indicator when exposed to the sample under test.

A paper tape treated with indicator dye (indicator paper) is dipped into the sample solution, and depending upon the pH of the solution, the tape will change colour and an approximate pH can be determined by comparison with a standard colour chart. This method does not have the accuracy of the potentiometric method but can provide a quick test.

6.2.4 Equilibrium pH

Equilibrium pH is the pH of the food at 25 °C when its acid is fully diffused throughout the food and successive pH readings as described above produce the same results.

The pH of an acidified product must be 4.6 or lower by the end of the processing. The small quantities of low-acid ingredients included in acidified foods should be at an equilibrium pH within 24 hours (see Section 6.7.3).

6.3 ACIDIFICATION OF FOODS

The natural buffering capacity of most foods naturally opposes acidification. Methods of acidification to reach an acceptable equilibrium pH level in the final product are many and include:

Example 6.4 Cold preservation of pickles by acids

Pickles were traditionally fermented, but more commonly are prepared from fresh by adding a weak organic acid, usually acetic acid, which has an excellent biocidal action. The amount of acid in the equilibrated product is a critical control point and should be measured analytically. This is usually expressed as % acetic acid content. Other acids may be used, but the required levels are higher than those for acetic acid.

- blanching of the food ingredients in acidified aqueous solutions;
- immersion of the blanched food in acid solutions;
- direct batch acidification, where a known amount of an acid solution is added to a specified amount of product;
- addition of a predetermined amount of an appropriate acid to individual containers during production. Liquid acids are generally more effective than solid or pelleted acids;
- addition of acid foods to low-acid foods in controlled proportions to conform to specific formulations.

The choice of method will depend on the food (Example 6.4). A Preservation Index (PI) is calculated from the acid content:

$$\text{PI} = \left[\text{acidity expressed as acetic acid} \times 100\right] / \left[100 - \text{total solids}\right]$$

For products with a PI value above 3.6%, very little or no heat treatment may be needed. However, a food with a PI <3.6% requires some heat treatment; generally, a process to achieve in excess of 10–15 minutes at 75 °C in the coldest slowest heating spot.

6.4 PROCESSING ACIDIFIED FOODS

The heat treatment necessary to achieve commercial sterility of an acidified food is significantly less than that necessary for processing low-acid foods, because the acidity of the final product will prevent bacterial spore outgrowth (see Table 6.4). The heat treatment is only required to kill the less heat-resistant moulds,

Table 6.4 Comparative heat resistance of bacteria important to thermally processed foods.

Class of food	Type of micro-organisms	Micro-organisms	Reference temperature	*D*-value (minutes)	*Z*-value (°C)
Low-acid food pH>4.6	Thermophiles	Flat-sour group *(B. stearothermophilus)*	121.1 °C	4–5	7.78–12.22
	Mesophiles (Spores)	Putrefactive anaerobes *C. botulinum* types A and B	121.1 °C	0.1–0.2	7.78–10
		C. sporogenes group (including PA 3679)	121.1 °C	0.1–1.5	7.78–10
Acid food and acidified food pH 4.0–4.6	Thermophiles	*B. coagulans* (facultatively mesophilic)	121.1 °C	0.01–0.07	7.78–10
	Mesophiles (Spores)	*Bacillus polymyxa* and *Bacillus macerans*	100 °C	0.1–0.5	6.67–8.89
		Butyric anaerobes *(C. pasteurianum)*	100 °C	0.1–0.5	6.67–8.89
		B. licheniformis	93.3 °C	4.5	15
Acid food and acidified food pH <4.0	Yeasts, moulds, and mesophilic nonspore-forming bacteria	*Lactobacillus* species, *Leuconostoc* species	65.6 °C	0.5–1.0	4.44–5.56

Source: Modified from FDA (2010).

yeasts, vegetative cells of bacteria, to inactivate enzymes, and cook the product where necessary. The heat treatment together with the acidification of the product is therefore incontrovertible critical factors within a HACCP plan. According to Codex Alimentarius (1993), when establishing a thermal process, the items listed below must be considered:

- pH of the product
- time to reach equilibrium pH
- product composition/formulation (including dimensions of solid ingredients)
- levels and types of preservatives
- water activity
- microbial flora (including *C. botulinum* and spoilage micro-organisms)
- container size and type
- organoleptic quality
- Product code or recipe identification

As with acid foods, properly acidified foods only need to undergo a pasteurisation procedure, which targets spoilage micro-organisms. During processing of acid foods, often a general rule of achieving 85 °C at the coldest spot in the container is used.

However, equilibration of the pH throughout all components of the food can take place during processing, so a hold time is necessary to enable equilibrium. It is therefore often better to calculate a lethality value (F) for the appropriate spoilage micro-organisms. The time and temperature of each scheduled process is dependent on the equilibrated product pH, as listed in Table 6.5 (see Examples 6.5 and 6.6).

Table 6.5 Table of suggested *F*-values for high acid and acidified products.

pH	*F*-value (minutes) T_{ref} = 200 °F (93.3 °C), z = 16 °F (8.9 °C)	Container centre temperature (°C)
<3.90	0.1	83.3 °C
3.91–4.10	1.0	85.0 °C
4.11–4.20	2.5	87.8 °C
4.21–4.30	5.0	90.6 °C
4.31–4.40	10.0	96.1 °C
4.41–4.50	20.0	98.9 °C

Source: Podolak (2010).

Example 6.5 Spaghetti in Tomato Sauce

Spaghetti (pasta) in Tomato Sauce is a popular canned food that is made to many different recipes. The amount of pasta (carbohydrate) plays an important role in the amount of heat that the product needs to produce a commercially sterile canned food. The pH of cooked pasta is between 5.6 to 6.1, whereas the pH of the tomato sauce usually varies from 3.9 to 4.6. Thus, the sauce recipe and the amount of carbohydrate will determine the final equilibrated pH. Studies have shown that the amount of carbohydrate, irrespective of the pH, also has an effect on the survival of micro-organisms (Sognefest et al. 1948). Table 6.6 may be used as a guide for the selection of sterilising values.

Example 6.6 Fresh packed whole pepper pickles (Downing 1996)

Jalepeno, sweet cherry, and hot cherry peppers are packed in a similar way to fresh pack dill pickles. The fresh peppers are blanched in hot water (~70–80 °C) for 3–6 minutes. They are packed into jars, which are then filled with hot brine, containing approximately 5% acetic acid and 9% salt. The jars are then capped and pasteurised to achieve 70–80 °C at the coldest spot in the container. The jars are cooled with cold water. Preservatives, like sodium benzoate, may be added to extend the shelf life of the product once the jar has been opened. Calcium chloride (0.2%) may be added to the brine to help the peppers retain their texture (calcium in the brine replaces calcium lost in the cell wall, which helps it retain its structure).

6.5 DESIGN OF PASTEURISATION PROCESSES

The pasteurisation process must be sufficient to eliminate the micro-organisms of concern so that they are either destroyed or cannot grow in the final product. It is important to consider the heat resistance of the micro-organisms that are of concern to the particular product for which the process is being designed. There are many factors that can influence the survival of micro-organisms in different recipe products (Campden BRI 1992, 2006):

- *Low moisture*: In semi-moist foods, the heat resistance of micro-organisms is generally increased, similar to in dry environments.
- *Fat*: Generally, the presence of high-fat content increases the heat resistance of micro-organisms due to the protective action of fat and the reduced moisture environment.
- *Water activity*: At reduced water activity levels, the heat resistance of micro-organisms is generally increased, probably due to the dry conditions. It must also be noted that the type of humectant used to reduce the water activity may result in a different *D*-value when compared to the same product with the same water activity, but achieved via a different humectant.
- *Salt or sugar levels*: Generally, the heat resistance of micro-organisms will be increased as the sugar or salt levels increase, again due to the drying effect.
- *pH*: Generally, the heat resistance of micro-organisms will be reduced at lower pH. Table 6.6 takes this into account when suggested levels of pasteurisation. The recommendations in the following subsections were taken from the Campden BRI Guidelines (1992, 2006), as were the data in Table 6.7.

It is difficult to state absolute pH groupings as micro-organisms behave differently in different products; the groups stated are typically what are used.

Table 6.6 Table of proposed process values (*F* value) in relation to pH and carbohydrate content of canned foods.

pH of product before heating	**Percentage carbohydrate in canned vegetable products**		
	3–6%	**9–12%**	**>15%**
4.5–5.0	0.5	1.0	2.0
5.0–6.0	3.0	4.5	6.0
6.0 and above	4.0	6.0	8.0

Source: Data from Sognefest et al. (1948).

Table 6.7 Table of limiting pH values below pH 4.5 for growth of micro-organisms.

Organism	Minimum pH for growth	Product/medium
Bacillus licheniformis	4.2	Tomato puree
Clostridium butyricum	4.0	Canned mung beans
B. polymyxa	3.8–4.0	Sucrose nutrient broth
B. macerans	3.8–4.0	Sucrose nutrient broth
C. pasteurianum	3.8 (a_w 0.98–0.99)	Canned pears
	3.8 (a_w 0.975)	Mango
B. coagulans	3.85	Tomato juice
C. barati	3.7	Canned mung beans
Lactobacillus brevis	3.5	Citrus fruit
Aspergillus spp.	2.0	Laboratory media
Fusarium spp.	2.0	Laboratory media
Penicillium spp.	2.0	Laboratory media

Source: Campden BRI (1992). © 1992, Campden BRI.

6.5.1 Medium acid range: pH 4.2–4.6

A lowered pH value will inhibit some micro-organisms. However, in this pH group, there is the risk of outgrowth and spoilage from acid-tolerant heat resistant spore-forming bacteria (e.g. *Bacillus* and *Clostridium* spp.), namely *B. macerans, B. polymyxa, B. thermoacidurans, C. tyrobutyricum, C. butyricum*, and *C. pasteurianum.*

6.5.2 Acid range: pH 3.5–4.2

Heat resistance is closely related to pH; i.e. micro-organisms are less heat resistant at lower pH values. The main spoilage organisms in this pH category are *C. pasteurianum, C butyricum*, and *C. tyrobutyricum*, as well as some of the acid-tolerant aerobic spore-forming bacteria such as *B. macerans* and *B. polymyxa.*

6.5.3 High acid range: pH below 3.5

The major risk in this category is from acid-tolerant micro-organisms such as lactic acid bacteria, yeasts, and moulds, some of which can produce relatively heat resistant ascospores (e.g. *Byssochlamys fulva* and *Byssochlamys nivea*). There is also a risk of spoilage from thermotolerant aciduric spore-forming bacteria (e.g. *Alicyclobacillus* spp.), although this organism will not grow in all foods.

A 6-fold reduction in the microbiological load is generally used when recommending processes for acidified foods. This is lower than the 12-fold recommendations used with low-acid foods. The consensus is that the lower log reductions are for two reasons: first, the processes are based on spoilage organisms and not the lethal toxin-producing *C. botulinum* and, second, there is a preservation hurdle of acidity present in the foods. The process calculations using *D*-value and log reductions are the same as for acid and high-acid foods, which are discussed in Chapter 5.

Some further processing guidelines (in addition to those in Table 6.5) for products with normal contamination loading are as follows (Campden BRI 2006) (Table 6.8):

Some further advice from NFPA 1968 is as follows:

- For products with pH between 4.3 and 4.6 (e.g. tomato-based products), the process should be equivalent to 10 minutes at 93.3 °C (200 °F).
- For products with pH between 4.0 and 4.3, the process should be equivalent to 5 minute at 93.3 °C (200 °F).
- For products with pH 3.7–4.2, the slowest heating spot in the container should achieve 85 °C (185 °F) for at least 5 minute or 95 °C (203 °F) for 30 seconds.
- For products with pH<3.5, the slowest heating spot in the container should achieve a minimum of 65 °C (149 °F) for 16.7 minutes or 70 °C (158 °F) for 2.1 minutes.

A summary of recommendations published in the FDAs 2010 draft guidelines for acidified foods can be found in Table 6.9.

Table 6.8 Thermal processes for acidified foods.

Finished equilibrium pH	Recommended process or equivalent	Micro-organisms of concern
pH<3.5	5 minutes at 70 °C; z-value 7.5 °C	Acidophilic yeasts, moulds, *Lactobacillus* sp
pH 3.5–4.0	5 minutes at 85 °C; z-value 8.3 °C	Acid tolerant spore-forming bacteria e.g. butyric anaerobes
pH 4.0–4.2	5 minutes at 95 °C; z-value 8.3 °C	Butyric anaerobes
pH 4.2–4.5	10 minutes at 95–100 °C; z-value 10 °C	

Table 6.9 Thermal process lethalities for acidified foods.

Finished equilibrium pH	Thermal process lethality[a]
3.1–3.2	$F_{5.56}/90.56$ = 0.1 minute
3.3–3.5	$F_{5.56}/90.56$ = 1 minute
3.5–4.0	$F_{5.56}/90.56$ = 16–23 minute
4.0–4.2	$F_{8.33}/93.33$ = 5 minute
4.3–4.6	$F_{8.33}/93.33$ = 10–23 minute
4.3–4.6	$F_{8.33}/100$ = 10 minute
4.4–4.5	$F_{10}/110$ = 1.6 minute

[a] Expressed as F_z/Ref, where *F* is the destruction time, *z* is the increase in temperature that results in a 90% reduction of the *D*-value (°C); and Ref is the reference temperature (°C). *Source:* FDA (2010).

6.6 HOT FILL AND HOLD PROCESSING

Many foods lend themselves to a hot-fill-hold process, with the hot product being used to sterilise the container. The entire container including the lid must be decontaminated to an acceptable microbial level. This is generally done by inverting or using a headspace pasteuriser for the container after filling and capping. The fill temperature of the product must relate to the target micro-organisms and the likeliness of post-filling introduction or survival. The minimum hot fill temperature is recommended to be 85–95 °C, to kill vegetative and heat-resistant pathogens (validated processes using temperatures as low as 75 °C may be acceptable). This is a similar process to the one used for the Everfresh worked example 5.7 in the previous chapter, which works for spores of yeast and mould as well as for aerobic pathogens.

It is important to bear in mind two key aspects of hot filling:

1. It is the temperature and time of the food in the container and at the container surfaces that is important i.e. after the initial cool on entering the container.
2. A hold period at the filling temperature, when in the container, is sometimes necessary for effective decontamination, before cooling may commence. This is common if the pH approaches 4.6.

A basic methodology is as follows:

- Food is heated (>85 °C),
- Hot food is filled into containers (often prewarmed),
- A closure or lid is applied.

Table 6.10 Hot fill and hold process and centre container temperature recommendations.

Product pH	Hot fill and hold process	Container centre temperature (°C)
3.0	$F_{71.11} = 0.75$	65.56
3.1	$F_{71.11} = 1.50$	67.22
3.2	$F_{71.11} = 2.50$	68.89
3.3	$F_{76.67} = 0.75$	71.11
3.4	$F_{76.67} = 1.50$	73.33
3.5	$F_{76.67} = 3$	75.00
3.6	$F_{76.67} = 6$	77.22
3.7	$F_{76.67} = 12$	79.44
3.8	$F_{76.67} = 24$	81.11

Source: Podolak (2010).

- The sealed container is inverted for an appropriate length of time to ensure pasteurisation of the container headspace. The length of hold time before cooling depends on the product, but for fill temperature of 90 °C a typical hold time of a minimum of 15–20 seconds is required to ensure inactivation of possible surface contaminating micro-organisms, and in some instances, 3–5 minutes may be more appropriate.
- The container can be put through a raining water tunnel to pasteurise the headspace and surfaces.
- After inversion, the container is turned right side up and allowed to cool before labelling (Table 6.10).

6.7 CRITICAL CONTROL POINTS IN THE PRODUCTION OF ACIDIFIED FOODS

A critical control point is a step in the process, which must be controlled to assure food safety. This is discussed in greater detail in Chapter 16.

6.7.1 Ingredients

Fresh vegetables and meat are highly perishable and good quality ingredients must be ensured. This includes receiving good quality ingredients and handling them in a hygienic fashion with adequate temperature and time controls. Ingredients such as herbs and spices are generally microbiologically stable because of their low water activity, but can harbour high loads of bacterial spores.

The size of the particulates in the formulation must be carefully controlled because this will significantly affect the time taken to reach an equilibrium pH in the final product. Materials such as peppers, tomatoes, and cucumbers may need to be sliced, chopped (to specific sizes), or pierced to provide intimate contact with the acidified brine.

6.7.2 Heat processing

An adequate thermal process must be delivered to ensure that the final product is commercially sterile. A process with a minimum of a 6-log reduction of the appropriate micro-organism must be used, and the appropriate temperature must be achieved for the particular container size, shape, and fill weight. This temperature must be held for the appropriate time in a retort or a hot fill at the appropriate fill temperature and hold time should be used for products not subjected to a post packaging heat treatment (Fig. 6.3).

Fig. 6.3 Pickled onions.

6.7.3 Post process equilibrated pH

Extreme care must be taken when acidifying low-acid foods so that they can be minimally processed without concerns over low-acid regions. The time taken for the pH of the product to equilibrate must be sufficiently short so that no micro-organisms have time to grow and produce toxin or cause spoilage.

The UK Department of Health Guideline (1994) states that pH equilibration of the product at pH 4.5 or below should take place within four hours of the end of the heat process. This guideline also advises that final equilibration of pH is not always achieved in this time, is not easily controlled, and may vary with different ingredients. However, it must be measured within a fixed time and suggests 24 hours from the end of the heat process.

The USA FDA (21CFR 114.3) regulations for acidified low-acid foods states that that the equilibrated pH of the product must be <4.6 by the end of heat processing.

The low-acid components in a product must be <4.6 within a period of 48 hours. A finished equilibrium pH of 4.6 or below to prevent botulism is always a critical control point for acidified foods. The pH of the brine or sauce must be sufficiently acid to result in an equilibrated pH that is less than or equal to 4.6. The critical control point for acidified foods is a finished equilibrium pH of 4.6 reached within 24 hours.

6.7.4 Container integrity

Container closures (caps, seals, or double seams) should be intact and free from food product on sealing surfaces in order to provide a proper hermetic seal. This should remain hermetic through the product shelf life.

6.7.5 pH during product shelf life

The pH of the product should be monitored during shelf life. Any increase in pH could indicate microbiological activity, or improper equilibration, with the risk of micro-organisms growing. These are measurements that should be carried out by the processing company.

References

Barron, F.H. and Fraser, A.M. (2013). *Acidified Foods: Food Safety Considerations for Food Processors*. InTech https://doi.org/10.5772/55161.

Campden BRI (1992). *Guidelines for Food Pasteurisation Treatments. Technical Manual No. 27*. Chipping Campden UK: Campden BRI.

Campden BRI (2006). *Food Pasteurisation Treatments. Guideline No. 51.* Chipping Campden UK: Campden BRI.

Codex Alimentarius (1993) *Recommended International Code of Hygienic Practice for Low And Acidified Low Acid Canned Foods.* CAC/RCP 23-1979, Rev. 2. FAO/WHO.

Downing, D.L. (1996). *A Complete Course in Canning and Related Processes.* Book III. Timonium MD: CTI Publications, Inc.

National Food Processors Association (NFPA) (1968). *Laboratory Manual for Food Canners and Processors.* Westport CT: The AVI Publishing Company, Inc.

Nielsen, S.S. (2003). *Food Analysis*, 3ee. New York: Kluwer Academic Publishers Group.

Podolak, R. (2010). *Acidified Foods: Principles of Processing Commercially Sterile Acid and Acidified Foods Products.* Washington DC: Food Products Association.

Sognefest, P., Hays, G.L., Wheaton, E., and Benjamin, H.A. (1948). Effect of pH on thermal process requirements of canned foods. *Food Research* 13: 400–416.

UK Department of Health (1994). *Guidelines for the Safe Production of Heat Preserved Foods.* London: HMSO.

USA Food and Drug Administration (2010). *Guidance for Industry: Acidified Foods (Draft Guidance).* Center for Food Safety and Applied Nutrition.

7 Heat Preserved Chilled Foods

Ready meals and prepared foods saw rapid growth in the UK and Europe during the 1980s and 1990s. This was in response to consumers spending less time preparing meals from scratch in the home and the availability of the microwave for reheating foods. The UK was one of the countries that led the way with innovations in chilled food products that were suitable for microwave reheating. There were many processing advances, including thermal processing systems and improved measurement capability, in retortable semi-rigid plastic packaging and in the chilled distribution chain. One of the reasons for the advances in chilled foods is the small geographical size of the UK that enables food products to be transported anywhere in the country within a few hours. Compare this to larger countries, such as Australia, USA, Russia, China, and Brazil, where the distribution network operates on a much larger scale, and the challenges in moving food are greater.

Chilled food products are vast in their complexity but tend to fall into three main categories; (i) products that do not receive a thermal process and have a very short shelf life (e.g. fruits, vegetables, meat, fish, sandwiches), (ii) those given a mild pasteurisation process with <10 days shelf life, and (iii) those given a more severe pasteurisation process with around 42 days shelf life (ACMSF 1992). This chapter considers only the products that receive a thermal process.

Those food products with up to 10 days chilled shelf life are, currently in the UK, required to receive a pasteurisation treatment at least equivalent to two minutes at 70 °C (ACMSF 1992; DoH 1989; FSA 2017). Microorganisms of concern with this group of foods are the aerobic pathogens that include *Listeria monocytogenes*, *Salmonella* spp. and *Escherichia coli*. These

Essentials of Thermal Processing, Second Edition. Gary Tucker and Susan Featherstone.

organisms exist only as vegetative cells and, as such, are relatively easy to kill by heat. The range of short shelf-life foods (<10 days stored at 3–8 °C) is vast and includes cooked poultry, ingredients for chilled foods, ready meals, plus a wide range of pies and pastries, and soups and sauces. The severity of the heat process given to these products must be sufficient to target the organisms that can grow over the storage time. The 10-day rule has been in place since 1992 when the Advisory Committee for the Microbiological Safety of Food (ACMSF) published guidance on the safety of chilled (3–8 °C) vacuum and modified atmosphere packaged foods with low oxygen headspace.

As the chilled shelf life is extended beyond 10 days, the possibility exists that psychrotrophic strains of *Clostridium botulinum* can grow. Controls must be put in place for *Clostridium botulinum* if chilled food products are given a shelf life of more than 10 days. *Sous-vide* was one of the first technologies to take chilled shelf life beyond 10 days. It originated in France as a mild thermal process (e.g. 70 °C for 40 minutes) for the manufacture of high-quality foods sold to the catering sectors in vacuum packs (Campden BRI 1992a). More recent *sous-vide* processes have targeted the psychrotrophic strains of *C. botulinum*, which has required the process severity to be increased beyond 70 °C for 40 minutes. A process equivalent to 10 minutes at 90 °C is now applied to achieve at least 6-log reductions in numbers of *C. botulinum* spores (FSA 2008, 2017). This will enable a shelf life of around 42 days to be achieved at a storage temperature in the range 3–8 °C.

This has been a growth sector in Europe, exploiting the ease of distribution provided by the long shelf life and the consumer move towards convenient meals that can be reheated in a microwave. Many of these food products are given an in-pack retort process to ensure both food and package are pasteurised. These heat preserved foods are sometimes known as REPFEDS (refrigerated products of extended durability). Since the name is ungainly, references to REPFEDS will be kept to a minimum.

7.1 UNDERSTANDING MICROORGANISM BEHAVIOUR

In order to understand how thermal processing is used to manufacture a heat preserved chilled food, it is important to know how microorganisms behave. This is best understood by adopting the HACCP approach. The HACCP context quantifies three main areas of biological risk, and these are as follows:

(i) introduction to the food;
(ii) growth within that food;
(iii) survival through the production stages.

A factory HACCP plan will consider introduction, growth, and survival at all stages of manufacture (Mortimore and Emond 2008) and put in place critical control points (CCPs) for their control. Temperature is the main preservation action, both from the storage of raw materials and finished products, but also from the processing perspective.

There are categories of temperature sensitivity of microorganisms that help to define the growth behaviour of microorganisms and also relate to the required severity of the thermal process. These were introduced in Chapters 1–5 but are outlined again here for their relevance to heat preserved chilled foods. Microorganisms are divided into the following categories according to their growth patterns at various temperatures. The first three categories are of greatest importance in chilled foods manufacture.

- Psychrophilic (cold loving), with growth temperature of 12–20 °C.
- Psychrotrophic (cold tolerant), with growth temperature of 20–30 °C.
- Mesophilic (medium range), with growth temperature between 30 and 45 °C.
- Thermophilic (heat loving), with growth temperature between 45 and 65 °C.
- Thermoduric (heat enduring), which can survive above 65 °C, but cannot reproduce at these temperatures.

Most of the pathogenic microorganisms fall within the mesophilic category, such as *Salmonella* spp., *L. monocytogenes* (LM) and *E. coli O157*, and so food production conditions are designed to minimise their growth and survival during manufacture. Note that LM tends towards psychrotrophic growth characteristics but is more typically mesophilic. These tendencies for organisms to grow or survive towards the limits of the above categories illustrate that these categories only provide a convenient set of divisions. Microorganisms have the ability to evolve within their short lifecycle stages, so can adapt to adverse conditions. Over a period of weeks or months, it is possible for a factory strain of a microorganism, for example *Listeria*, to evolve and develop some resistance to dilute cleaning chemicals or the ability to multiply several degrees colder than its preferred growth range. This presents a challenge to chilled food factories and requires good standards of hygiene coupled with temperature control, and otherwise, micro-organism colonies can evolve and become established.

It is also important to understand the four stages in bacterial growth, of which the first two, namely the lag phase and log phase, are most important during chilled food manufacture:

(i) *Lag Phase*: during which the bacteria acclimatise to their environment, which can be several hours long.
(ii) *Log Phase*: during which reproduction occurs logarithmically for the first few hours, where conditions for growth are ideal, and it is during this period that most toxin production starts.
(iii) *Stationary Phase*: during which the bacteria's reproduction rate is cancelled by the death rate.
(iv) *Mortality or Decline Phase*: during which exhausted nutrient levels or the level of toxic metabolites in the environment prevents reproduction, with the result that the bacteria gradually die off.

The lag phase is critical in chilled foods production because it allows the food manufacturer to complete the processing and assembly stages for a food without microorganisms germinating from spores or increasing in cell numbers. A chilled food manufacturer will usually prepare the food in a clean environment held at low temperatures (8–10 °C) to control and suppress growth. Preparation is usually done in a low care environment where standards of hygiene are high but do not need to be at the levels required with high care environments. The latter are essential when foods are assembled without any further process, for example a thermal process step, being given to the foods.

Thermal processing conditions of hold temperature and hold time are calculated from knowledge of the intended storage conditions, which dictate which microorganisms can grow. This is a form of commercial sterility, although many thermal processing specialists do not recognise the definition when applied to the chilled foods sector.

7.1.1 Pathogenic microorganisms relevant to chilled foods

A common theme with pasteurisation treatments is the design of a 6-log kill process. For example, if the initial loading of psychrotrophic *C. botulinum* spores is 10^2/g in an unprocessed food, then a 6-log process will reduce this to 10^{-4}/g in the processed food.

This is a very small number of surviving cells, which is best described as a probability of a spore surviving the process rather than an absolute number. In this case, the probability is 1 spore in 10 000 g of food.

As mentioned earlier, there are two main categories of chilled foods in which a 6-log heat process is applied to extend the shelf life:

(i) Food products sold chilled with up to 10 days shelf life are currently, in the UK, required to receive a pasteurisation treatment at least equivalent to two minutes at 70 °C (DoH 1989). These products include cooked poultry, meat as an ingredient for ready meals, some ready meals, plus a wide range of pies, pastries, soups, and sauces for the high-quality end of the market. Microorganisms of concern with this group of foods are referred to as aerobic pathogens, which include *L. monocytogenes*, *Salmonella* spp. and *E. coli* (Campden BRI 1992b). These organisms exist only as vegetative cells and as such are relatively easy to kill by heat.

(ii) As shelf life is extended beyond 10 days, the possibility exists that psychrotrophic strains of *C. botulinum* can grow. A process equivalent to 10 minutes at 90 °C is designed to achieve at least a 6-log reduction in numbers of *C. botulinum* spores (Campden BRI 1992b; FSA 2008). The precise shelf life will depend on spoilage organisms in the raw materials.

The above processes rely on chilled storage at 3–8 °C to prevent growth of microorganisms that are more heat tolerant than either the aerobic pathogens (*L. monocytogenes*, *Salmonella* spp., and *E. coli*) or psychrotrophic *C. botulinum*. With respect to the latter, there are other control mechanisms that can be applied. These include pH <5.0, salt (NaCl) $>3.5\%$, or a_w <0.97 throughout the food. The adoption of hurdle technologies, as described in Chapter 3, can be used to reduce the thermal process severity by using a combination of these mechanisms to prevent growth/toxin formation. Challenge testing should be carried out to determine whether *C. botulinum* can grow in the food over the storage time.

The following section provides more information on two of the main pathogenic microorganisms of relevance to extended life chilled foods. The first, *C. botulinum*, is the target organism for the 10 minutes at 90 °C process, and the second, *Bacillus cereus*, will become important as shelf lives are extended upwards of around 30 days.

7.1.1.1 *Clostridium botulinum*

C. botulinum is a ubiquitous bacterium which is found in soil, fresh water and in the gastrointestinal tract of animals and fish (Marth 1998). There are seven types of *C. botulinum* (A–G) separated by the properties of the toxin; of these seven toxin types, A, B and E are known to cause human disease, and type F has occasionally been implicated. Type E is often associated with fish and aquatic environments (ACMSF 1992). As well as different toxin types, *C. botulinum* is also classed into groups, with groups I and II responsible for the largest number of foodborne botulism cases. Spores of group I are proteolytic and have a higher heat resistance; typically, these spores are controlled by an $F_0 3$ process (described in Chapter 3). Organisms in group I are not of concern to chilled foods, as the minimum growth temperature is 10–12 °C (Peck 2006). Group II organisms are non-proteolytic and have a lower minimum growth temperature of 2.5–3.0 °C and therefore are of concern in the chilled food sector (Peck 2006).

The occurrence of *C. botulinum* in raw foods is varied and may be dependent upon geography, for example, in Europe high levels are reported in fish from Scandinavia (Lindström et al. 2006). Non-proteolytic *C. botulinum* outbreaks are associated most frequently with processed fish (Peck 2006). The heat resistance of the different types of *C. botulinum* are given in Table 7.1, which shows that these values can vary depending on the type and the substrate used for heating. Foods containing lysozyme and

Table 7.1 Heat resistance of psychrotrophic *C. botulinum*.

Psychrotrophic strain	Heating substrate	Temp (°C)	*D* value (minutes)	*z* value (°C)	Reference
C. botulinum type B	Cod	90	1.10	9	Gaze et al. (1991)
C. botulinum type E	Tuna in oil	82.2	6.6	6.1	Bohrer et al. (1973)
C. botulinum type F	Crab meat	82.2	1.16	6.38	Lynt et al. (1979)
C. botulinum type B	Crab meat	94.4	2.9	8.5	Peterson et al. (1997)
C. botulinum type B	Phosphate buffer with added lysozyme	90	18.1	7.6	Peck et al. (1993)
C. botulinum type E	Phosphate buffer with added lysozyme	90	12.6	8.3	Peck et al. (1993)
C. botulinum	—	90	1.5	—	ECFF (1996)

Worked example 7.1 6-log process for psychrotrophic *C. botulinum*

Measured D_{90} value of 1.1 min.

Target 6 log reductions in psychrotrophic *C. botulinum*

The equation to calculate the target *F*-value is:

$$F = D_T.\log\left(\frac{N_o}{N}\right)$$

Hence, the target *F*-value is $6 \times 1.1 = 6.6$ minute. This was rounded up to 10 minute, which was a decision taken by the ACMSF committee (ACMSF 1992), and so the actual log reductions of psychrotrophic *C. botulinum* spores achieved by 10 minute at 90 °C is just greater than 9.

recovery medium with added lysozyme have been shown to increase the heat resistance of non-proteolytic *C. botulinum* (Peck et al. 1993). This issue has stimulated many debates amongst thermal microbiologists.

The heat resistance values accepted as the target values for the 10 minutes at 90 °C process are those of Gaze et al. (1991). Worked example 7.1 presents how the process was derived.

7.1.1.2 *Bacillus cereus*

Bacillus cereus is a ubiquitous spore-forming bacterium, which is found in soil and many foods such as rice, milk, dairy products, spices, and vegetables, and therefore can be present in cooked chilled foods (Guinebretiere et al. 2003). The organism can cause two different types of food poisoning; emetic and diarrhoeal, depending on the type of enterotoxin produced (ACMSF 1992). In the emetic form of poisoning, symptoms are nausea and vomiting and this is often associated with rice dishes (Altayar and Sutherland 2006). Generally, this organism causes food poisoning at levels of 10^5–10^8 cells per g in the contaminated food but has caused limited outbreaks at levels of 10^3–10^4 per g (EFSA 2005).

Some strains of the organism are psychrotrophic and are capable of growth at 4 °C, which makes them of concern for refrigerated pasteurised foods (Choma et al. 2000). This organism can display high levels of heat resistance, as shown in Table 7.2, often higher than those used to establish the current process

Table 7.2 Heat resistance of *B. cereus* spores.

Organism	Heating substrate	Temp (°C)	*D* value (minutes)	*z* value (°C)	Reference
B. cereus (emetic and diarrhoeal strains)	Buffer, pH 7	85	32.1–106 (mean 45.4)	6.9–13.9 (mean 8.9)	Johnson et al. (1982)
B. cereus spores psychrotrophic strain	Trypticase soy broth, pH 6.0	90	5.0–7.1	–	Membre et al. (2006)
B. cereus	Buffer, pH 7.0	90	71	9.8	Goepfert et al. (1972)
B. cereus	Buffer, pH 7.0	100	8.0	10.5	Molin and Smygg (1967)
B. cereus vegetative cells	Phosphate buffer	90	0.33–0.50	–	Kamat et al. (1989)
B. cereus spores	Phosphate buffer	90	8–15	–	Kamat et al. (1989)
B. cereus	Buffer, pH 7	85	220	9.8	Goepfert et al. (1972)
B. cereus spores, psychrotrophic strains	Sterile distilled water	90	<1.5	–	Choma et al. (2000)
B. cereus spores, unable to growth at ≤10°C	Sterile distilled water	90	0.9–5.9	–	Choma et al. (2000)
B. cereus spores, isolated from Spanish rice	Sterile distilled water	90	3.23–23.26	–	Sarrías et al. (2002)
B. cereus spores, two strains	Honey	105	1.7–2.1	–	Etoa and Adegoke (1996)

recommendations (see above for *C. botulinum*). Therefore, the risks of food poisoning are controlled by selection and/or treatment of raw ingredients and a reduction in shelf life. For example, if a particular ingredient, such as a dried spice, has a high level of *Bacillus* spp., this ingredient can be heat treated before being added to the recipe dish.

Psychrotrophic strains of *C. botulinum* and *B. cereus* are the two organisms that could result in food poisoning incidents with extended shelf-life foods. Most other known pathogens either cannot grow at chilled conditions or are less heat resistant.

7.1.2 Microorganisms likely to be found in chilled foods

Apart from food poisoning risks, there is a commercial requirement to manufacture food products that retain their appeal to consumers throughout the duration of shelf life. It is therefore important that the relevant food spoilage organisms are also taken into account when establishing the thermal process. Spoilage organisms are key to shelf-life determination and are sometimes ignored because their effects on the food are to cause it to spoil but not to produce toxins. However, the number of types of spoilage organism is substantially greater than for pathogenic organisms.

One recent piece of work by Tucker et al. (2008) used a fish curry ready meal as an example product in which the microorganisms in each raw material were enumerated. The curry was a prawn makhani, chosen because it contained many of the raw materials known as those that can introduce high microbial loading to food recipes. Microbiological tests included:

- *APC*: aerobic plate count;
- *Lactics*: lactic acid bacteria;
- *PSRC*: presumptive sulphite reducing Clostridia;
- *ANPC*: anaerobic plate count;
- *ASC*: aerobic spore count;
- *ANSC*: anaerobic spore count.

Results from the microbiological tests on the raw ingredients showed high levels in the fresh coriander with levels of 5.5×10^6 and 9.8×10^5 cfu/g for the APC and ANPC, respectively (Table 7.3). In general, the herbs and spices contributed a high proportion of the microbiological loading for this product. These data are also shown graphically in Fig. 7.1.

Results in Table 7.3 and Fig. 7.1 demonstrate that many of the herbs and spices in recipe dishes contain high numbers of microorganisms. Levels of spoilage organisms were considerably higher than for pathogenic organisms, with the latter being isolated by the ANPC and ANSC tests. Spores of psychrotrophic *C. botulinum* (ANSC) were only found in appreciable numbers in the spices, with garlic puree and cardamom having the highest levels (15000 and 1200 cfu/g, respectively). When diluted into the curry recipe as a small percentage of the total, the spore numbers in the product prior to the thermal process are low. This gives confidence in the 6-log recommendation for establishing a 10 minutes at 90 °C process. However, it does require that preparation conditions for a recipe dish are controlled to maintain these low numbers, otherwise the chances of surviving *C. botulinum* spores increase.

Table 7.3 Levels of microorganisms in the raw ingredients of a prawn makhani curry (cfu/g).

Ingredient	APC	Lactics	PSRC	ANPC	ASC	ANSC
Rice	1.20E+03	5.00E+00	<5	5.90E+02	<5	5.00E+00
Prawns	2.30E+03	2.27E+02	<5	1.90E+02	3.00E+01	6.50E+01
Tomatoes	5.00E+00	<5	<5	<5	<5	<5
Cream	<5	<5	<5	2.00E+01	<5	<5
Honey	2.18E+02	4.00E+01	1.00E+01	7.50E+01	1.85E+02	5.00E+01
Tomato puree	<5	<5	<5	<5	<5	<5
Butter	5.00E+00	<5	<5	<5	<5	<5
Starch	1.00E+01	<5	<5	<5	<5	1.00E+01
Sunflower oil	5.00E+00	<5	<5	5.00E+00	<5	<5
Lemon juice	<5	<5	<5	<5	<5	<5
Ginger puree	<5	<5	<5	<5	<5	<5
Fresh Coriander	5.50E+06	1.20E+03	1.50E+01	9.80E+05	6.00E+02	5.00E+01
Paprika	8.00E+01	<5	<5	1.00E+01	1.90E+02	3.00E+01
Salt	<5	<5	<5	<5	<5	<5
Garlic puree	7.80E+04	7.00E+03	2.00E+01	1.90E+04	5.10E+04	1.50E+04
Ground Coriander	5.09E+02	1.00E+02	<5	1.81E+02	1.20E+02	1.40E+02
Curry powder	2.91E+02	<5	<5	5.00E+00	9.65E+02	1.00E+01
Cumin	1.10E+02	1.50E+01	<5	5.50E+01	1.60E+02	1.00E+01
Chilli	8.00E+01	1.00E+01	1.00E+01	<5	7.50E+01	1.00E+01
Cardamom	5.10E+03	4.45E+02	4.00E+01	4.50E+03	1.30E+03	1.20E+03

7.2 METHODS OF MANUFACTURE

Production methods for chilled foods are many and varied but can be categorised into those where the thermal process step occurs: (i) prior to packaging, for example, using heat exchangers, cooking vessels, and ovens; or (ii) after packaging, for example, in a retort. Applying the thermal process before packing offers most variety in terms of production methods and packaging formats. It is common for factories making short shelf-life meals to be divided into high and low care (or risk) areas. Equipment for achieving the thermal processing step for both in-pack and in-line systems is described in more detail in Chapter 8.

Packaging materials for heat preserved chilled foods do not need to provide the strength or protection demanded by fully sterilised ambient stable foods. There are several reasons for this:

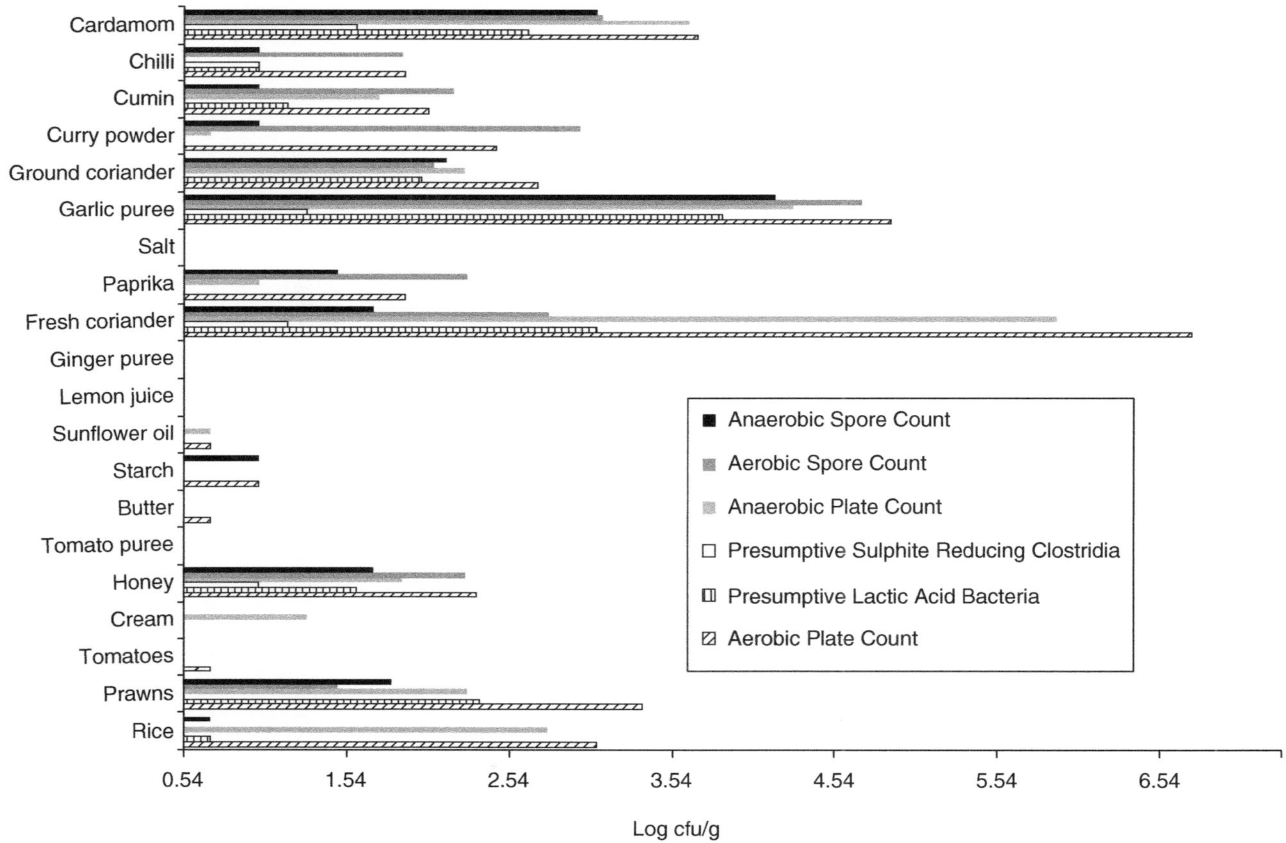

Fig. 7.1 Levels of microorganisms in the raw ingredients of a prawn makhani curry (cfu/g). For a colour version of this figure, please see the colour plate section.

- Processing temperatures are in the pasteurisation region (e.g. 70–105 °C) rather than sterilisation (e.g. 112–132 °C);
- pressure differential between inside and outside of the package is much lower;
- storage life is a few weeks compared with many months;
- storage temperature is 5–8 °C and so gas transfer occurs at slower rates.

Hence, the packaging is typically in barrier plastics, either pouches, bowls, or trays, but with some paperboard cartons used for soups, sauces, and vegetables. One advantage of a plastic package compared with one containing metal is the microwavable nature of that package. This provides convenience for the consumer. A single type of plastic, for example polypropylene, is unlikely to be suitable for retorting or will not provide the barrier properties over the shelf life. Multi-layered materials are required to provide barrier protection against ingress of water, oxygen, and light. Details on the package construction and properties are not appropriate for this book and can be found in *Food Packaging Technology* by Coles et al. (2003). These multi-layered packages are less strong than the metal can and require greater caution during the processing steps to avoid container damage. This is particularly important for retorted foods in which retort overpressure is required to maintain the container shape during the process. This is considered in one of the following sections.

One very important consideration when choosing the packaging materials is to avoid metal cans for chilled foods. These should not be used for this type of food because of the risks of consumers confusing how the cans should be stored. Consumers expect certain types of containers, including metal cans and glass jars, to be stored in ambient conditions. This is highlighted in the recent botulism outbreaks in the example in Box 7.1.

7.2.1 Thermal process step applied prior to packaging

This is the method for chilled food manufacture that was used when the industry sector first gained popularity soon after refrigeration was available to households. The first food products were relatively simple in their construction, with most being commodity items such as milk and fruit juice. During the 1980s, the UK was instrumental in pioneering a more ambitious chilled foods market in which the concepts of convenience food were started. Foods such as soups, sauces, and ready meals were often prepared in large kettles and hot filled into packages. These were then

Box 7.1 Food poisoning example: Canned chilled carrot juice

In the USA in 2006/7, Bolthouse Farms manufactured a chilled carrot juice in a metal can, and this initiated four separate *C. botulinum* poisoning incidents over a few months. The product was intended for storage in the refrigerator but was mistakenly put into store cupboards because it was similar in appearance to other ambient stored cans.

In September 2006, three people living in Georgia developed foodborne botulism that was traced to the carrot juice from a single can. Soon after, there was an additional case in Florida and two more in separate incidents in Ontario, Canada. At least one person died and two remain on ventilators today.

In June 2007, the FDA modified its guidance for refrigerated low-acid juices to recommend adding a validated juice treatment method, such as acidification or appropriate thermal treatment, to decrease the risk of *C. botulinum* contamination, should any breaches in refrigeration occur. No further incidents are reported for this or other products (FDA 2007).

Source: Based on FDA (2007).

chilled or frozen in the sealed packages using air blast refrigeration. The heat process in the kettles is adequate to knock out aerobic pathogens in the food by achieving two minutes at 70 °C. Heat from hot filling provides a surface pasteurisation effect, although difficult to quantify, and this is sufficient to control yeast and mould on the packaging surfaces. Chilled food prepared in this way has a shelf life of up to 10 days.

Not all components are cooked immediately before assembly. Some are cooked and held in chilled storage until ready for packing. This includes both the protein and carbohydrate components. A significant reliance is placed on the integrity of the chilled storage chain to prevent growth of any contamination that enters the package during assembly and closing. Microorganisms, for example LM, are known to grow down to temperatures close to 0 °C but this is in ideal laboratory conditions and the growth rates are very low. The 10-day storage time is unlikely to provide enough time for the psychrotrophic microorganisms to grow in the chilled conditions. Therefore, this process is reliable and only tends to fail if the chilled storage is broken.

Methods to validate the pasteurisation achieved during cooking and filling have developed in recent years. These are described in Chapter 11.

When companies demand longer shelf life, it is necessary to increase the thermal process severity for the cooking process and to ensure the hot filling achieves at the very minimum a yeast and mould process. The driving force is the retailer's desires to reduce waste on its premises and provide the consumer with more days of useable shelf life in the fridge. The FSA advice (FSA 2008) for the extended process is to achieve 10 minutes at 90 °C during cooking (T_{ref} 90 °C, z 9 °C), so that 6-log reductions in psychrotrophic strains of *C. botulinum* are achieved. This organism can grow down to low temperatures (e.g. 3–8 °C), albeit slowly, but over a shelf life that can be around 42 days, the chances of it growing need to be considered.

One option used by some companies to increase the security of the packed food is to hot fill into the package. Evidence from commercial products that are hot and cold filled suggests that hot filling provides a few more days of shelf life. Hot filling needs to take place in clean conditions so that bacterial contamination is prevented and the risks are only from air-borne yeast and mould spores. These are easier to kill by heat. The hot fill process should achieve at least two minutes at 70 °C on the packaging surfaces (T_{ref} 70 °C, z 7.5 °C). Contamination from bacterial spores is likely to survive the hot filling process and so attention to filler cleaning and start-up procedures, and filler room hygiene is essential.

Filling after processing incurs a higher risk than when processing a filled and closed container. Factories that manufacture REPFEDs will usually enclose the filling environment in a high care or high-risk area (Campden BRI 1992a). Standards of hygiene are stricter than in a low care environment where the food product receives its cooking process. If microbiological contamination gets into a food before or during cooking, the thermal process will destroy the contaminating microorganisms. This assurance is not present during filling. The next section outlines some of the practices involved in assembling product within high care environments.

7.2.1.1 Low care–high care factories

Ready meals of high quality are being manufactured using cooked protein (e.g. chicken, meat or fish) and raw vegetable components that are cooked and cooled separately. The thermal processing steps (cooking) take place in a low care environment and are followed by packing into the sealed final package in a high care environment. Strict control of hygiene in the high care area is essential in order that microorganisms are not introduced to the food. Once packaged, there is no further preservation hurdle to prevent or slow down microorganism growth apart from chilled storage.

It is increasingly common for the cooked protein to be transported frozen from countries local to its production, for example, Thailand, Brazil, and China. The minimum process for cooked protein in a short shelf life product is to achieve at least two minutes at 70 °C at the core (T_{ref} 70 °C, z 7.5 °C). As with in-pack processes, there is a need to validate the heat treatment and to control its implementation in a factory using HACCP approaches. Methods for validating these processes often require techniques such as microbiological or biochemical TTIs, because of the continuous cooking and cooling equipment that is used. Chapter 11 provides details of these techniques.

The advantage of the assembly method for manufacture of multi-component prepared meals is in the quality of protein and vegetable components that can be achieved. Many vegetables, such as peppers and spring onions, easily lose their texture when cooked. Careful attention to their critical dimensions and cooking method is required so that quality is optimised. This is best achieved by cooking the vegetables separate to the protein components because the protein can often withstand higher levels of cooking. By assembling a meal with all components cooked separately and chilled to 5–8 °C, and maintaining the chilled shelf life for only a few days, it is possible to produce textures that cannot be achieved by cooking the complete meal.

This method of chilled food manufacture is more complicated than the few paragraphs above suggest. It often uses modified and controlled atmospheres to suppress microorganism growth, applies new technologies such as pulsed light and other surface treatments, and places emphasis on hurdles to microbial growth. It is debatable whether it should be included in a book on thermal processing, but for completeness and because of the cooked protein components, it is included. Tucker (2008) gives further information on the above techniques in his book on *Food Biodeterioration and Preservation*.

7.2.2 Thermal process step applied after packaging

Retorting after filling is now the preferred method for extending the shelf life of heat preserved chilled foods. It is more secure than hot filling because the retort process pasteurises both the packaging and the food. Ready meals have benefited from this to the extent that the majority of UK chilled ready meals (in trays) are now manufactured by retorting the filled package. The shelf life can go considerably beyond 10 days, although there is no recommendation on how many days are achievable. This decision is the responsibility of the manufacturer.

The retorting process is carried out in a similar way to that for canned foods; however, the difference is in the need to operate the process using an overpressure that counteracts expansion of the packages. If the food is a ready meal in a plastic tray, the tray will be made from flexible plastics (e.g. polypropylene or CPET) and the lid will be a thin film (e.g. polyester). These trays are delicate and distort easily at small pressure differentials. One of the first retort manufacturers to seize on this opportunity was the French company Lagarde in partnership with the UK company Holmach Ltd. They engineered the steam/air retorting system with air overpressure so that it could be applied to processing barrier plastic trays. Many of the early heat preserved chilled food products on UK retailer shelves were thermally processed in Lagarde steam/air retorts. Any retort system with air overpressure facility can be used to process meals in barrier plastic trays.

Processing a delicate plastic tray requires a higher level of retort control than with a stronger metal can or glass jar. The main control issue is to maintain an overpressure in the retort that is sufficient to force the tray back to its original size and shape but not stress the seals with excessive force. Figure 7.2 illustrates the changes in pack deflection that occur during processing, as measured using an Ellab deflection sensor positioned with the sensing tip just touching the upper pack surface. In this example, the package was a ready meal tray, the retort used steam/air as the heating media, and the test product was water. Figure 7.2 shows the tray increase in volume during the venting period of the retort, with the deflection brought under control at the end of the vent period by increasing the retort pressure to counteract the expansion. Overpressure was held steady during the hold period and for further time into cooling while the product was hot and internal pack pressure was high. The example in Fig. 7.2 is not intended to be an ideal deflection curve but to illustrate how deflection changes during the retorting cycle.

Further complications can arise because of differences between filled packs in the air content or pressure inside the filled packs. This can create inconsistency in pack deflection across a retort load. This effect is complicated with changes in fill temperature for products filled hot (for viscosity purposes) in which the cooked batches cool during the time taken to fill the packs for a retort

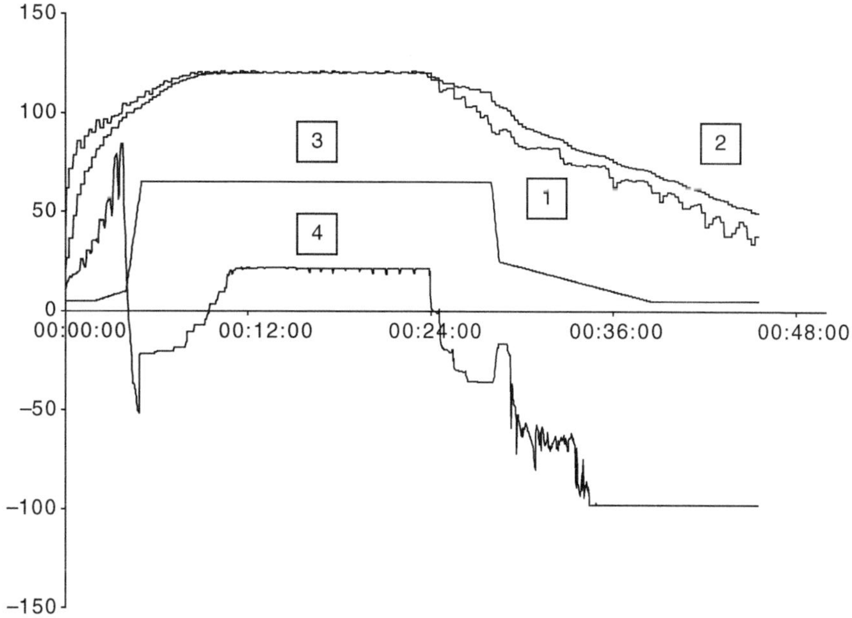

Fig. 7.2 Graph showing changes in deflection in the upper surface of a flexible package processed in an overpressure retort (1 – retort temperature, 2 – product temperature, 3 – retort pressure, 4 – deflection sensor reading).

load. The first packs filled are at a higher temperature and so they draw a higher vacuum level inside the packs as the product cools compared with the cooler packs filled at the end of the batch. Fill temperature differences may only be 20 °C, but this is sufficient to generate pressure differences across the retort load.

Figure 7.3 illustrates improved pack deflection control by reducing the air headspace and increasing the retort overpressure. Less gas inside the package and higher overpressure forces the tray lid closer to its original position. This avoids the damaging changes in lid deflection that are detrimental to pack integrity.

7.2.2.1 Caution with latent heat for frozen protein

As with ready-meal assembly in high care environments, retorted foods are also now using cooked protein imported in frozen form. This is economic, easier to handle than fresh protein, can be stored for long periods, and does not suffer from protein cookout that damages the appearance and texture of sauces. However, there is one major disadvantage of using frozen protein in a retorted ready meal, and this is the issue of the latent heat of freezing that must be overcome before the product temperature can rise from 0 °C.

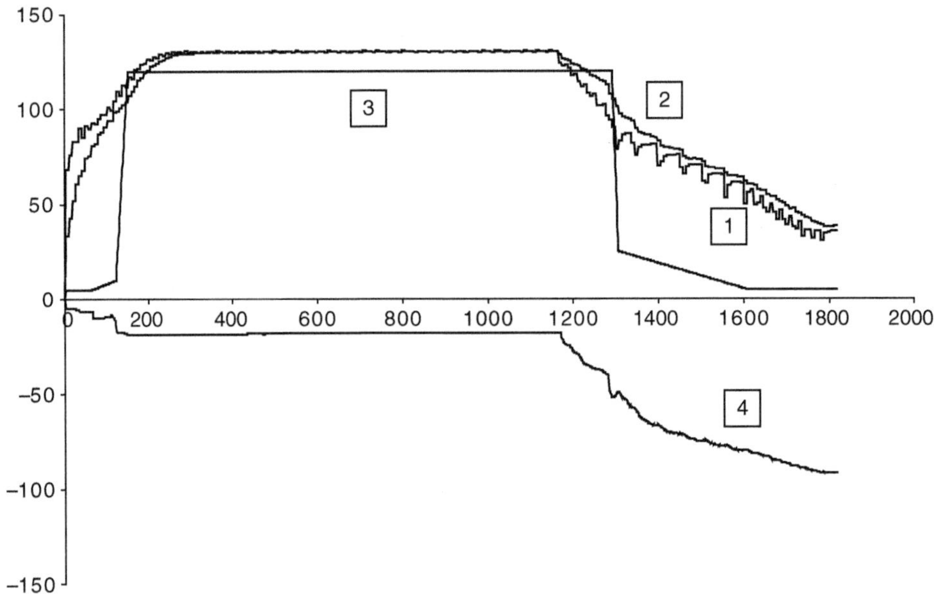

Fig. 7.3 Graph showing changes in deflection in the upper surface of a flexible package processed in an overpressure retort; with lower headspace and high retort pressure than for Fig. 7.2 (1 – retort temperature, 2 – product temperature, 3 – retort pressure, 4 – deflection sensor reading).

Latent heat is an energy-intensive property that occurs as water and dissolved substances (e.g. salts, acids, colours) change their state from liquid to solid. To reverse this process, as is required in retorting a ready meal containing frozen protein, the same high quantities of energy are required but in reverse. Worked example 7.2 demonstrates the proportion of energy required with frozen protein.

The worked example of a ready meal with 30–40% frozen protein is commonly experienced with this industry. Companies with canning backgrounds are aware of the issues with frozen ingredients. Guidance is available that suggests using defrosted materials or restricting the proportion of frozen material to a low level (e.g. 5–10%) that will defrost before the container enters the retort (Campden BRI 1977; IFTPS 1995).

Volumes of frozen material used in ready meals can be high, as in worked example 7.2. Defrosting this material prior to filling or within the filled packs before retorting is a challenge. Substantial factory space is required to defrost these materials under careful temperature control. There are food safety concerns around recontamination and microorganism growth when allowing a lengthy defrosting period for the frozen components. Hence, it is common to fill the frozen components into the packs and retort while still frozen.

Worked example 7.2 Energy required to thaw frozen chicken

The ready meal is a 400 g chicken curry with 150 g of chicken. The retorting process raises the meal temperature from 0 to 90 °C. Assume the frozen protein (150 g) is at 0 °C, and the sauce (250 g) is also at 0 °C but is not frozen.

The energy required to thaw the frozen chicken is given by:

$$Q = m \times L$$

where Q is the energy required (J), m is the mass of frozen chicken (kg), and L is the latent heat for frozen chicken (assume a value for water) at 33.5×10^4 J × kg^{-1}:

$$Q = 0.15 \times 33.5 \times 10^4 = 50\ 250\ \text{J}$$

So 50 250 J are required to thaw the chicken, with the chicken temperature remaining at 0 °C. The energy required to raise the temperature of the chicken and sauce is given by:

$$Q = m \times C_p \times \Delta T$$

where Q is the energy required, known as sensible heat (J), m is the mass of chicken or sauce (kg), and C_p is the specific heat capacity for chicken or sauce (assume a value for water) at 4186 J kg^{-1} xx K^{-1}:

$$Q = 0.4 \times 4186 \times 90 = 150\ 696\ \text{J}$$

So 150 696 J are required to raise the thawed chicken and sauce from 0 to 90 °C. The total energy to process the ready meal (excluding packaging) is the sum of both the latent and sensible heat components, which is 200 946 J. Latent heat is 25% of the total energy; which is a significant amount.

Heat penetration testing requires probes to be inserted through drilled holes into frozen lumps of protein. Great care is required when taking the experimental data to ensure the probe remains in position throughout the process. There is a lot of pack movement that takes place while the instrumented packs are placed into the retort but also with the overpressure profile. It is good practice to check the probe has remained at the centre of the protein component once the packs are removed.

Most processors use the General Method for their calculations of hold time. This avoids the complications that arise from the thawing step at the start of the process. Analysis of the product time–temperature data shows an extended lag period while the frozen components defrost before increasing in temperature towards the retort hold temperature. Other calculation methods must be used with caution. The Ball Method was not set up for frozen material so would incur inaccuracies with j-value analysis. Numerical approaches also require the heating (f_h) and *j*-values to be calculated from semi-logarithmic plots but use the *j*-value in a different way to the Ball Method. Data analysis will require broken heating methods with high f_h and *j*-values at the start while the material remains frozen.

References

Advisory Committee on the Microbiological Safety of Food (1992). *Report on Vacuum Packaging and Associated Processes.* London: HMSO.

Altayar, M. and Sutherland, A.D. (2006). *Bacillus cereus* is common in the environment but emetic toxin producing isolates are rare. Journal of Applied Microbiology 100: 7–14.

Bohrer, C.W., Denny, C.B., and Yao, M.G. (1973). *Thermal Destruction of Type E Clostridium botulinum. Final Report on RF 4603.* Washington DC: National Canners Association Research Foundation.

Campden BRI (1977). *Guidelines to the Establishment of Scheduled Heat Processes for Low-Acid Foods. Technical Manual No. 3.* Chipping Campden UK: Campden BRI.

Campden BRI (1992a). *The Microbiological Safety of Sous-Vide Processing. Technical Manual No. 39.* Chipping Campden UK: Campden BRI.

Campden BRI (1992b). *Pasteurisation Heat Treatments. Technical Manual No. 27.* Chipping Campden UK: Campden BRI.

Choma, C., Guinebretière, M.H., Carlin, F. et al. (2000). Prevalence, characterization and growth of *Bacillus cereus* in commercial cooked chilled foods containing vegetables. Journal of Applied Microbiology 88: 617–625.

Coles, R., McDowell, G., and Kirwan, M.J. (2003). *Food Packaging Technology, Chapter 7* (ed. R. Coles). Oxford: Blackwell Publishing Ltd.

DoH (1989). *Chilled and Frozen Guidelines on Cook-Chill and Cook-Freeze Catering Systems.* London: The Stationery Office.

DoH (1994). *Guidelines for the Safe Production of Heat Preserved Foods.* London: The Stationery Office.

ECFF (1996). *Guidelines for Good Hygienic Practice in the Manufacture of Chilled Foods.* Kettering, Northants, UK: European Chilled Food Federation.

EFSA (2005). Opinion of the scientific panel on biological hazards on *Bacillus cereus* and other *Bacillus* spp in foodstuffs. The EFSA Journal 175: 1–48.

Etoa, F.X. and Adegoke, G.O. (1996). Evolution of thermal resistance of spores of two *Bacillus* species in pepper and honey. Sciences des Aliments 16: 49–59.

FDA (2007). *Guidance for Industry Refrigerated Carrot Juice and Other Refrigerated Low-Acid Juices*. College Park: Office of Plant and Dairy Foods http://www.cfsan.fda.gov/guidance.html.

FSA (2008) *Food Standards Agency Guidance on the Safety and Shelf-life of Vacuum and Modified Atmosphere Packed Chilled Foods with Respect to Non-proteolytic* Clostridium botulinum. FSA website. www.food.gov.uk

FSA (2017) *Guidance on the safety and shelf-life of vacuum and modified atmosphere packed chilled foods with respect to non-proteolytic Clostridium botulinum*. www.food.gov.uk

Gaze, J.E., Brown, C.D., and Gaskell, D.E. (1991). *Growth of Clostridium botulinum Non-Proteolytic Type B and Type E in 'Sous Vide' Products Stored at 2–15°. Campden BRI Technical Memorandum No. 635*. Chipping Campden UK: Campden BRI.

Goepfert, J.M., Spira, W.M., and Kim, H.U. (1972). *Journal of Milk and Food Technology*, **35**, 213–227. Cited in. In: *Microbial Survival in the Environment* (eds. E. Mitscherlich and E.H. Marth). Berlin-Heidelberg-New York-Tokyo: Springer-Verlag.

Guinebretiere, M.H., Girardin, H., Dargaignaratz, C. et al. (2003). Contamination flows of *Bacillus cereus* and spore-forming aerobic bacteria in a cooked, pasteurised and chilled zucchini purée processing line. International Journal of Food Microbiology 82: 223–232.

IFTPS (1995). Protocol for carrying out heat penetration studies. www.iftps.org

Johnson, K.M., Nelson, C.L., and Busta, F.F. (1982). Germination and heat resistance of *Bacillus cereus* spores from strains associated with diarrhoeal and emetic foodborne illnesses. Journal of Food Science 47: 1268–1271.

Kamat, A.S., Nerkar, D.P., and Nair, P.M. (1989). *Bacillus cereus* in some Indian foods, incidence and antibiotic, heat and radiation resistance. Journal of Food Safety 10: 31–41.

Lindström, M., Kiviniemi, K., and Korkeala, H. (2006). Hazard and control of group II (non-proteolytic) *Clostridium botulinum* in modern food processing. International Journal of Food Microbiology 108: 92–104.

Lynt, R.K., Kautter, D.A., and Solomon, H.M. (1979). Heat resistance of non-proteolytic *Clostridium botulinum* type F in phosphate buffer and crabmeat. Journal of Food Science 44: 108.

Marth, E.H. (1998). Extended shelf-life refrigerated foods: microbiological quality and safety. Food Technology 52 (2): 57–62.

Membre, J.M., Amézquita, A., Bassett, J. et al. (2006). A probabilistic modeling approach in thermal inactivation: estimation of postprocess *Bacillus cereus* spore prevalence and concentration. Journal of Food Protection 69 (1): 118–129.

Molin, G. and Smygg, B.G. (1967) Quoted from J.M. Goepfert, W.M. Spira and H.U. Kim (eds) (1972). *Journal of Milk and Food Technology*, 35, 213–227. Cited in: E. Mitscherlich and E.H. Marth (eds), *Microbial Survival in the Environment*, Springer-Verlag, Berlin-Heidelberg-New York-Tokyo.

Mortimore, S. and Emond, S. (2008). Principles of HACCP: The importance of HACCP systems in food manufacturing. In: *Food Biodeterioration and Preservation, Chapter 2* (ed. G.S. Tucker). Oxford: Blackwell Publishing Ltd.

Peck, M.W. (2006). *Clostridium botulinum* and the safety of minimally heated, chilled foods: an emerging issue? Journal of Applied Microbiology 101: 556–570.

Peck, M.W., Fairbairn, D.A., and Lund, B.M. (1993). Heat-resistance of spores of non-proteolytic *Clostridium botulinum* estimated on medium containing lysozyme. Letters in Applied Microbiology 16: 126–131.

Peterson, M.E., Pelroy, G.A., Poysky, F.T. et al. (1997). Heat-pasteurization process for inactivation of non-proteolytic types of *Clostridium botulinum* in pickled Dungeness crabmeat. Journal of Food Protection 60: 928–934.

Sarrías, J.A., Valero, M., and Salmerón, M.C. (2002). Enumeration, isolation and characterisation of *Bacillus cereus* strains from Spanish raw rice. Food Microbiology 19: 589–595.

Tucker, G.S. (ed.) (2008). *Food Biodeterioration and Preservation.* Oxford: Blackwell Publishing Ltd.

Tucker, G.S., Shaw, H., and Gaze, J. (2008). *Exploring the Potential to Reduce Heat Processes for Refrigerated Extended Life Products. R&D Report 286.* Chipping Campden UK: Campden BRI.

8 Processing Systems

8.1 IN-PACK PROCESSING: RETORT SYSTEMS

In the traditional in-pack process, the packs are filled and hermetically sealed before being thermally processed in a retort. Care must be taken to ensure that the heat penetrates to the slowest heating point in the can, so that no part of the food is left under-processed. A metal can is the ideal package from a processor's perspective because, relative to other packaging media, it offers high production speeds, pack size flexibility, and high compression strength to withstand physical abuse during processing and distribution. However, many of the recent retorts have arisen because of the desire to process more sensitive pack types. After heating, the food needs to be cooled, and it is vital that no post-process contamination occurs through the package seals or seams.

Batch retorts (Fig. 8.1) operate with a variety of heating media, which includes condensing steam, mixtures of steam and air, water immersion, or water droplets either sprayed or rained onto the packs (Britt 2008; May 2000). These offer considerable flexibility for many combinations of food type and package. By applying an air overpressure, above that of the saturated steam pressure, the package shape can be maintained through the process so that stress on the seals is reduced. This allows delicate packages such as pouches and trays to be processed, for glass jars it prevents the lids from being forced off, and it ensures that plastic packs retain their shape and size.

The baskets (or crates) within a batch retort can be rotated in order to induce mixing inside the food by end-over-end agitation of the packs. This increases the rate of heat transfer to the thermal

Essentials of Thermal Processing, Second Edition. Gary Tucker and Susan Featherstone.

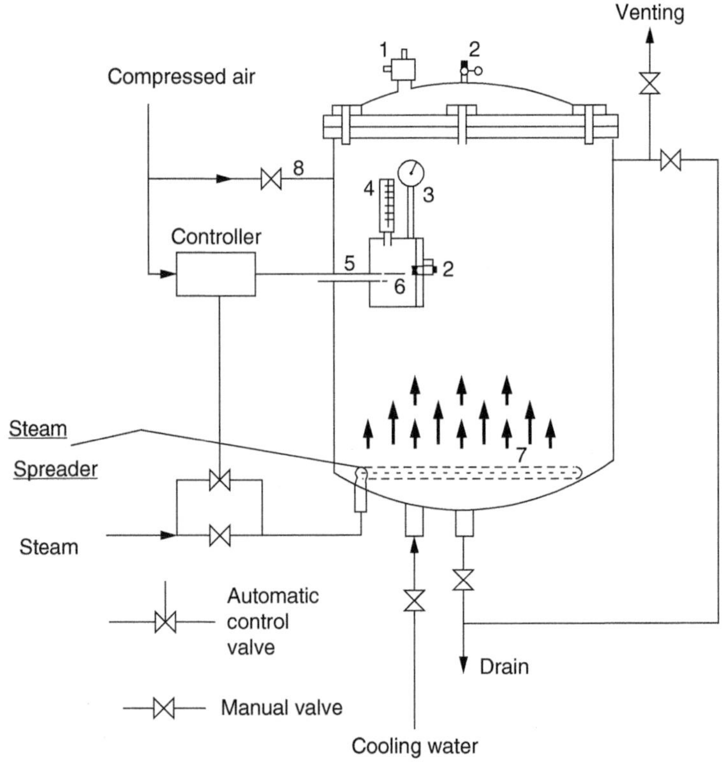

Fig. 8.1 Batch condensing steam retort.

centre (i.e. slowest heating point) of the pack. Typical rotation speeds can vary between 2 and 30 rpm, depending on the strength of the pack and the convective nature of the food inside the pack. For example, a plastic pouch containing rice would be rotated slowly (e.g. 2–5 rpm) so that the delicate pack and its contents are not damaged; however, the rotation is sufficient to reduce the process times to an extent that economic gains are made and measurable quality benefits are achieved.

8.1.1 Condensing steam retorts

Steam retorts are historically the method of choice for processing foods packaged in metals cans (Fig. 8.1). They were the system used when the canning industry was established over 100 years ago. These retorts are vented at start-up to eliminate air pockets in the retort that can result in low temperature regions and reduced heat transfer. Very high surface heat transfer coefficients associated with the condensing steam process ensure uniform heating throughout the vessel.

Nowadays, most companies in Europe and the USA have replaced their steam retorts with a retort type that can apply an overpressure. This enables a wider variety of container types to be processed.

8.1.2 Crateless retorts

These consist of a single can conveyor and a bank of batch processing vessels (Fig. 8.2) that are arranged so the flow of filled cans is almost continuous. At the start of an individual process, the retort is initially filled with water to cushion the cans as they enter the retort directly from the conveyor. When an individual retort is full, as determined by a can counter, the conveyor diverts the continuous flow of cans to the next retort in the bank. Simultaneously, the door is closed on the filled retort, a bottom vent is opened, and steam enters the vessel through a circular distribution spreader at the top of the retort, rapidly evacuating the remaining cushion water together with any air present in the retort. Cans are then heated in condensing steam for the design process time. At the end of heating, the retort is flooded with cooling water, and the product is pressure cooled for a specified length of time. At the end of the pressure cooling step, a door located in the bottom of the vessel is opened, and cans are discharged onto a conveyor located in a cooling canal. The processed cans move through the cooling canal and then on to the labelling area.

The concept is an interesting one but with some drawbacks. On the positive side, energy efficiency is improved over the traditional steam retort by eliminating the need for a venting time. Also, the

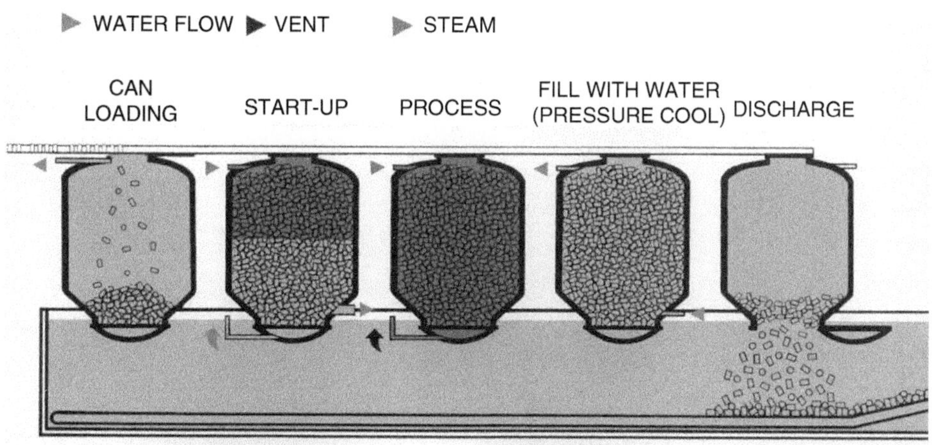

Fig. 8.2 Crateless retort showing the sequence in operating a bank of retorts.

flow of cans through a bank of crateless retorts can be almost continuous and therefore is efficient. However, the method of filling the retorts by dropping cans into water is known to lead to a small level of can damage. Orientation of the cans in a crateless retort is totally random, as the falling cans settle into a random packing pattern. This results in cans being processed in different orientations and nesting of shallow profile cans, which reduces the heat penetration effectiveness.

Most crateless retorting systems have been removed over the last 20 years and replaced with modern overpressure retorts. The following section describes some of the most important designs for overpressure retorts.

8.1.3 Water immersion retorts

One of the first types of water immersion retort was the Stock Rotomat (now Satori Stocktec GmbH, Germany), as shown in Fig. 8.3. During a retort cycle, processing water is preheated in the upper vessel and then released at the start of the process to fill the processing or lower vessel. This reduces the come-up time that would otherwise be lengthy in heating a large volume of water to

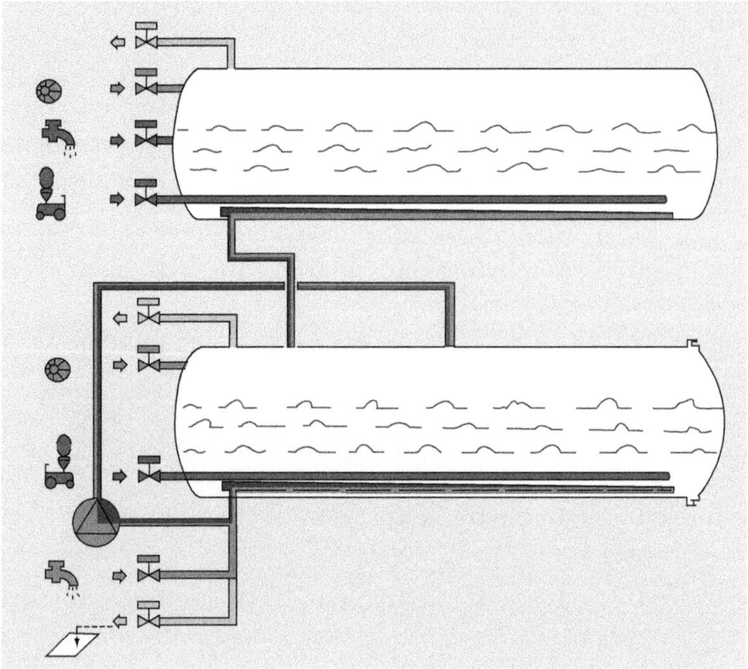

Fig. 8.3 Water immersion retort showing the two vessels. Courtesy of Holmach Ltd.

processing temperature. When the containers are fully immersed, water is pumped from the base of the lower vessel through an external steam injection heat exchanger and back into the top of the same vessel. At the end of the hold period, cold water is injected into the top of the processing vessel while hot water is pumped from the base of the processing vessel back into the upper vessel to be reheated and used in the next cycle. Overpressure is established by introducing steam or compressed air to the retort during processing.

The major disadvantage of the water immersion system is the need to operate with two vessels, which makes this system considerably more expensive than the alternatives. As a result, their popularity has gone down over the last 15 years. One distinct advantage of immersion is the buoyancy provided by the water that allows glass jars to be processed using quite high end-over-end rotation speeds (e.g. up to 20 rpm). This can result in high-quality sauce products where product movement is used to increase heat penetration rates.

8.1.4 Water spray and cascade

These retorts operate by circulating water from the base of the retort through an external heat exchanger and then distributing it inside the retort, either through spray nozzles (Fig. 8.4) or as a cascade generated by flow through a perforated plate (Fig. 8.5).

Fig. 8.4 SuperAgi TM concept for a water spray retort. Courtesy of JBT FoodTech, formerly known as FMC FoodTech.

Fig. 8.5 Water cascade retort showing the powerful water shower.

The resulting spray or cascade of heated water is then directed through the load to heat individual containers. Overpressure is established by introducing steam or compressed air to the retort during processing. Several designs are in commercial use.

Barriquand (now Steriflow SAS) pioneered the Steriflow system, in which a perforated plate provides a powerful cascade of droplets over the containers. This system has been used for cylindrical metal cans and allows cans with easy-open ends to be processed by operating with an overpressure towards the end of the heating phase. Heating and cooling of the process water uses a tubular heat exchanger. Several incidents of canned food spoilage are known to have resulted from leaks in the heat exchanger that allowed untreated cooling water to contaminate the sterilised process water. This is one point of weakness of this system. Another is the introduction of top-up water into the retort during cooling. Most of the older retorts have since been modified to prevent cooling water being introduced during cooling.

FMC (now JBT FoodTech) developed the Surdry sprayed water system that uses a series of spray nozzles to direct water into the crates of food containers. This is a recently introduced retort but has quickly gained popularity because of its flexibility and excellent temperature distribution. The latest version has a ring of spray nozzles that rotates (SuperAgi retort) around the food containers so that the temperature distribution is improved further.

8.1.5 Steam/air retorts

These retorts utilise condensing steam to supply the heat energy for heating and compressed air for overpressure (Fig. 8.6). Processes start with a short vent cycle to provide a large initial flow of steam into the retort and to vent out most of the air. After venting, steam and air are added independently to obtain the desired operating conditions. A high-speed circulation fan pulls the steam/air mixture through the load and then forces it to the back along the length of the retort, either through an annular space between the rotating cage and the retort shell, or through side plenum chambers in a static system.

At the end of the hold period, the cooling cycle commences with a pre-cool when cooling water is sprayed slowly into the circulating steam/air mixture to collapse the steam in a controlled manner. Air is also added to maintain the retort pressure and ensure package integrity. After the pre-cool, the circulation fan is stopped and water accumulated in the base of the retort is pumped through an external heat exchanger to cool the water. The circulation pump returns the cooled water back into the processing vessel via a row of spray nozzles at the top of the retort. In rotary processes, rotation commences with steam-on and continues through the completion of the cooling cycle. With no buoyancy forces to protect the containers from damage, rotation speeds are limited to a maximum of around 15 rpm.

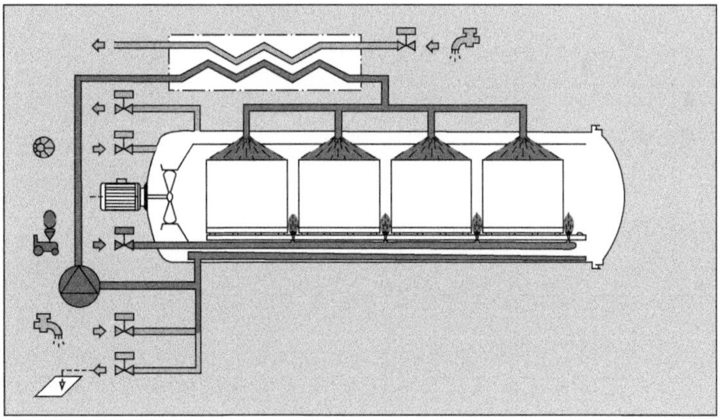

Fig. 8.6 Steam/air retort. Courtesy of Holmach Ltd.

8.1.6 Shaka retorts

This is a unique concept and a recent addition to the options with batch retorts. Its point of difference is that it uses high frequency longitudinal agitation instead of end-over-end or axial rotation.

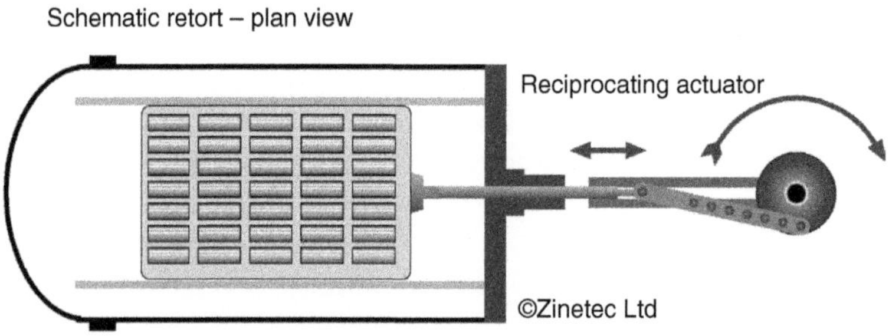

Fig. 8.7 Shaka retort principle showing the reciprocating basket movement. © Zinetic Ltd.

This is the mode of agitation that an individual would use if asked to agitate a cylindrical metal can of food product. The principle is that the headspace will be forced through the liquid or semi-solid food during one lateral agitation cycle. Reductions in process time are claimed to be significant, which are highlighted by heating factors (f_h values) that drop from 30 to 40 minutes in a static process to 2–3 minutes in a Shaka process (for a standard can, 73×115 mm). This has benefits in reducing energy use and process times, and in doing so increasing production efficiency. Figure 8.7 shows the Shaka retort principle.

All the above retorts operate in a batch mode in which the containers are loaded into crates that are loaded into the retort. A more efficient means of operating a sterilisation process is to continuously load the system. Continuous retorting systems come in two main types: reel and spiral and hydrostatic. Both uses the ability of the metal can to roll along a pathway as the means of propelling the cans into the system.

8.1.7 Reel and spiral retorts

In these systems, cans enter and exit the processing vessel through mechanical pressure locks (Fig. 8.8). Once in the vessel, cans move through a spiral track mounted on a reel that is rotating inside a horizontal cylindrical shell. In one revolution of the reel, cans roll by gravity along the bottom part of the arc (⊠90–120°), which provides most of the product mixing within the can. This is known as fast axial rotation (FAR) and delivers a very rapid rate of heat penetration to the can centre. The cans are essentially static as they pass through the remainder of the arc (⊠240–270°).

Systems are usually configured as a number of vessels with different functions connected in series. For example, a pre-heat shell, heating shell and cooling shell could make up a complete

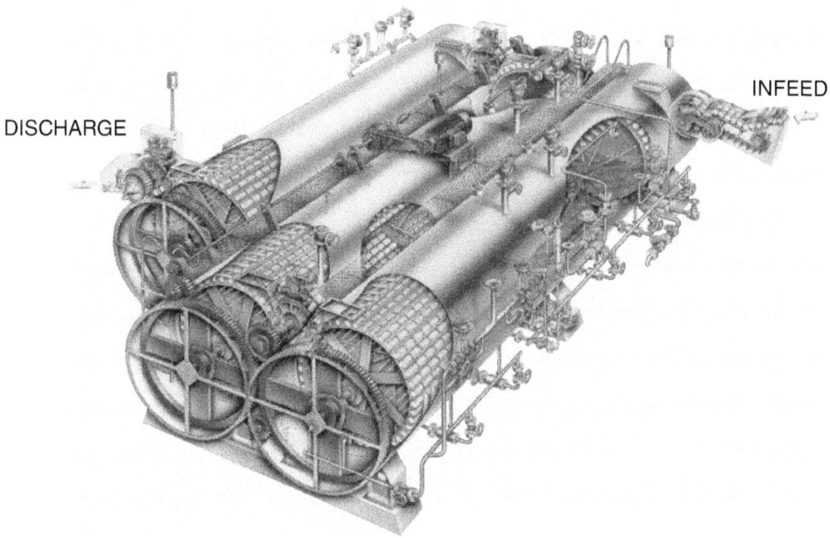

Fig. 8.8 Reel and spiral retort. Courtesy of JBT FoodTech.

system. Soups, sauces, and foods that can move within the can are processed in reel and spiral cooker–coolers. These will benefit from agitation. It is estimated that almost 50% of the worldwide production of foods in metal cans is produced using reel & spiral retorts.

8.1.8 Hydrostatic retorts

A hydrostatic retort (Fig. 8.9) does not invoke such dramatic rotation but instead carries the cans on carrier bars through various chambers. The only rotations are half-turns as the cans move between the chambers. It uses the pressure generated by the height of water in the entry and exit legs to create a pressure in the heating dome equal to the saturated steam pressure at the processing temperature. During processing, containers undergo a pre-heat treatment as they move through the entry leg, are exposed to condensing steam in the heating dome, are cooled as they move through the exit leg, through a cooling water spray, and are then discharged. Typical chamber temperatures are, for example, pre-heat at 80–90 °C, sterilisation at 120–130 °C, pre-cool at 80–90 °C, and final cooling at 40 °C.

The length and speed of the conveyor or chain carrying the containers determine the processing time in the steam dome. Because of their large size, venting of these retorts is time and energy intensive. This, combined with a high capital investment,

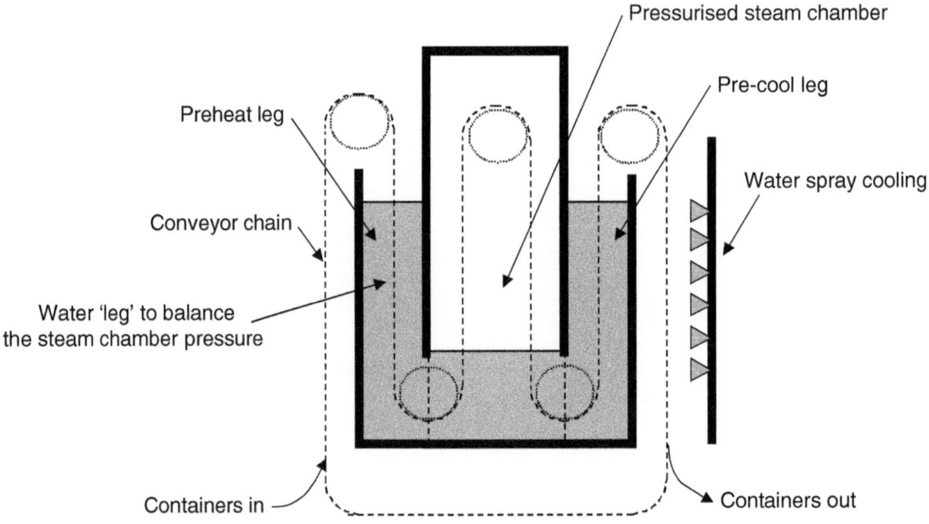

Fig. 8.9 Hydrostatic retort showing the movement of cans through different chambers.

requires that these retorts generally run 24 hours per day, 7 days per week. Hydrostats are used for high viscosity foods where rotational forces cannot be utilised to increase heat transfer rates, for example in solid petfoods and meat products.

8.2 IN-LINE PROCESSING: HEAT EXCHANGERS

In-line processing of liquid foods, as compared to in-container processing, offers many economic advantages (Emond 2000; Tucker and Bolmstedt 1999). It is claimed that by minimising the exposure of food to the adverse effects of high temperatures and long processing times, commercial benefits can be realised in improved product quality, increased safety, and increased plant throughput. A classic example of this is with UHT milk, where an in-pack sterilisation process usually results in excessive browning reactions taking place and an undesirable product. By operating an HTST process (high temperature, short time), it is possible to reduce these browning effects and produce milk that has a more acceptable appearance and taste to the consumer (Dairy Processing Handbook 2003). A heat exchanger such as a plate pack or narrow bore tube will be the core of this process and will deliver a rapid rise and fall in milk temperature. The heat exchanger part of the HTST process for milk is not now a difficult technical challenge, with the exception of minimising the fouling that builds up over the runtime and reduces heat transfer efficiency.

A challenge in the design of heat transfer equipment for the liquid food industry are the so-called prepared food products, such as tomato products, soups and sauces, and dessert products. These products are normally of high viscosity and of complex composition. In each case, the optimal heat processing equipment has to be chosen in order to retain particulate integrity, flavour, and colour of the end product. There is not one heat exchanger that will process all of these types of products from fibrous fruit juices to soups with particulates. These are the product types for which it is necessary to understand how their flow behaviour interacts with the heat exchanger choices.

A typical continuous processing line consists of preparation modules, the actual heat exchangers, and a filling machine. The preparation modules are used mainly for formulated food products, for example, vanilla puddings and salsa sauces, where the product is prepared and exits this section ready to be pasteurised or sterilised. The processing modules comprise the heat exchangers where the product flow is counter-current to the heating or cooling media. Filling systems will either operate hot or cold, or require a balance tank or aseptic storage tank that acts as the buffer between the processing and filling modules.

With respect to flow behaviour or rheology, most of the formulated products are typically non-Newtonian, showing in many cases quite extraordinary behaviour. Few foods flow as Newtonian liquids, apart from milk and thin fruit juices. Most food products are shear-thinning and some also display time-dependent properties caused by the complex macromolecular structures that are used in their formulation. These properties must be considered when designing a processing system, in order that excessive damage is avoided to the delicate structures. Depending on the rheological properties of the product and the possible presence of particulates, the design and choice of equipment can vary significantly from case to case. Various additives, for example thickeners and stabilisers, often also change the physical and rheological properties of the product. The next section provides a brief introduction to some of the complexities with flow behaviour of foods.

8.2.1 Flow behaviour

In the design of heat exchangers and choice of heat exchanger configurations, the flow behaviour of the product has to be taken into consideration (Barnes 2000). The flow behaviour will affect, for instance, the residence time distribution in the pipework and hence the design of heat exchangers and holding cells to obtain the sufficient thermal treatment. There are two types of flow that will be experienced by a flowing food; these are laminar and turbulent, but with a transitional region between the two types as the flow changes from one to the other. It is important to know which flow regime is present in all parts of a continuous process so the residence times can be calculated correctly.

The basic difference between laminar (streamline) and turbulent flow is well known, as is the effect on the velocity profile from heating or cooling of the product. For example, the maximum velocity in laminar flow, originating from the parabolic velocity profile, is theoretically twice the mean velocity, and in turbulent flow, it is around 20% higher than the mean velocity (Fig. 8.10). Laminar flow is assumed to occur up to Reynolds numbers of around 2100, whereas turbulent flow occurs at greater than 10000 (Equation (8.1)). The region in between 2100 and 10000 is referred to as the transitional region, because the flow regime is changing from laminar to turbulent. This is a region that equipment designers will try to avoid because of the uncertainties in flow behaviour and the key relationship between fastest and mean velocity. Should a heat exchanger be operated under transitional flow, then it would be safe to assume that the fastest liquid along the pipe centre could be travelling twice as fast as the mean velocity. The US FDA takes a Reynolds number of 4000 as the division between laminar and turbulent flow and assumes there is no transitional region. This is a simplification of what happens in practice, but it does make process calculations much clearer.

For viscous products, the flow conditions are nearly always laminar. A tomato paste steriliser, for instance, operates at Reynolds numbers (Re) around 1 (see Worked example 8.1). For milk and juice products, flow conditions are almost always turbulent (see Worked example 8.2):

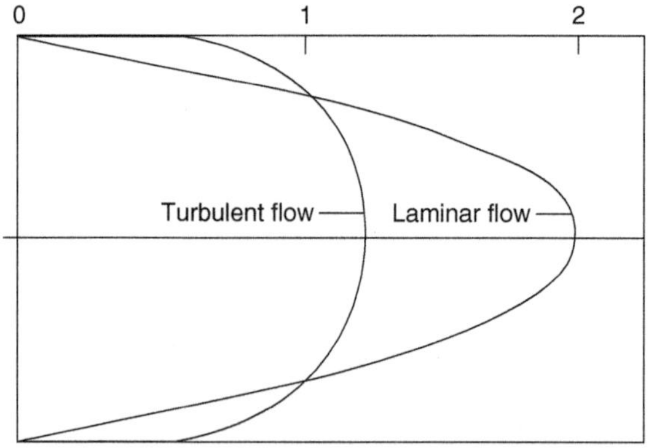

Fig. 8.10 Velocity profiles for turbulent and laminar (streamline) flow showing the ratio of maximum velocity to average velocity.

$$\text{Re} = \frac{d_h \times \rho \times \text{v}}{\mu} \quad (8.1)$$

where d_h is hydraulic diameter of the processing system, m, μ is dynamic viscosity of liquid, Pa s, ρ is density of liquid, kg m^{-3}, and v is mean velocity in the processing system, m s^{-1}.

Worked example 8.1 Reynolds number for a tomato ketchup

Density = 1000 kg m^{-3}
Velocity in tube = 0.8 m s^{-1}
Tube diameter = 0.075 m
Viscosity = 760 cP (0.76 Pa.s)

$$\text{Re} = \frac{0.075 \times 1000 \times 0.8}{0.76} = 78.9$$

Flow conditions are laminar.
Assume maximum centreline velocity is 2x the mean velocity in the holding tube.

Worked example 8.2 Reynolds number for a fruit juice

Density = 1000 kg m^{-3}
Velocity in tube = 1.2 m s^{-1}
Tube diameter = 0.075 m
Viscosity = 5 cP (0.005 Pa s)

$$\text{Re} = \frac{0.075 \times 1000 \times 1.2}{0.005} = 18000$$

Flow conditions are turbulent.
Assume maximum centreline velocity is 1.2x the mean velocity in the holding tube.

Equation (8.1) is for Newtonian liquids that have a viscosity independent of shear rate. Most, if not all, formulated foods contain macromolecular thickening agents in which the viscosity depends on the shear applied in the processing system. These non-Newtonian foods are certain to flow under laminar conditions because of their high viscosity values, and they also display velocity profiles that are still more complex. As the degree of non-Newtonian behaviour increases, the velocity profile increases in flatness across the pipe cross section. This means in practice that the maximum velocity decreases from its Newtonian value of twice the mean velocity. However, there are few commercial operations that do not apply the factor 2 when calculating holding tube length, irrespective of the measured flow behaviour index.

Further complications to the velocity profile arise when additives, for example xanthan or gellan gum, are used. These have viscoelastic properties and are beneficial for enhancing the particulate carrying properties of a sauce. The so-called yield value, normally a measure of the product's willingness to flow by itself, for example from a storage tank or bottle of ketchup, is also a measure of the particulate carrying abilities. A significant yield value, typical of paste-like products, also adds to the flatness of the velocity profile and hence further increases the deviation from the parabolic shape.

The above descriptions of flow behaviour and Reynolds number are very much a simplification of what is a highly complex and fascinating subject. However, they serve as an introduction to the subject and are the minimum information required for designing a holding tube in a continuous flow process.

8.2.2 Choice of heat exchanger

The choice of optimal heat exchanger depends to a great deal on the flow conditions. Fluids with low viscosities and no particulates are preferably treated in a plate heat exchanger (Fig. 8.11). This should be the first choice of exchanger because of its low cost and high heat transfer rate. For fruit juices with pulp and fibres of up to 5 mm length, special types of plates are available with more open channels that allow the fibres to pass through unhindered. In addition, even with high viscosity foods, the plate heat exchanger can be utilised as long as the pressures developed are not too high and the rheological behaviour does not indicate that a yield stress is present. Foods that contain yield stresses can experience maldistribution of flow between the plate gaps, and in some instances, this can lead to flow stagnation and blockages.

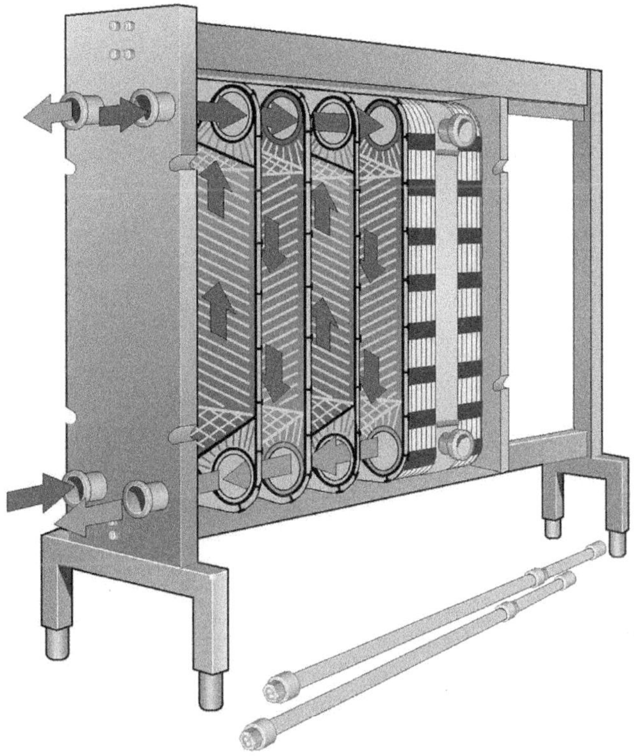

Fig. 8.11 Plate type heat exchanger. Courtesy of Tetra Pak.

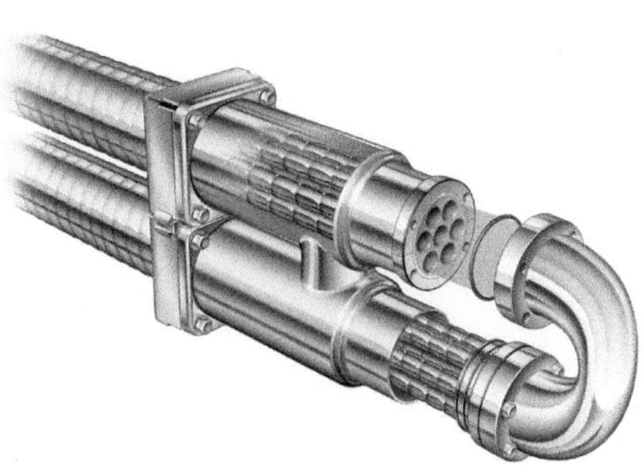

Fig. 8.12 Multi-tube type tubular heat exchanger. Courtesy of Tetra Pak.

For fruit juices with fibres of up to 15 mm length and for relatively water-like foods, a multi-tube tubular heat exchanger is preferably used (Fig. 8.12). Also, fluids of moderate to high viscosity with only small particulates (<5 mm) will flow through a multi-tube heat exchanger without problems. Despite being less thermally efficient than a plate heat exchanger, multi-tubes can be configured in various options of the number of tubes in parallel and of tube diameter. This gives them a high degree of flexibility.

A common application for the multi-tube exchanger is milk sterilisation, despite the heat transfer disadvantages of tubes compared with plates. Tube bundles, containing large numbers of small diameter tubes in parallel, offer adequate thermal efficiency, with the benefit of cleanability without dismantling. A plate heat exchanger would need to be taken apart to remove the fouling deposits that build-up over several hours of processing. This increases the downtime and increases the chances of contamination being introduced with poorly fitted gaskets.

Most tubular heat exchangers now use corrugations on the shell and tubes to enhance heat transfer with the heating and cooling media, typically water. Design of the corrugations will be specific to the company supplying the heat exchanger but they will each create the same effect, which is to generate turbulence in the water flows. This reduces the resistance to heat transfer caused by boundary layers that can be set up adjacent to the tube walls. In effect, it ensures that the media, whether it is heating water or cooling water, does not restrict the heat transfer performance of the exchanger. The limit to performance is therefore within the tubes.

The length of most commercial tubular systems has been standardised at 6 m; therefore, tubular heat exchangers are long and thin in terms of their geometry. This makes them suitable for placement next to factory walls or even above head height. The need for access to the tubes is rare because the tubes themselves are designed to be cleaned in place and access will only be required to the connecting pipes and equipment. These can be positioned at ground level where access is easy.

One of the greatest challenges to heat exchanger design is when the food fluids are significantly viscoelastic, i.e. exhibit a large yield value, often in combination with a high viscosity that is shear dependent. For these fluids, there is a risk of mal-distribution across the inner tubes of a multi-tube heat exchanger. In the worst case, the product flow will stop in some of the tubes causing overcooking of parts of the product and also cleaning problems where the food burns onto the inner tube surfaces. Examples of such products are hot break tomato pastes or stiff dessert puddings, where the multi-tube is not suitable and concentric tubes are a preferred choice (Fig. 8.13).

Concentric tubes have only one product channel, which eliminates the risk of mal-distribution. Here, the product flows in

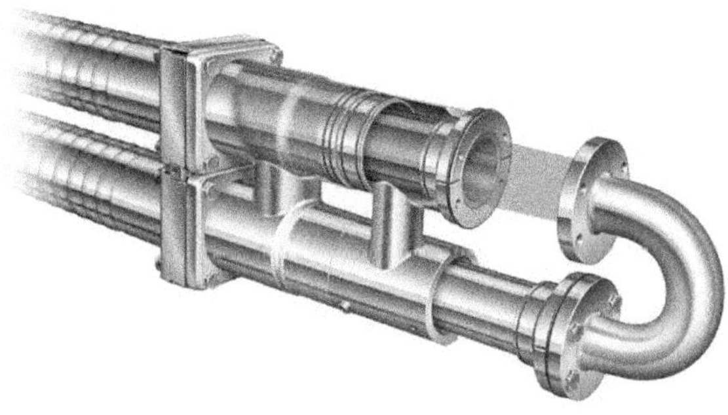

Fig. 8.13 Concentric channel type heat exchanger. Courtesy of Tetra Pak.

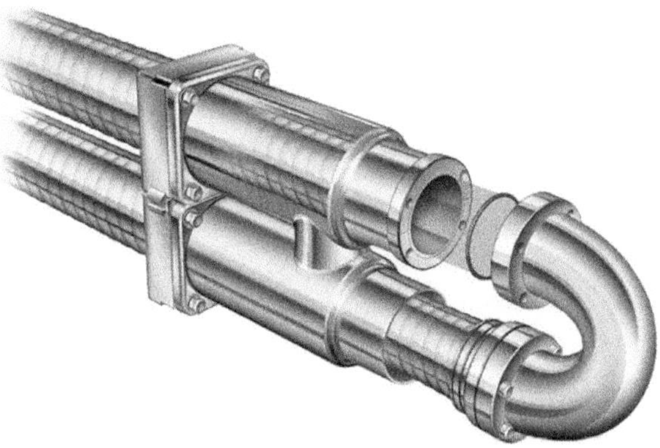

Fig. 8.14 Mono-tube type tubular heat exchanger. Courtesy of Tetra Pak.

a gap between two concentric tubes with the media on both sides, therefore increasing the heat transfer efficiency. Particulate products can be processed in wide gap modules, with particulate sizes up to 5–6 mm. Concentric tubes are a common choice of heat exchanger for tomato ketchup manufacture because this product possesses a yield stress that is an essential feature of the ketchup.

If large particulates are present in the food product, the mono-tube is probably the optimal choice (Fig. 8.14). The drawback with a mono-tube compared to a concentric tube or multi-tube is reduced thermal efficiency due to the thicker product layer. However, the particulates present in the product will to a great extent work as ‘internal mixers’ and will promote heat transfer. This makes design of mono-tube exchangers difficult, unless prior knowledge on the heat transfer behaviour of that food is available. The limiting design criterion for mono-tubes is the heat treatment given to the particulates, because of the need for heat to conduct into the particulate. The need for sufficient contact time between the carrier liquid and the particulates can be an advantage of the tubular concept, in which the food product has sufficient residence time in the exchanger to equilibrate towards a uniform temperature.

If none of the tubular types are suitable, a scraped surface heat exchanger is the last option (Fig. 8.15). In principle, a scraped surface heat exchanger is a mono-tube equipped with a rotating internal scraper. The scraper keeps the heating surface free from any deposits and also promotes turbulence. This type of heat exchanger is ideal for products of very high viscosity, possibly also containing large particulates, and especially for products that can foul the heated or cooled surfaces. Unlike tubes that usually operate with water as the media, scraped surface heat exchangers often use steam for greater heat transfer efficiency. There are drawbacks with scraped surface heat exchangers in that the costs are high for purchasing a system and ongoing maintenance is higher than with tubular heat exchangers because they have moving parts that wear.

One common application for scraped surface heat exchangers is where cooled surfaces, without the scraping action, are likely to foul quickly as high viscosity layers develop on the surfaces or hard fats deposit. These will:

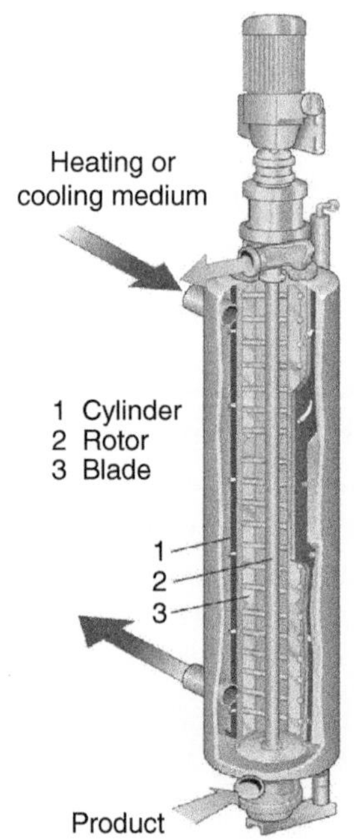

Fig. 8.15 Scraped surface heat exchanger. Courtesy of Tetra Pak.

- insulate the cold surfaces from the product to be cooled and so reduce thermal efficiency; and
- reduce the effective cross-sectional area for flow, which will increase the velocity and again reduce thermal efficiency.

Continuous removal of these layers is essential for this application.

There are a small number of novel designs with scraping actions that do not conform with the typical heat exchangers described above. For example, UNICUS® is a tubular heat exchanger with a lateral scraped surface action, as is the ViscoLine dynamic unit. Both have applications for products such as fruit and vegetable purées, pulps and concentrates, ketchup, dairy desserts, chocolate, and UHT milk. Attempts have also been made to construct scraped plate systems so that the exchangers can benefit from very high surface areas and scraped surfaces kept free of fouling. The OCTATOR from Nova 2K is one example.

8.2.3 Maximising product recovery

Food products have a high ingredient cost and so efforts are made to recover as much of the product as possible from the processing systems (Tucker and Shaw 2001). Pigging systems for tubular heat exchangers are in widespread use in the food industry because of the high costs and high volume of product held up in the tubes. In a food application for a cook-in sauce product, this can be up to several tonnes of product. The simplest pigging system operates using air to displace the product. Complex geometries can be pigged using air at high pressure; however, high viscosity materials often require pressures too high for air systems. There is also a significant risk that the air will core through the centre of the flow channels where the material viscosity in a cooling application is at its lowest.

Pigs are usually made of plastic- or plastic-coated spheres that are pushed through the tubes by water or air pressure. Clearance between the pig and tube surfaces is only a few millimetres. They can go around 90° bends but cannot deform and pass through the gaps experienced in plate or scraped surface heat exchangers. These exchangers require a different type of pigging system such as a cheap material with similar physical properties to the food, for example a gelled starch. A recent development in deformable pigging options is the ice pig. It has an advantage over a starch pig in that the ice moves through the system as a solid plug with the boundary between ice and surfaces lubricated by a thin water layer. Recovery of expensive product is likely to be greater with the ice pig compared with a starch pig.

8.3 NEW THERMAL TECHNOLOGIES

Food manufacturers are continually looking for new ways to produce food with enhanced flavour and nutritional characteristics. Traditional thermal processes using retorts or heat exchangers tend to reduce the vitamin content of food and can affect its texture, flavour, and appearance (Holdsworth 1997). There are many new and emerging technologies that make great claims for their beneficial effects on food quality; however, few have reached commercialisation. Of these, only ohmic heating, high pressure, and microwave processing have a role to play in niche thermal processing markets.

High pressure processing was originally considered in the 1890s, but it was not until the 1970s that Japanese food companies started to develop its commercial potential. Pressures of several thousand atmospheres (500–600 MPa) are used to kill microorganisms, but there is little evidence that high pressure is effective on spores or enzymes. Thus, chilled storage or high acidity is essential hurdles in preventing microbial growth. Jams were the first products to be produced in Japan, and the process is now being used commercially in Europe and the USA. Product types now include meats, fish, and shellfish, and heat sensitive sauces such as avocado dips, fruit juices, and smoothies.

High pressure sterilisation is still some way off, although there are patents that demonstrate possibilities for applying the technology to certain foods, including petfoods. The killing effect on *Clostridium botulinum* spores cannot be obtained with high pressure alone and so this effect will rely on thermal destruction. Most commercial equipment is batch, using vessels similar in appearance to medium-sized horizontal retorts.

Ohmic heating achieves its preservation action via thermal effects, but instead of applying external heat to a food as with in-pack or heat exchangers, an electric current is applied directly to the food. The electrical resistance of the food to the current causes it to heat it up in a similar way to a light bulb filament. The advantage is that much shorter heating times can be applied than would otherwise be possible, and so the food maintains more of its nutritional and flavour characteristics. The limitation is that ohmic cooling, or some other means of affecting rapid cooling, cannot be applied and so cooling relies on traditional methods that are slow in comparison with ohmic heating.

Foods containing large particulates are suited to ohmic heating because the electrical properties of the particulate and carrier liquid can be designed so that the particulate heats preferentially.

The only commercial ohmic heater in operation in the UK (at the time of writing) is used to pasteurise fruit preparations, in which good particle definition is a key requirement. Most commercial ohmic applications are for high acid foods to avoid the requirement for aseptic filling. The technology has yet to find its niche market, and the increase in applications has been slow. Nevertheless, for a high acid food with large particulates, ohmic heating confers significant quality benefits when compared with conventional heat exchangers. One application that has yet to be explored fully is the use of ohmic heating for achieving the fill temperatures prior to hot filling into jars, cans, or plastic packages. Hot filled products can be designed so they receive no further process and cooling occurs within the filled package. This is gentle on the particulates.

Microwave processing, like ohmic heating, destroys micro-organisms via thermal effects. Frequencies of 950 and 2450 Hz are used to excite polar molecules, which produces thermal energy and increases temperature. On the continent, a small number of microwave pasteurisation units are in operation, primarily manufacturing pasta products in transparent plastic trays. Heat generated by the microwaves pasteurises the food and the package together, and the products are sold under chilled storage to achieve extended shelf-lives. Benefits of rapid heating can result in improved quality for foods that are sensitive to thermal degradation. The technology has not received widespread adoption because of the high capital costs of the equipment and the wide distribution in temperatures across a package.

Microwave sterilisation has not developed much because of the need for air overpressure to maintain the shape of the flexible packages during processing. This creates complications with continuous systems in that transfer valves are required between the chambers.

References

Barnes, H. (2000). *A Handbook of Elementary Rheology*. Institute of Non-Newtonian Fluid Mechanics, University of Wales.

Britt, I.J. (2008). Thermal processing. In: *Food Biodeterioration and Preservation* (ed. G. Tucker). Oxford: Blackwell Publishing Ltd.

Dairy Processing Handbook (2003) 2e. Tetra Pak, Lund, Sweden.

Emond, S.P. (2000). Continuous heat processing. In: *Thermal Technologies in Food Processing* (ed. P. Richardson). Cambridge: Woodhead Publishing Ltd.

Holdsworth, S.D. (1997). *Thermal Processing of Packaged Foods*. London: Blackie Academic and Professional.

May, N.S. (2000). Retort technology. In: *Thermal Technologies in Food Processing* (ed. P. Richardson). Cambridge: Woodhead Publishing.

Tucker, G.S. and Bolmstedt, U. (1999). Gently does it. *Liquid Foods International* 3 (3): 15–16.

Tucker, G.S. and Shaw, G.H. (2001). *Heat Recovery in Tubular Heat Exchangers for Medium Viscosity Food Products. R&D Report No.140.* Campden BRI, Chipping Campden, UK.

9 Cook Values and Optimisation of Thermal Processes

Thermal processing of foods has a dual purpose; first to commercially sterilise the product so it is safe to eat and second to cook it to an acceptable level. Unlike sterilisation, which has a precise objective, the term cooking is defined in many ways. For example, it can refer to the softening of the product texture as with vegetables or meat, retention of specific nutrients or vitamins, or a breakdown in a colour pigment.

Thermal process times and temperatures are usually chosen so that a target F-value or P-value is achieved and often to achieve a fast throughput in the factory. However, there are many products that require the cooking criteria to be taken into account, for example white sauces that turn brown easily if overcooked or foods containing delicate particulates such as fruit pieces. The methods used to establish process times and temperatures require additional measurements and calculations.

This chapter addresses the challenges of optimising a specific quality attribute without negatively impacting on the microbial process safety. Tables of data are given on the different quality parameters appropriate to heat preserved foods and how these can be mathematically analysed. Examples are then presented to illustrate how process times and temperatures can be adjusted so that maximum quantities of a nutrient, such as a vitamin, are retained.

9.1 MATHEMATICAL ANALYSIS OF COOKING

Cooking processes can be considered in a similar mathematical way to microbiological destruction, by assuming they follow first-order reaction kinetics. This is not appropriate to all cooking reactions but sufficient reactions fall into this category that the analysis has merit.

Essentials of Thermal Processing, Second Edition. Gary Tucker and Susan Featherstone.

The first-order analysis becomes less accurate if chemical reactions occur during processing, for example when interactions take place between compounds naturally present in the foods. Reaction orders are then based on the concentration of the reacting chemical species, and this can result in more complex orders. These can still be treated mathematically but are more complicated than with first-order systems (see Chapter 4 for explanation of first-order kinetics). The analysis that follows is appropriate for cooking reactions that can be approximated by first-order kinetics.

Compared with microorganism destruction, cooking rates have a different temperature coefficient of reaction, or z-value. Often, the rate of change (z-value) with temperature for a cooking process, such as softening of vegetable texture, is in the range 25–35 °C. The rate of microbial kill in sterilisation processes increases 10-fold for every 10 °C, whereas the rate of change in many cooking processes is much lower over the same temperature change. For sterilisation, this means that very little microbial kill is taking place at low temperatures, but this changes drastically as the temperature rises; for every 10 °C rise, the rate of kill increases tenfold. For example, with a vegetable canning process, the rate of softening of the vegetables (e.g. z 30 °C) occurs over a wider temperature range; for every 30 °C rise, the rate of softening increases tenfold. Advantage is taken of this fact in the sterilisation of milk, which turns brown all too easily if heated for too long. In continuous flow heat exchangers, milk is processed at ultra-high temperatures for short times (e.g. 135–140 °C for 3–6 seconds) with minimal browning. Compare this with an in-pack process that has to operate at lower process temperatures (118–122 °C) for longer hold times (10–20 minutes), and the result is an end product with more browning.

Process optimisation is the task of judiciously selecting a process to give the desired degree of cooking but at the same time ensuring an adequate process value to take account of microbial destruction. There are many cooking attributes that can be used to optimise a process, and it depends on the specific food product in the choice of which attribute is most relevant. Some of those cooking reactions reported to show first-order kinetics are as follows:

- colour breakdown, e.g. reds and greens in canned vegetables;
- browning of white sauces and milk;
- meat, fish, and vegetable texturisation;
- development of off-odours during processing;
- vitamin or nutrient destruction;
- enzyme inactivation.

Mathematical analysis of cooking follows a similar mechanism to that for the destruction of microorganisms by heat, as given in Chapter 4. The principle equations that govern this analysis are described in the following section.

There is, however, a further and major difference in the way that cooking is assessed compared with sterilisation, or pasteurization. When establishing the process times and temperatures for a packaged food, whether in-pack or in-line processed, it is the process at the point of lowest lethality that is of interest. Efforts are made to understand where this position is located within the pack or holding tube and to place temperature or microbiological sensors at these locations. However, the consumer will eat the entire contents of a packaged food, and therefore, it is important to understand how the extent of cooking differs across the pack. This lends itself to calculations for average cook values as opposed to sterilisation or pasteurization values that are established at the point of lowest lethality. This will be revisited later in this chapter.

9.1.1 Cooking equations and kinetic data

The example equations presented here are for a sterilisation process, although the principles are equally applicable to pasteurisation. Thus, the symbols are for the unique sterilisation F_0-value for the destruction of mesophilic *Clostridium botulinum* spores.

The sterilisation value of a thermal process (T_{ref} = 121.1 °C, $z = 10$ °C) is given by the integrated form in Equation (9.1):

$$F_0 = \int_0^t 10^{\frac{(T-121.1)}{10}} dt \tag{9.1}$$

If the product temperature (T) is constant during the process, Equation (9.1) reduces to a simplified form of the equation [Equation (9.2)]. In thermal processing, the nearest example to this is found for continuous flow processing of low viscosity products such as milk or fruit juice. Heat transfer is very effective with thin liquids and so the liquid food increases in temperature very rapidly in the heat exchanger. Most of the lethal kill occurs during the time spent (t) by the product at a fixed temperature (T) in the holding tube. Equation (9.1) can be reduced down to Equation (9.2):

$$F_0 = t \times 10^{\left(\frac{T-121.1}{10}\right)} \tag{9.2}$$

In a similar way to the sterilisation calculation, the cook value (C-value) can also be represented by an equation for first-order reaction kinetics. This is given by Equation (9.3), in which the times and temperatures are the same as those used in Equation (9.2). The difference between Equations (9.2) and (9.3) is in the z-value for the cooking reaction rate, which is often higher when dealing with a quality attribute, for example colour degradation, texture, or vitamin retention. In Equation (9.3), a z-value of 33 °C is chosen to illustrate the differences, which is a value representative of the destruction of thiamine (Table 9.1):

$$C = t \times 10^{\left(\frac{T-121.1)}{33}\right)} \tag{9.3}$$

Tables 9.1–9.5 present a selection of data on cooking attributes that can be used in process optimization studies. Data in the Tables 9.1–9.3 were taken from review sources including Holdsworth (1997) and Ling et al. (2015). All reaction rates are approximated by first-order kinetics. The tables are not intended as an exhaustive list of cooking attributes, but instead to illustrate the range of kinetic data associated with cooking analysis. Decimal reduction times (D-values) are given in minutes, which is the conventional unit, and the kinetic factors (z-values) in degrees Celsius.

The D-values are given at the reference temperatures used in the original research papers (D_2 and T_2). Equation (9.4) can be used to convert the D-value to a different reference temperature (D_1 and T_1). It is essential when comparing D-values to use the same reference temperature.

$$\log(D_1) = \log(D_2) + \left(\frac{T_2 - T_1}{z}\right) \tag{9.4}$$

Table 9.1 lists the destruction of vitamins and shows two important characteristics; first, that the D-values are much higher than those for microorganism destruction (e.g. the *C. botulinum* $D_{121.1}$ value is 0.21 minutes) and, second, that the z-values are often higher (e.g. the *C. botulinum* z-value is 10 °C). The fact that vitamin D-values are high indicates that these vitamins are less sensitive to temperature change than commonly thought. For example, when fruit is cooked, the ascorbic acid or vitamin C content is reduced considerably, but this occurs through several mechanisms and its breakdown by heat is less significant than either leaching from the fruit into the cook-water or by oxidation with oxygen from air.

Table 9.1 Kinetic factors for vitamin destruction.

Heat sensitive vitamin	Temperature (°C)	D_T-value (min)	z-value (°C)
Vitamin A (beta carotene)			
Beef liver	103–127	$D_{122} = 40$	23.0
Carrot juice	104–132	$D_{104} = 393$	25.5
Vitamin B1 (thiamine)			
Buffer	109–150	$D_{109} = 158$	24.0
Carrots	109–150	$D_{150} = 13.8$	22.0
Spinach	109–150	$D_{150} = 101.7$	22.0
Pea puree	121.1	$D_{121.1} = 167$	31.3
Lamb puree	109–150	$D_{122} = 11.8$	25.0
Pork luncheon meat	100–127	$D_{127} = 105$	35.0
Vitamin B6 (pyridoxine)			
Cauliflower	106–138	$D_{121} = 400$	43.0
Pantothenic acid			
Beef puree pH 5.4	118–143	$D_{121.1} = 2{,}300$	35.8
Beef puree pH 7.0	118–143	$D_{121.1} = 2{,}250$	19.3
Folic acid			
Apple juice	100–140	$D_{140}=1{,}667$	31.0
Vitamin C (ascorbic acid)			
Peas	110–132	$D_{121.1} = 833$	18.0
Spinach	70–100	$D_{100} = 432$	74.4
Orange juice (ohmic heated)	100–125	$D_{90} = 15$	20.7
Orange juice (microwave heated)	100–125	$D_{100} = 46$	19.2
Orange juice (water bath heated)	100–125	$D_{100} = 13$	24.4
Cupuacu nectar	60–99	$D_{80} = 177$	36.7
Grapefruit juice (11.2 °Brix)	20–92	$D_{75} = 1{,}354$	49.3
Grapefruit juice (31.2 °Brix)	20–92	$D_{75} = 1{,}228$	45.0
Lime	20–92	$D_{75} = 1{,}186$	35.8
Lemon	20–92	$D_{75} = 949$	44.6
Tangerine	20–92	$D_{75} = 771$	46.5
Grapefruit	20–92	$D_{75} = 1{,}276$	36.5
Orange juice (12.5 °Brix)	70–98	$D_{75} = 24{,}110$	19.0
Orange juice (12.5 °Brix)	70–98	$D_{75} = 10{,}447$	24.9
Tomato juice	70–90	$D_{90} = 14.0$	57.9
Anthocyanin			
Raspberries	100–150	$D_{110} = 112$	37.4
Purple potato	100–150	$D_{120} = 32$	47.9
Carotenoids (β-carotene)			
Citrus juice	70–90	$D_{90} = 435$	22.9
Tomato juice	70–90	$D_{90} = 54.7$	40.0
Lycopene			
Tomato juice	70–90	$D_{90} = 52.5$	126.5

Sources: Holdsworth (1997) and Ling et al. (2015).

Table 9.2 Kinetic factors for enzyme destruction.

Heat sensitive enzyme	Temperature (°C)	D_T-value (min)	z-value (°C)
Peroxidase			
Horseradish	60–160	$D_{120} = 13.8$	27.8
Potato puree	100–140	$D_{120} = 1.2$	35.0
Catalase			
Spinach	60	$D_{60} = 1.0$	8.3
Lipoxygenase			
Pea/soya	50–80	$D_{77} = 12.0$	3.4
Pectinesterase			
Guava syrup pH 4.0	74–95	$D_{96} = 0.6$	16.5
Polyphenol oxidase			
Potato	80–110	$D_{89} = 16.7$	7.8

Sources: Holdsworth (1997) and Ling et al. (2015).

Table 9.2 shows that enzymes display a wider variation in both *D*- and *z*-value than for the vitamins in Table 9.1. Much of the research on measurement of these values was done for the frozen foods industry, in which blanching is a key process step. Blanching serves the main purpose of inactivating enzymes that would otherwise break the food down during the freezing process and, also, perhaps more importantly, during the time taken during preparation of the food by the consumer. Some enzymes, for example peroxidase from sources such as horseradish, are very heat stable and can cause issues with low levels of residual activity following mild pasteurisation processes. It is very unlikely that enzymes will remain active after sterilisation processes, with the possible exception of low water activity systems.

Table 9.3 contains data for three of the main pigments of relevance to thermally processed foods, and these are green, red, and brown. Green pigments are important for canned vegetables, where they readily change colour to an undesirable grey shade. For example, canned peas and beans need careful processing regimes, which includes some degree of pH control to avoid undesirable colour changes. Red pigments from berries give fruit juices a characteristic bright colour but can turn brown if the processing regime is too severe. For example, strawberries turn from their fresh bright red colour towards a brown colour when canned, as well as undergoing an undesirable structure breakdown. Milk products are known to turn brown easily when heated. One of the most studied colour reactions is that of browning in dairy products (Burton 1955; Kessler and Fink 1986). These studies have defined the UHT sector in producing sterilised milk products of a higher

Table 9.3 Kinetic factors for pigment destruction.

Heat sensitive pigment	Temperature (°C)	D_T-value (min)	z-value (°C)
Green (chlorophylls)			
Green beans	80–148	$D_{121.1} = 21$	38.8
Peas	80–148	$D_{121.1} = 25$	39.4
Red			
Raspberry juice	78–108	$D_{108} = 117$	30.4
Grapes	76.7–121	$D_{121} = 127$	54.7
Browning reactions			
Chestnut paste darkening	105–128	$D_{121.1} = 2\,350$	24.6
Milk, hydromethyl furfural	105–160	$D_{130} = 0.2$	26.7

Sources: Holdsworth (1997) and Ling et al. (2015).

Table 9.4 Kinetic factors for colour changes.

Heat sensitive pigment	Temperature (°C)	D_T-value (min)	z-value (°C)	Reference
Tomato paste				
a		$D_{85} = 245$	62.4	Barreiro et al. (1997)
b		$D_{85} = 1\,735$	29.8	
Peach puree				
L		$D_{122.5} = 794$	28.0	Avila and Silva (1999)
b		$D_{122.5} = 576$	27.5	
Asparagus				
b		$D_{85} = 3500.2$	45.2	Lau et al. (2000)
Pineapple puree				
L		$D_{110} = 1\,097$	21.7	Chutintrasni and Noomhorn (2007)
b		$D_{110} = 378$	25.7	
Guava (seedless)				
L		$D_{85} = 26$	20.0	Ganjloo et al. (2011)
a		$D_{85} = 378$	27.7	

quality than can be produced by in-pack methods. Milk and dairy products are treated at ultra-high temperatures (135–140 °C) for short times (3–6 seconds) so the sterilisation occurs in an instant but cooking reactions occur at a reduced rate.

Table 9.4 outlines further colour change data taken using instrumental colour measurements. This contrasts with the data in Table 9.3 that were derived from sensory panel scores for the colour changes before and after processing. Instrumental measurements used the Hunter L, a, b scale where 'L' is a measure of brightness (0–50 is dark, 51–100 is light), 'a' represents the red/green split (+ is green, – is red), and 'b' represents the yellow/blue

Table 9.5 Kinetic factors for texture changes.

Textural parameter	Temperature (°C)	D_T-value (min)	z-value (°C)	Reference
Hardness				
Green beans	90–122	D_{100} = 295.2	29.0	Van Loey et al. (1995)
Peas	90-122	D_{100} = 300.0	30.2	
Pink Salmon				
Tenderising	100-131.1	D_{111} = 121	28.1	Kong et al. (2007)
Toughening	100-131.1	D_{121} = 1 001	39.1	

split (+ is yellow, – is blue). The colour of the fruit or vegetable dictates which L, a, or b measurement is appropriate.

Table 9.5 presents data for hardness of vegetables and tenderization and toughening of fish. Vegetables and fish are popular canned products that both suffer from textural changes caused by the thermal process. The *z*-values for softening, the reverse of hardness, of either fish or vegetables is high and typically in the range 28–30 °C.

Tables 9.1–9.5 illustrate that microbial destruction and cooking reactions take place at different rates. The example that follows explains how the competition between sterilisation and quality works in practice, and how optimal regions for thermal processing can be calculated.

9.1.2 Competition between sterilisation and cooking

The fact that *z*-values for microbiological safety and for quality attributes are often different by more than 20 °C implies that the rate of these reactions with temperature increase occurs quite differently. Table 9.6 shows the calculated F_0 and *C*-values for a unit time interval (1 minute), calculated from Equations. (9.2) and (9.3). In effect, Table 9.6 presents the lethal rates at fixed process temperatures. The *z*-value for sterilisation is 10 °C, and the chosen value of 33 °C for cooking is appropriate to a cooking criterion such as thiamine destruction.

The F_0 values in Table 9.6 increase by a factor of 10 for every 10 °C rise in temperature, whereas the *C*-value approximately doubles for the same temperature increase. The effect of this difference can be seen more clearly if a plot of log time versus temperature is made (Fig. 9.1). Each of the two lines in Fig. 9.1 represents a line of constant *F*-value or *C*-value. What this means is that at any point along either line, the *F*-value or *C*-value within

Table 9.6 Effect on F_0 and C-values of increasing the product temperature.

Temperature (°C)	F_0 value (minutes)	C-value (minutes)
100	7.76×10^{-3}	0.23
105	2.45×10^{-2}	0.32
110	7.76×10^{-2}	0.46
115	2.45×10^{-1}	0.65
120	7.76×10^{-1}	0.92
121.1	1.00	1.00
125	2.45	1.30
130	7.76	1.85
135	24.54	2.62
140	77.62	3.71
145	245.4	5.25
150	776.2	7.42

Values represent the lethal rate for 1 minute at these temperatures.

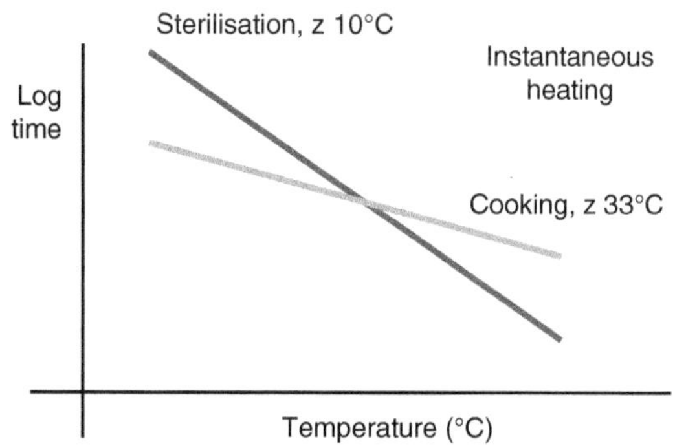

Fig. 9.1 Lines of constant F_0 and C-value for foods instantaneously heated.

the product is the same. For example, the F-value line could represent $F_0 3$ at any combination of process temperature and process time, and so the line indicates a division between a process that achieves a minimum botulinum cook and one that does not. Above the line is in excess of $F_0 3$, and below line is a sub-lethal process.

Data in Fig. 9.1 are for an imaginary product that heats instantaneously. This could apply to the heating of a thin layer of material whereby temperature rise is instantaneous, for example in the processing of a fruit juice in a plate heat exchanger. When the heat penetration depth of the material is considered, as in containers of food, the relationships become curved lines (Fig. 9.2). Data for Fig. 9.2 were calculated for a spherical food particle with radius 1 cm, but a similar shape of graph applies to any in-container processed food product.

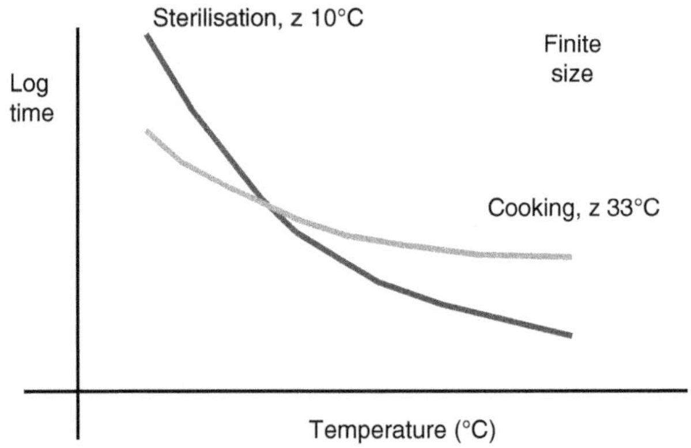

Fig. 9.2 Lines of constant F_0 and *C*-value for foods of finite size, calculated for a spherical food particle with radius 1 cm.

Despite the curvature in the lines, the regions for optimising the process times and temperatures still exist.

9.1.3 Optimisation of temperature/time in processing

Figures 9.1 or 9.2 can be used to highlight four processing regions in which the process temperature and time can be selected. This is the starting position for a process optimisation study that considers the cooking criteria when choosing the process temperature. It is essential that a minimum safety criterion or processing target is established, which is represented in Figs. 9.1 and 9.2 by the sterilisation lines. Process conditions must be chosen so that the scheduled process is above this line. This constrains the regions available for selecting the process time and temperature combinations to the two regions above the sterilisation line.

Worked example 9.1 shows how a process can be optimised to retain the maximum quantity of the vitamin thiamine.

The above worked example is for a high viscosity conduction heating product in which minimal product movement takes place in the container. Hence, product towards the edges of the container will be exposed to much higher temperatures than at the centre, which can result in high levels of quality loss. Not all products will show optimum process temperatures between 115 and 121 °C. It will depend on the *z*-value for the quality attribute of interest and on the rate of heating for the product. In general, fast heating convection products tend to show optimum processing temperatures higher than with conduction products. This is because the product can physically move within the container, which acts to even out the effects of exposure to time and temperature.

Worked example 9.1 Quality optimisation for a vitamin with *z*-value of 33 °C (e.g. thiamine)

One cooking criteria that is often used in thermal processing is the retention of vitamins following the process. This is a good example to illustrate how a process can be optimised to retain the highest percentage of a specific vitamin. Thiamine typically has a *z* value of 25–33 °C, much higher than for *C botulinum* spores at 10 °C, and so it is reasonable to expect the rates of destruction of thiamine and *C. botulinum* spores to occur at very different rates. When choosing the process temperature, it can be useful to consider what impact this has on the product quality. Table 9.5 presents some calculated options of process temperature and hold time that each result in a minimum botulinum cook for a canned pea puree product. These were calculated using the Campden BRI CTemp programme based on a set of heat penetration data for pea puree at 121.1 °C retort temperature.

Each of the above combinations of temperature and time were used as input parameters in the CTemp programme, so that cook values (*C*-value) could be calculated for a *z* value of 33 °C using Equation (9.3). Conversion of the *C*-values to a number that represents the retention of thiamine uses another of the fundamental equations in thermal processing (Table 9.7). This equation relates the number of log reductions in the quality attribute (in this case thiamine) to a measure of the heat stability of thiamine, namely the *D*-value, as given in Equation (9.5):

$$C = D_T \times \log\left(\frac{N_0}{N}\right) \tag{9.5}$$

where D_T is the decimal reduction of thiamine, minutes, N_0 is the initial thiamine concentration, mg per can, and N is the final thiamine concentration, mg per can.

By rearranging the equation so the log reduction ratio is inverted (N/N_0) and multiplying by 100 to bring the answer to a percentage, it is possible to calculate the % retention of thiamine in each can. Equation (9.6) presents the rearranged equation:

$$\frac{N}{N_0} = \frac{100}{10^{(C/D_T)}} \tag{9.6}$$

Table 9.7 Process temperatures and hold times that result in a minimum botulinum cook ($F_0 3$) for 73 × 115 mm canned pea puree product.

Process temperature (°C)	Process time (minutes)
105	205
110	122
115	90
121.1	72
125	65
130	58

Equation (9.6) gives the percentage retention of thiamine directly and does not require either the initial or final concentration to be known. For thiamine retention, Table 9.1 gives a *D*-value of 166 minutes at 121.1 °C. The calculated *C*-values for each temperature and time combination are substituted into Equation (9.6) to result in different percentage retention for thiamine.

Figure 9.3 presents the percentage retention of thiamine in the pea puree (conduction pack) for each of the equivalent *C. botulinum* processes. The optimised process requires a retort temperature of 118 °C, in which around 55% of the thiamine is retained in the processed puree. Figure 9.3 also highlights the disadvantage of in-can HTST processing from a quality perspective.

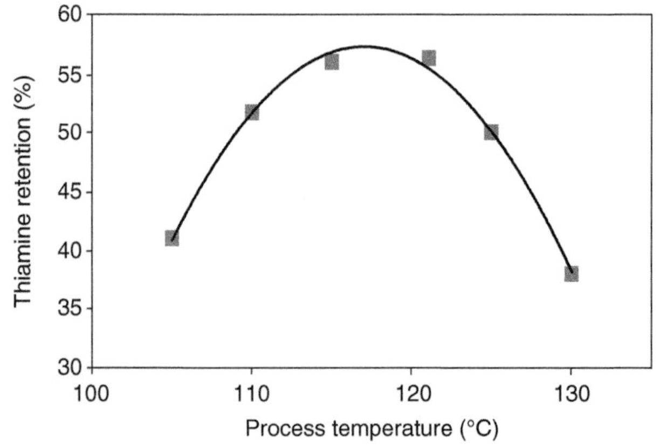

Fig. 9.3 Percentage thiamine retention for pea puree processed at various temperatures and times showing a process temperature optimum at 118 °C.

Retention of vitamins is only one criterion that can be applied to an optimisation calculation. Loss of colour, flavour changes, and enzyme stability may also be limiting factors in selecting the optimal process temperatures for a product. Tables 9.1–9.5 present selected kinetic data that can be used in the calculations for *C*-values and retention of those properties. Worked example 9.2 shows how a process can be optimised to minimize the amount of browning developed during the retorting process and worked example 9.3 is for canned oranges in which vitamin C retention is important.

Worked example 9.2 Quality optimisation for a white sauce

White sauces are highly susceptible to browning through the Maillard reactions. This example is for a glass jar containing a four-cheese sauce, processed in a full water immersion process at 15 rpm end-over-end rotation. For safety, the target F_0 is 6 minutes at the end of heating. Calculations were done using the CTemp program to determine the processing temperature between 110 and 125 °C that resulted in the lowest cook-value. Each combination of time and temperature achieved $F_0$6 at end of heating. Cook values were based on browning criteria of D_{130} = 12.0 seconds, z = 26.7 °C, as used by the milk industry. An assumption was made that the rotation process mixed the product effectively so the calculations for the centre were representative of the bulk. Heating factor (f_h) for the product was 27 minutes, and *j*-value was 1.12. Figure 9.4 shows that there is a minimum cook value at 120.4 °C.

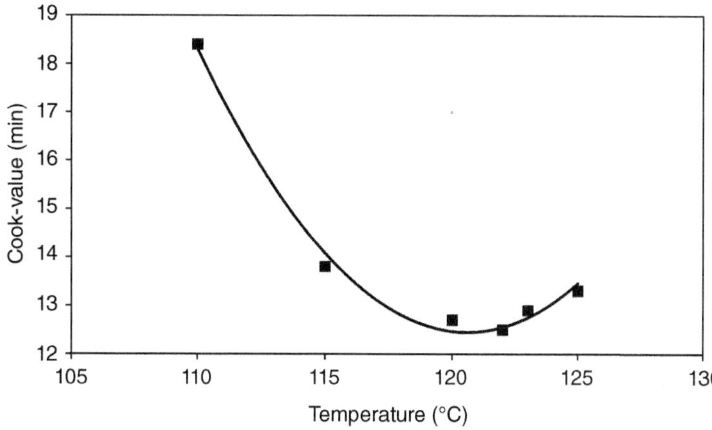

Fig. 9.4 Calculating the retort temperature to minimise the browning reactions in a cheese sauce.

Worked example 9.3 Optimisation for retention of vitamin C in canned oranges

The most important quality attributes for a canned fruit such as oranges include maximising the retention of vitamin C and minimising softening of the segments. Table 9.8 presents some calculated options of process temperature and hold time that each result in a P-value of 10 minutes at 85 °C (T_{ref} 85 °C, z 10 °C) at end of heating. The high levels of acidity in the oranges enabled a mild pasteurisation process to be applied. The calculations for *C*-value assumed the oranges heated uniformly in the can, which is a reasonable assumption for a product with a low viscosity liquid (orange juice) including small particulates (orange segments). Calculations again used the Campden BRI CTemp program based on a set of heat penetration data (f_h 15 minutes) at a process temperature of 85 °C. *C*-values for vitamin C loss were calculated for a z value of 19 °C, taken from Table 9.1. These *C*-values were calculated at the end of the cooling period. Conversion of *C*-values to the retention of vitamin C used Equation (9.6).

Figure 9.5 presents the end of cooling *C*-values for vitamin C and Fig. 9.6 the calculated retention of vitamin C as a

Table 9.8 Process temperatures and times that achieve *P*-value of 10 minutes at end of heating, together with *C*-values end of heating and end of cooling (EOH, EOC, respectively) and percentage of vitamin C retention.

Process time (mins)	Process temp (°C)	C-value EOH for vitamin C (minutes)	C-value EOC for vitamin C (minutes)	Retention of vitamin C (%)
56	80	68.4	76.2	64.6
33	85	46.0	60.5	70.7
24	90	37.0	56.2	72.5
19	95	31.0	52.9	73.9
17	100	27.0	60.7	70.6
15	105	24.0	65.2	68.8

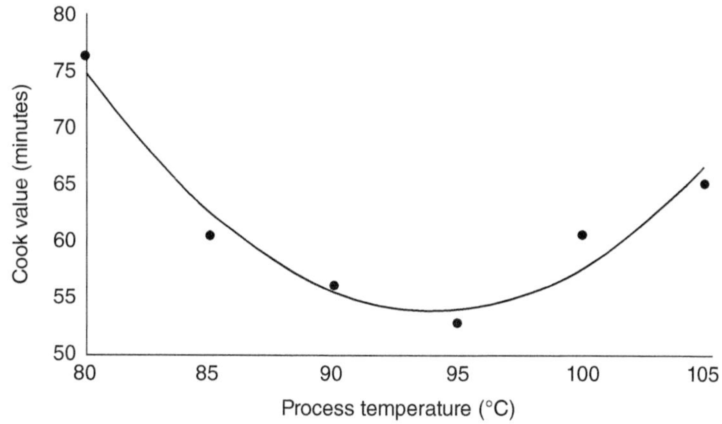

Fig. 9.5 End of cooling *C*-values for vitamin C in canned oranges, for process temperatures between 80 and 105 °C.

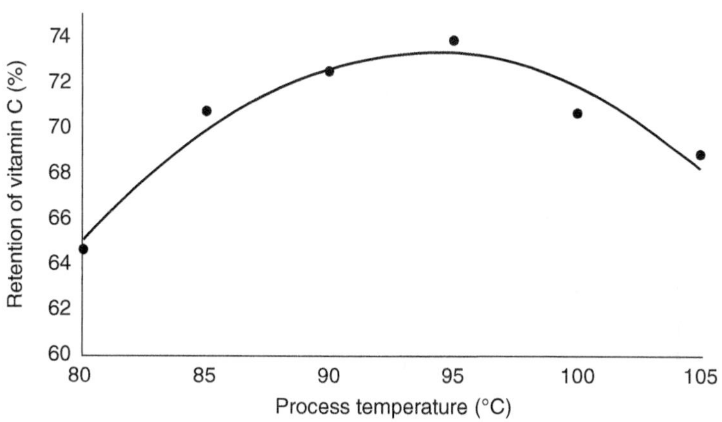

Fig. 9.6 Percentage retention of vitamin C in canned oranges, for process temperatures between 80 and 105 °C.

percentage of the original value. Note that it is not necessary to know the starting amount of vitamin C to calculate the percentage retained. In this example, the optimum process temperature to retain vitamin C is 94 °C and the optimum retention of vitamin C is 74%.

In most thermally processed foods, the heat is damaging to the quality and there is a need to minimize the impact of the process on product quality. However, there are some products in which a process in excess of the safety target must be applied so that the product quality is acceptable; such as softening of tough vegetable skins or fish bones. One example of this is with canned peas. If the minimum botulinum cook ($F_0$3) is given at 126.7 °C in an agitating cooker, this results in a product with tough skins, residual enzyme activity, and unstable colour and flavour. An F_0 value of 8 minutes is required (at 126.7 °C) to inactivate the enzyme peroxidase, which equates to a *C*-value of 30 minutes. At higher process temperatures than 126.7 °C, in order to have the same *C*-value effect on peroxidase, the F_0 value must be higher than 8 minutes.

Another quality attribute of commercial importance to the fish industry is bone softening. Fish species such as mackerel and sardines are canned whole and the process has to be sufficient to soften the bones. This is achieved by lowering the retort temperature to between 112 and 116 °C, depending on can size and heat penetration rates, and processing for long times. It is not possible to produce an acceptable bone-in fish product using high retort temperatures and short times. The effect will be hard and sharp bones.

9.2 SETTING PROCESS TARGETS

Chapter 11 discusses the thought processes behind establishing safe process times and temperatures. These revolve around the assumptions behind worst case scenarios for the product and process. It is imperative that thermally processed foods are safe at the point of consumption. The present chapter on 'cooking' introduces new challenges and puts this into context, since processors of thermally processed foods need to balance safety and quality. The competition between safety and quality should be considered when carrying out the process establishment studies and during subsequent calculations.

Food products suffer from degradation by heat that causes the products to take on inferior quality attributes. Data presented in Tables 9.1–9.5 show that the quality aspects include vitamin retention, colour changes, enzymatic reactions, and texture changes. Consumers will not repeat purchase poor-quality products. It is important to understand how the balance between safety and quality can be achieved in practice. Some thoughts on this follow in the next section.

9.2.1 How to select processing conditions without excess quality damage

When establishing process times, there are many worst-case scenarios that can be considered for each product type and processing system. For example, pieces of vegetable clumping together in a can, overfilled vegetables in a can, low product initial temperature, can location in a retort cold point, short come up time, and so the list goes on. The danger is that they can all get combined into a worst-case scenario when carrying out the heat penetration studies; to the disadvantage of the product quality. It is likely that this is unrealistic. Evidence from record sheets should be used to make an informed decision that allows some of the factors to be positioned towards the worst-case but not by using the individual worst-case conditions.

Products that are processed using end-over-end or axial rotation need further consideration. The worst case in many canned products is to overfill and to use very little or no headspace. This severely limits the action of the headspace bubble as it moves within the product during rotation. It is the movement of the air bubble, which was the headspace when filling the container, that mixes the product during processing and forces it to heat faster. A container closed without a headspace will heat up significantly slowly because of the limited mixing, making the processing time much longer than realistically necessary. Headspace for products that are processed with rotation is a critical factor and the level of control must be determined from recorded evidence so that a realistic headspace can be used for that product in that establishment.

Worked example 9.4 discusses the factors that need to be considered when setting safe times and temperatures for a canned soup with vegetables. Assigning worst-case levels for each of these factors will result in excessively long process times and low-quality product.

There are some products which have, for example, a delicate flavour, a crunchy texture or a colour that is easily damaged by heat. Vegetables such as asparagus, beetroot, and sweetcorn and fruits such as raspberries, strawberries, plums, apricots, and peaches fall into these categories. The quality of these products will benefit from a process that is close to the target for food

Worked example 9.4 Deciding on worst case heat penetration factors for canned soup with vegetables

The target process for a canned soup with vegetables is to achieve $F_0$6 at the slowest heating point by the end of the heating phase. This is a static process. Factors to consider when carrying out heat penetration testing include the following:

- largest vegetable pieces
- clumping together of vegetables
- overfill of solids
- lowest IT
- highest starch concentration or soup viscosity
- overfill giving lowest headspace
- short come up time (if not programmable)

The above list is intended as an example, and there could be more worst-case factors that can be chosen. In addition, the heat penetration testing will be carried out after the retort cold point has been identified through temperature distribution tests. A combination of the worst-case factors will result in a significant extension to the holding time to ensure the worst case can achieve $F_0$6.

During normal production runs the cans when the above factors are at normal levels, the cans will achieve F_0 values much higher than those from the HP tests. F_0 values could range upwards from 25 minutes. This will impact negatively on the product quality, with a softening of all the vegetables, poorer coloration of vegetables such as carrots and peas, and reduced vitamin levels. This combination of factors is damaging and will result in a process time and/or temperature that is too severe for the product. It is necessary to apply evidence-based judgements when considering the conditions that can combine to give the lowest level of process in a container of food; for example, look at the fill records for solids and thickening agents and choose a realistic worst case, similarly look at what is actually measured as the lowest IT and avoid choosing room temperature, even though that would realistically be the actual worst case for a hot fill product. This also applies to continuous processing using heat exchangers and holding tubes.

Worked example 9.5 Setting optimised processes close to $F_0 3$

Retorted (sterilised) asparagus spears in brine are a product that softens and discolours easily. For food safety, the product requires an $F_0 3$ end of heating but attention to the extent of cooking is important. The example is for a 73×116 mm can processed in a jumble crate retort with initial product temperature of 48 °C and 4 minutes retort come-up time. A series of CTemp simulations were carried out to determine the process time for temperatures from 105 to 125 °C. Heating factor (f_h) was 8.2 minutes. The factory retort operated at a retort temperature of 118 °C. Criteria for calculating *C*-values were taken from Table 9.4, which were colour changes for asparagus with D_{85} of 3.500 minutes and a z of 45.2 °C.

Table 9.9 presents the results of the CTemp calculations for these process times and temperatures, together with the *C*-values at the can centre and at the can surface. This provides a simplified method to assess the impact of the process on product cooking at the centre and the surface. Proportionally, there is more product around the outer edges of the can than at the centre, and therefore, *C*-value data at the surface are of interest.

These data show that higher process temperatures give lower *C*-values resulting in reduced colour loss at both centre and surface. The choice of process temperature, however,

Table 9.9 Process temperatures and times that achieve $F_0 3$ end of heating, together with *C*-values end of heating and end of cooling (EOH, EOC, respectively).

Process time (mins)	Process temp (°C)	*C*-value EOH at can centre (minutes)	*C*-value EOC at can surface (minutes)
129	105	56.8	58.1
46	110	25.8	27.5
20	115	13.8	16.1
14	118	10.7	13.5
11	120	9.0	12.0
8	125	7.7	11.5

is also based on other parameters. A short process time is not always possible due to steam pressure limits and will go out of control quickly. In this example, a more consistent process and product quality can be achieved by operating at a slightly lower temperature. The difference between the *C*-values at centre and surface for the operating RT of 118 °C was 10.7 and 13.5 minutes, respectively. This would give consistent product quality throughout the can.

safety, whether it is a minimum botulinum process or a pasteurisation process for acidic products. One of the factors that helps ensure process safety is achieved is the contribution to lethality from cooling. This is not usually taken into account when establishing process conditions unless there is good justification to do so. However, the impact of cooling on product quality can be considerable, particularly for conduction heating foods such as beetroot. The confidence in using cooling lethality is that a conduction heating product, by its nature, cannot cool quickly. There will always be a significant contribution to the overall F_0 from cooling. It is useful to determine the range of cooling F_0 or P values achieved during the process establishment trials to quantify the contribution. These should be highlighted in the report.

It was mentioned earlier in Section 9.1.3 that convection products tend to require higher process temperatures than conduction products to achieve optimum cooking criteria. Worked example 9.5 is for a vegetable product (asparagus spears in brine) that heat rapidly. The texture and colour of the spears can be damaged if the F_0 levels are much higher than $F_0 3$. Therefore, the process conditions are chosen to achieve the minimum botulinum cook at the end of heating but without incurring significantly more F_0 during cooling.

This chapter discussed the need to consider product quality when establishing safe processing times. Estimating the effect of the process on product quality factors must be achieved by thorough product knowledge and understanding of the process/product interactions. There are calculation tools to estimate cook-values that provide information on the effect of the process on specific quality factors. Modelling programmes can be used to calculate the effect of different process temperature and hold times to achieve the same lethal effect (F_0 or P-value) but with different quality impact (*C*-value). This enables process times and temperatures to be chosen that maximise or minimise the desired quality attribute. The ultimate aim is to optimise the process settings to achieve an improved product quality.

References

Ávila, I.M.L.B. and Silva, C.L.M. (1999). Modelling kinetics of thermal degradation of colour in peach puree. *Journal of Food Engineering* 39 (2): 161–166.

Barreiro, J.A., Milano, M., and Sandoval, A.J. (1997). Kinetics of colour change of double concentrated tomato paste during thermal treatment. *Journal of Food Engineering* 33 (3–4): 359–371.

Burton, H. (1955). Colour changes in heated and unheated milk I. The browning of milk on heating. *Journal of Dairy Research* 21: 194–203.

Chutintrasi, B. and Noomhorn, A. (2007). Color degradation kinetics of pineapple puree during thermal processing. *LWT – Food Science and Technology* 40: 300–306.

Ganjloo, A., Rahman, R.A., Osman, A. et al. (2011). Kinetics of crude peroxidase inactivation and color changes of thermally treated seedless guava (Psidium guajava L.). *Food and Bioprocess Technology* 4 (8): 1442–1449.

Holdsworth, S.D. (1997). *Thermal Processing of Packaged Foods*. London: Blackie Academic and Professional.

Kessler, H.G. and Fink, R. (1986). Changes in heated and stored milk with an interpretation by reaction kinetics. *Journal of Food Science* 51: 1105–1111.

Kong, F., Tang, J., Rasco, B., and Crapo, C. (2007). Kinetics of salmon quality changes during thermal processing. *Journal of Food Engineering* 83 (4): 510–520.

Lau, M., Tang, J., and Swanson, B.G. (2000). Kinetics of textural and color changes in green asparagus during thermal treatment. *Journal of Food Engineering* 45 (4): 231–236.

Ling, B., Tang, J., Kong, F. et al. (2015). Food kinetics of food quality changes during thermal processing: a review. *Food and Bioprocess Technology* 8: 343–358.

Van Loey, A., Fransis, A., Hendrickx, M. et al. (1995). Kinetics of quality changes of green peas and white beans during thermal processing. *Journal of Food Engineering* 24: 77–86.

10 Process Validation: Temperature and Heat Distribution

10.1 TEMPERATURE DISTRIBUTION

Any in-container processing system (retort, autoclave, or sterilizer) will contain regions in which the temperature of the heating medium is lower than that measured by the master temperature indicator (MTI). It is critical that food containers in these colder regions absorb sufficient heat to achieve the target process value. The location of these cold spots should be determined by performing TD tests throughout the system (Adams and Hardt-English 1990). Figure 10.1 illustrates an example of where the cold regions are likely to be found for steam and cascading water systems.

The concepts for TD testing are simple; however, the practicalities of making the measurements are fraught with difficulty (Campden BRI 1977; IFTPS 1992; NFPA 1985). For example, regions of low temperature may exist within a retort because of the flow restrictions imposed by the closely packed containers. Flexible pouches are notorious for restricting media flow because of their desire to expand to fit the available space. To position probes within a batch of food containers requires thermocouple wires to be trailed between the containers. This can open up the flow channels and can affect the TD results. These issues are overcome with remote logging systems (Ball Datatrace and Ellab Tracksense), but the downside of remote loggers is that results are not available until the loggers are removed and downloaded.

One cautionary word here that further complicates TD testing is that a uniform TD throughout the retort does not necessarily imply that heat transfer to the containers is also uniform (Tung et al. 1990).

Essentials of Thermal Processing, Second Edition. Gary Tucker and Susan Featherstone.

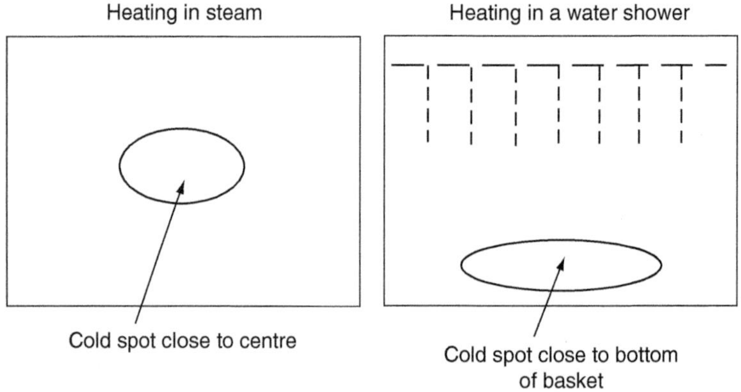

Fig. 10.1 Likely retort cold spots for steam heating and cascading water heating.

This relates to the poor heat transfer properties of air compared with steam or water, and is discussed later in this chapter.

However, uniformity in temperature is the minimum that should be studied and an additional HD study is advisable if there are concerns about air entrapment or heat transfer coefficient reductions throughout a container load (Campden BRI 1997b). Rotary retorts using excessive air overpressure are one example where the potential exists for air to collect at the crate centres. This occurs as a consequence of centrifugal effects of the heating media. This tends to be an issue only at high rotation speeds, perhaps upwards of 20 rpm, and is likely to be more significant when the heating media is more dense than air, for example with a water immersion, spray, or cascading retort. A heat distribution test will need to use containers filled with a consistent material that heats at a similar rate to the product.

A TD test should be carried out when a retort is commissioned, with intermittent re-testing required as factors change that could affect the retort performance. A suggested frequency for re-testing in the UK is every three years according the Department of Health guidelines (DoH 1994). Retorts require, as a minimum, re-testing in the event of engineering work likely to affect the TD of the retort, which could include:

- relocation of the retort or installation of another retort that uses the same services;
- modification to the steam, water, or air supply;
- replacement of the key components, e.g. pumps and valves;
- repair or modification to water circulation or steam injection systems within the retort;
- if there are any doubts about the performance of the retort.

The Good Manufacturing Practice guidelines for TD tests in batch retorts, as defined in DoH (1994), are as follows:

> In steady-state operation, the temperature spread across the sterilizing vessel should ideally be 1 °C or less. However, when this degree of control is not achievable due to design or characteristics of the equipment, any deviation from the limit should be allowed for in the scheduled process.

10.1.1 Temperature measurement systems

Temperature measurement systems are usually the first choice for most companies because of their ease of use, the greater quantity of information generated during the tests, and the relative expense when compared with the alternatives. The methodology for TD and HP testing evolved in the early part of the 1900s where mercury-in-glass thermometers were the available technology. These tests were time-consuming and difficult to carry out, so PC methods were established that enabled results to be transferred from one set of conditions to another. The Ball method was one of the first models to be derived and is still in regular use today, that is, despite the advances in temperature measurement technology.

Modern data probes and loggers are typically multi-channel systems with digital outputs allowing data to be recorded directly to computers for display and to maintain permanent records. Thermocouples based on type *T* (copper/constantan) with PTFE insulation are most common because they are inexpensive, accurate over the desired temperature range, and respond rapidly to changing temperature. For example, Fig. 10.2 shows a multi-channel system from Ellab that can be connected to 16 type *T* temperature sensors.

Other types of temperature measurement are based on a change in electrical resistance with temperature, such as thermistors and platinum resistance thermometers (PT100). These are most commonly used in data loggers where the logging unit is a self-contained unit (e.g. Ball Datatrace, TMI Orion, or Ellab Tracksense, as in Fig. 10.3). These are also referred to as resistance temperature detectors or RTDs.

Calibration of a temperature sensor against a traceable instrument is essential, otherwise the user does not have confidence in the numbers obtained. This is usually done each time the instrument is used in a set of TD or HP trials. The method recommended by IFTPS (2014) and Bee and Park (1978) is to use the master temperature indicator on the retort (MTI), which itself must be calibrated at no less than six-monthly intervals. It is worth noting

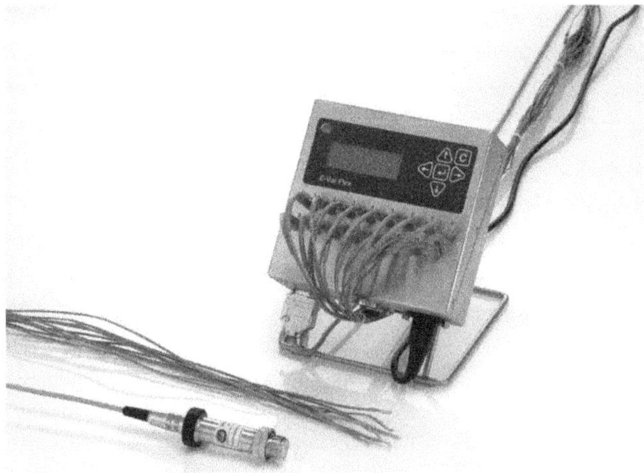

Fig. 10.2 Multi-channel system from Ellab that can be connected to 16 type *T* temperature sensors.

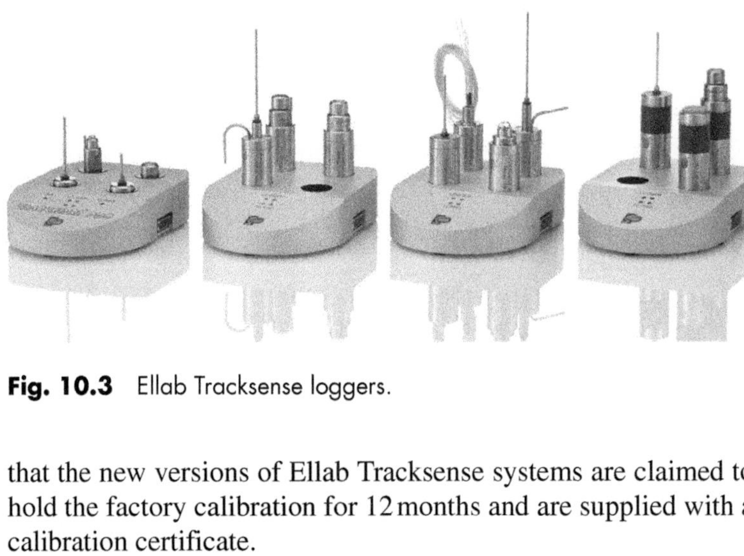

Fig. 10.3 Ellab Tracksense loggers.

that the new versions of Ellab Tracksense systems are claimed to hold the factory calibration for 12 months and are supplied with a calibration certificate.

The use of time and temperature measurements to validate thermal processes is likely to remain the most widely used method. Advances in technology are providing more computing power to analyse these results and increase the accuracy in defining the calculated process values. Historic canning processes were evaluated using lethal rates at time steps of one minute, because of the capabilities of data recorders at that time, but modern process values can be estimated from temperature measurements taken at much reduced frequencies, for example every second. The increased data storage capabilities also allow for more temperature probes to be used in each TD or HP test, with multiple loggers linked together in the software.

The alternative to temperature systems is to use microbiological or biochemical log reduction methods. These are required where trailing wires and data loggers can interfere with the measurements. Their relevance is more to HP data collection so will be detailed in Chapter 11.

10.2 HEAT DISTRIBUTION

During retorting of food products, whether it is sterilization or pasteurization, knowledge of the rate of heat transfer, and the factors that affect it is important. The tests that are used to evaluate heat transfer are known as heat distribution (IFTPS 2008) or heat transfer distribution and should be considered when evaluating the cold point of a retort. The one exception is with condensing steam retorts because the rate of heat transfer is very high, often considered to be infinite at the surface. This means the container surface temperature reaches that of the steam temperature in an instant. Thermal resistance caused by boundary layers does not affect the condensation of steam at a cold surface. This, however, is not the situation with modes of heating other than condensing steam, in which heat transfer is less effective. This includes water immersion, water spray or droplet, and steam/air modes. Each of these systems relies on convective heating in which effective movement of the heating medium across and onto the surface of the packs is required. This ensures the heat transfer conditions are suitable to transfer heat to the pack surface; the first step in enabling heat to penetrate to the pack centre. In order to understand heat transfer to retorted food packs, there are some theoretical considerations that describe the ways that heat flows to a solid body.

Heat will always flow from regions of high temperature (e.g. steam in a retort) to low temperature (e.g. a can) and never the other way around. The rate at which heat flows is described by a series of equations, starting with the governing equation for heat flow (Equation 10.1). This calculates the total amount of energy (Q) available for heat transfer (e.g. in the water) or that can be absorbed by the body (e.g. a can). It is calculated from the thermal characteristics of either the heating medium or the product. This tends to be the limiting or governing equation for heat transfer because it tells us the maximum quantity of heat that can be transferred.

$$Q = m.C_p.\Delta T \quad (10.1)$$

where m is the mass of the product (kg), C_p is the specific heat capacity of the product, defined as the heat required to raise a unit mass of the product by 1 °C ($J\,kg^{-1}\,K^{-1}$) and ΔT is the change in temperature of the medium or product (°C).

There are three main heat transfer media used in retorting systems: steam, water, and air. Air is used for control of the overpressure required to prevent damage to food containers (e.g. to prevent jar lids from coming off or flexible packs expanding) but is more common in baking, refrigeration, and freezing processes where dry heat is required. Avoiding too much air in the heating or cooling medium is critical for in-pack processes because of the poor nature of air as a heat transfer medium. By far the best heat transfer medium in terms of the amount of energy that can be transferred to a surface is steam. This is because of the energy released as the steam condenses onto a colder surface, known as the latent heat.

Steam contains two sources of energy:

(i) Sensible heat – The heat gained when the water is raised from ambient to its boiling point in the boiler.

(ii) Latent heat – The energy required to convert the water at its boiling point into steam at the same temperature, with an equivalent amount of energy released as the steam condenses back to water on a cooler surface.

To illustrate the significant differences in energy between sensible and latent heat, data from steam tables can be used. For example, at 30 psi absolute pressure (approx. 2 bar) saturated steam has a total energy content of 2709 kJ/kg. The sensible heat content of this is 509 kJ/kg and the latent heat portion is 2200 kJ/kg. As can be seen, the latent heat portion constitutes around 80% of the available energy and this is released during condensation of steam on the surfaces of containers.

In addition, the steam may be superheated to help prevent condensation within steam distribution systems. This ensures that steam supplied to retorts does not contain too many water droplets because of the physical damage these can cause to the supply pipework, valves, and fittings from water droplets moving at high velocity. Superheating steam involves increasing its energy content without increasing the pressure. This results in the steam temperature increasing above that for saturated steam. Data at a steam pressure of 30 psi abs, but with 10 °C of superheat, shows that the extra available energy is 23 kJ/kg, giving a total energy content of 2732 kJ/kg. The latent heat portion is still around 80% of the total heat available, which informs that it is condensation of steam that is of most importance with steam heat transfer. The few degrees of superheat have limited effect on the energy but are sufficient to ensure delivery of dry steam to the retort.

The high energy associated with steam condensation is not suitable for all types of retorted pack material. It can cause damage to glass jars through differential expansion rates between outside and inside surfaces, resulting in cracking. Hot water at the same temperature as steam transfers less heat therefore is less likely to set up temperature gradients so quickly. The potential heat shock to glass jars is greatly reduced if they are processed in water-based systems.

10.2.1 Modes of heat transfer

Before discussing the merits of different heat transfer media in greater detail it is necessary to know about the three main modes of heat transfer; radiation, conduction, and convection. These transfer their heat in different ways.

10.2.1.1 Radiation

All bodies emit and absorb radiated energy in the form of electromagnetic waves. If emitted radiated energy falls on an object it may be partially reflected, transmitted, or absorbed. Only the portion that is absorbed will appear as heat in the object (see Fig. 10.4). Radiant heat gets absorbed within the object so has the benefit that faster rates of heat transfer can be achieved in achieving a temperature at the thermal centre of that object.

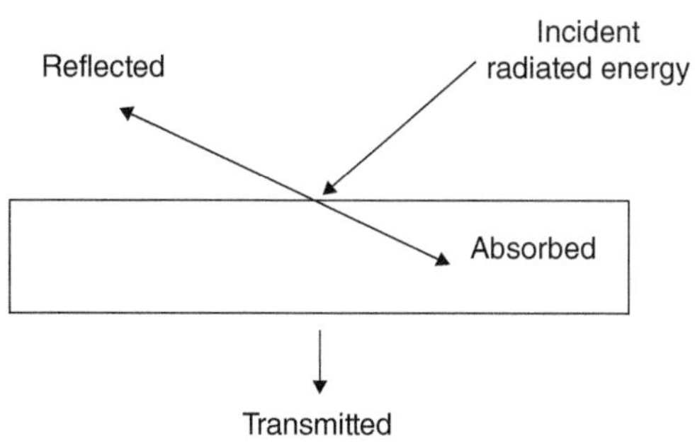

Fig. 10.4 Radiant heat transferred to an object.

The rate at which heat is emitted is dependent on the temperature of the body, with a hotter body radiating at a greater rate than one at lower temperature. The rate of heat transfer by radiation (Q) from a body is given by:

$$Q = \varepsilon.A.k.T^4 \tag{10.2}$$

where ε is the emissivity of the object (dimensionless), A is the area available for heat transfer (m^2), k is the Stefan Boltzmann constant, with value of 5.67×10^{-8} ($W.m^{-2}K^{-4}$) and T is the surface temperature of the object in degrees Kelvin (K).

The surface properties will also affect the rate at which radiation is emitted with black matt surfaces emitting at a greater rate than white shiny surfaces. The effect of surface properties is taken into account by the inclusion of the emissivity term (ε in Equation 10.2).

Radiation is the only mode of heat transfer that does not require a medium to transfer the heat from the emitter to the receiver; that is radiation can take place in a vacuum. Heat transfer by radiation is used in high-temperature baking ovens but it not significant in retorting situations because of the lower temperatures involved.

10.2.1.2 Conduction

Conduction describes the mode of heat transfer that occurs in a solid material. Heat will flow from high to low temperature by means of vibrational energy. Molecules within the material that receive thermal energy will vibrate more, 'bumping' into the molecules next to them and hence the energy is passed from molecule to molecule through the material. The quantity of heat transferred by conduction can be described by Equation (10.3) and illustrated in Fig. 10.5:

$$Q = \frac{k.A.\Delta T}{x} \tag{10.3}$$

where Q is the quantity of heat transferred (W), A is the surface area available for heat transfer (m^2), ΔT is the temperature gradient within the product, $T_1 - T_2$ (°C) and x is the thickness or length of the product over which transfer of heat is occurring (m).

Heat transfer by conduction is common in the food industry, taking place whenever a food or package is in contact with a surface at different temperature. It relies on molecular vibration and

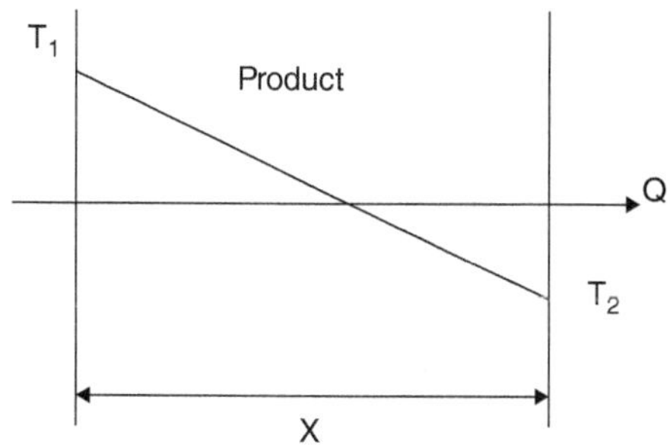

Fig. 10.5 Conduction heat (Q) transferred from the product surface at temperature (T_1) to a distance into the product (x) at temperature T_2.

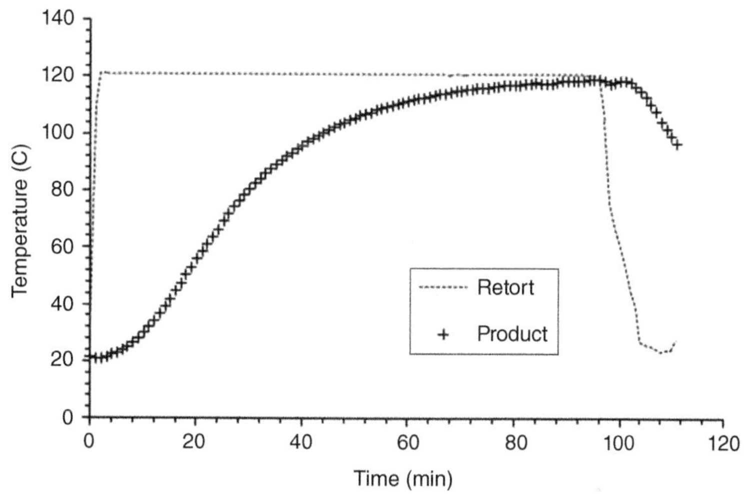

Fig. 10.6 Slow rise in temperature typical of a conduction-heating product; a canned meat processed in a steam retort.

so takes place at a slow rate of heat transfer. Many canned foods behave essentially as conduction packs, for example meat products such as corned beef, tightly packed vegetables, and some pet foods. Figure 10.6 shows a typical time–temperature curve for a conduction heating pack. There is an initial lag period where the temperature at the centre of the pack remains constant because of the time required to slowly conduct heat from the surface through the food product. For conduction heating products the point of slowest heating will be at the geometric centre. This is the location for the probe used to take heat penetration data.

According to heat transfer theory, the approach of the centre temperature to the heating medium temperature will be logarithmic. Theoretically, the centre temperature will never reach the heating medium temperature. However, in practice, the industry works to one decimal place because of temperature sensor calibration and accuracy limits, which means that retort and product temperatures appear to reach the same values. For conduction heating packs this can take considerable time. Figure 10.6 illustrates an example in which the centre temperature got to within 1 °C of the retort temperature.

10.2.1.3 Convection

Convection describes the mode of heat transfer that occurs in a fluid (liquid or a gas). There are two types of convection; natural and forced. Both are important in thermal processing; however, the advantages that can be realized by using forced convection in retorts is substantial.

Natural convection is caused by density differences within the fluid. Hot fluid, with a lower density will rise up the edge of the container where it is hottest, and colder fluid, with a higher density will fall down the centre of the container where it is coldest. This movement of fluid, known as convection currents, is the mechanism by which heat is transferred. Despite natural convection transferring heat more effectively than conduction, the rate of heat transfer can be increased by forcing the convection currents.

To enhance the rates of heat transfer, external forces are used to aid this movement of fluid and this is known as forced convection (see Fig. 10.7). Food packs are agitated using three different modes of agitation; end-over-end as with batch retorts, axial as

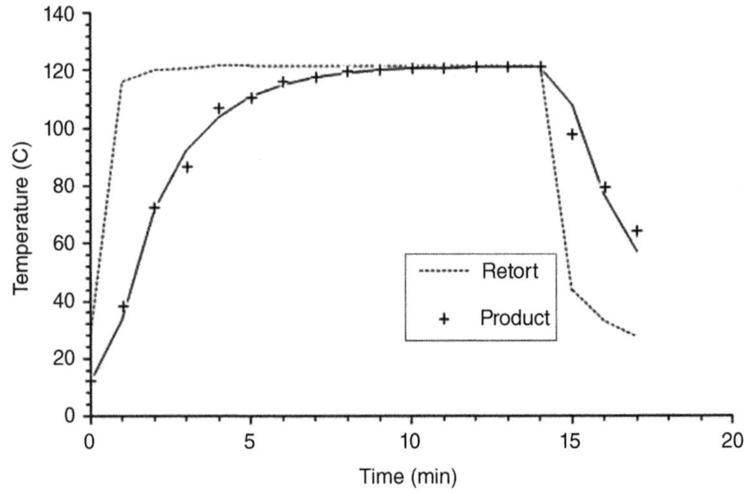

Fig. 10.7 Rapid rise in temperature typical of a convection-heating product; canned sliced carrot processed in a steam retort.

with reel and spiral cooker-coolers, and more recently reciprocal agitation with the Shaka retort. The enhancement of internal pack agitation results in higher rates of heat transfer and reduced thermal processing times.

Defining an equation for convection is less straightforward than with radiation or conduction, because of the complexities with the physical fluid movements. However, the Equation (10.4) applies and can be used to understand the benefits of different heat transfer media, packaging types, and food viscosities:

$$Q = U.A.\Delta T \tag{10.4}$$

where U is the overall heat transfer coefficient ($W.m^{-2}.K^{-1}$), A is the area available for heat transfer to occur (m^2) and ΔT is the temperature gradient acting as the driving force for heat transfer (°C).

This equation describes how the heat is transferred from the heating medium to the surface of the pack; the first step in transferring heat to the thermal centre. U describes the overall heat transfer coefficient, which can be made up of several different components. For a packaged food product heated in a retort, there are three components of the heat transfer that must be considered when optimizing heat transfer:

a. from the heating media to the pack surface,
b. through the wall of the packaging material,
c. from the inside wall to the inner surface of the product.

The heat transfer coefficient (U) can be calculated from each of these separate resistances to heat transfer, using Equation (10.5):

$$\frac{1}{U} = \frac{1}{h_o} + \frac{x}{k} + \frac{1}{h_i} \tag{10.5}$$

Where h_o is the heat transfer coefficient from the processing media to the outer pack surface ($W.m^{-2}.K^{-1}$), x is the thickness of the packaging material (m), k is the thermal conductivity of the packaging material ($W.m^{-1}.K^{-1}$), and h_i is the heat transfer coefficient from the inner pack surface to the food product ($W.m^{-2}.K^{-1}$).

When optimizing a thermal process, to deliver heat effectively to the thermal centre of a pack, it is important to consider each of the components in Equation (10.5). The section that follows looks at each of the three 'resistance' terms in sequence.

The external heat transfer coefficient (h_o) allows a comparison of the relative effectiveness of different heat transfer media, as shown in Table 10.1. Data in this table were extracted from various sources. Of interest is that air is a poor heat transfer media, even the use of a high-speed fan to increase air velocity over the surfaces does not make air as efficient as static water. Condensing steam is exceptionally efficient at transferring heat, whether it is by droplet or film condensation, both are an order of magnitude higher than the best of the other media.

The next term in Equation (10.5) accounts for heat transfer through the packaging material (*x/k*). Typical packaging values are given in Table 10.2, again extracted from various sources including on line resources such as the Engineering Toolbox (www.Engineeringtoolbox.com). Values for wall thickness and thermal conductivity will vary depending on the package design and supplier; however, it is the average values that are relevant to the comparison with Tables 10.1 and 10.3. This comparison will be made when the last of the three terms in Equation (10.5) is evaluated.

The final term in Equation (10.5) is the most complex and difficult to estimate. This is because of the laminar boundary layers that build up at the surfaces and offer resistance to heat transfer. Data is

Table 10.1 Approximate values for the external heat transfer coefficient (h_o) and resistance term, for the common heat transfer media used in thermal processing.

Heat transfer media	Heat transfer coefficient (h_o) ($W.m^{-2}.K^{-1}$)	Resistance term ($1/h_o$) ($(W.m^{-2}.K^{-1})^{-1}$)
Static air	10–15	$6.7–10.0 \times 10^{-2}$
Moving air	20–300	$3.3–50.0 \times 10^{-3}$
Static water	400–700	$1.4–25.0 \times 10^{-3}$
Moving water	700–3000	$3.3–14.3 \times 10^{-4}$
Water droplets or spray	1000–5000	$2.0–10.0 \times 10^{-4}$
Condensing steam	10,000–50,000	$2.0–10.0 \times 10^{-5}$

Table 10.2 Approximate values for the *x/k* term in Equation (10.5) and resistance term, for different food packaging materials.

Packaging material	Typical wall thickness, *x* (m)	Thermal conductivity, *k* ($W.m^{-1}.K^{-1}$)	Resistance term, *x/k* ($(W.m^{-2}.K^{-1})^{-1}$)
Glass	2.5	0.9	2.6×10^{-3}
Plastic (polypropylene)	1.0	0.2	5.0×10^{-3}
Stainless steel	0.2	16.0	1.3×10^{-5}

Table 10.3 Approximate values for the internal heat transfer coefficient (h_i) and resistance term, for liquids transferring heat to surfaces.

Heat transfer media	Heat transfer coefficient (h_o) (W.m^{-2}.K^{-1})	Resistance term ($1/h_o$) (W.m^{-2}.K^{-1})$^{-1}$
Gases in natural convection	5–37	2.7–20.0 × 10^{-2}
Water in natural convection	100–1200	8.3–100.0 × 10^{-4}
Oil in natural convection	50–350	2.9–20.0 × 10^{-3}
Static particulate in sauce	27–140	7.1–37.0 × 10^{-3}

Source: Data extracted from Thermopedia web site (www.thermopedia.com) and McKenna and Tucker (1991).

scarce and varies considerably. For example, there are various correlations derived for heating of particulates in flowing foods (McKenna and Tucker 1991) and used in the Campden BRI PCTemp model for heating of particulates. These suggest the internal heat transfer coefficient can be as low as 27 W.m^{-2}.K^{-1} for static liquids foods but range upwards from 1000 W.m^{-2}.K^{-1} as the velocity over the surface increases. These values are given in Table 10.3, which also includes data from other sources such as the Thermopedia web site (www.thermopedia.com). Food products retorted in packages are likely to be higher in viscosity than the oil given in Table 10.3 so it is reasonable to infer that the internal heat transfer coefficient will be lower. A low value closer to 27 W.m^{-2}.K^{-1}, as for static particulates in a sauce is likely to be closer to that for a packaged food.

If the data is compared in the above three tables it is clear that the greatest resistances to heat transfer are (i) at the internal surfaces of the package, and (ii) when air surrounds the package. This explains why rotary processing of packages is much preferred to static retorting. For heat transfer with media that could include air, it explains why it is essential that the heat transfer media (steam, water droplet/spray, water immersion) is in contact with the pack surfaces at all times.

Heat distribution testing is intended to measure the differences in heat transfer at selected positions in a retort. The methods used in doing this will be discussed later in Section 10.3.

10.2.1.4 Broken heating or mixed heating

Not all food products heat in a simple manner. Some food products exhibit both conduction and convection heating during processing, depending on the viscosity changes that take place in the product. The most common examples are with products containing starch thickeners that are added as dry powders to the batch mixer. Many soups and sauces are formulated in this way. Food products containing dried pulses in a sauce (e.g. beans in tomato sauce) are known to exhibit a complex heating mode similar to that with a starch. As heating progresses, starch contained within the pulse leaches out into the sauce. Initially, these products are relatively low in viscosity so heat rapidly by convection until the starch gelatinization temperature is reached. This causes a sudden increase in viscosity and the remaining heating is through much slower conduction, as shown in Fig. 10.8. Designing a product formulation and preparation method to heat fast initially makes good use of the thermally efficient convective heating stage. This allows energy to get to the thermal centre quickly, which reduces the processing time required to achieve sterility. A sauce made by gelatinizing the starch thickeners during mixing will require considerably longer to process. This is less efficient with energy consumption but is also likely to have a damaging impact on the product quality because of the longer process time.

The other type of viscosity change experienced with thermally processed foods is when high-viscosity products reduce in viscosity (get thinner) as their temperature increases. Initial heating rates will be mostly conduction but with a small amount of convection. As the sauce temperature increases, its viscosity will decrease and the proportion of convection will increase. Heat transfer rates will increase as a result of this. When heat penetration data are plotted

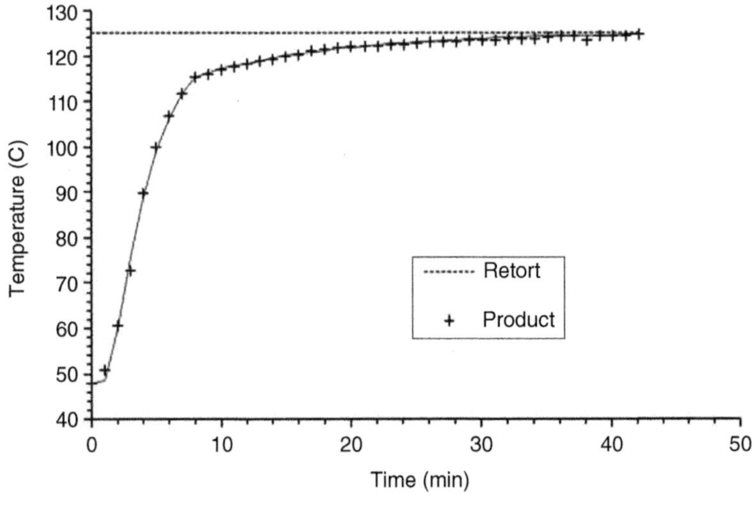

Fig. 10.8 Time–temperature plot for a broken-heating product, canned meat chunks in gravy processed in a steam retort, showing rapid convection at start and slower conduction at end.

on a semi-log graph it can be difficult to decide where the break-in heating, from conduction to convection, took place. It tends to happen over a range of temperatures rather than a well-defined change as with a gelatinization change. This presents a challenge to thermal process calculations.

Both of the above changes in heating mode are known as broken heating and require extra care when establishing thermal processes. It is important to understand what factors affect the time at which the break-in heating occurs, whether it is a starch gelatinization change, more gradual sauce thinning, or even a breakdown of a natural starch caused by temperature or shear. Analysis of process deviations is particularly difficult and should only be attempted if the physical changes within the product are understood completely. Changing the time–temperature and/or the agitation rate of a food pack will cause the viscosity changes to take place at different times in the process. The most dangerous to food safety is if gelatinization takes place earlier than predicted so that more of the process occurs with the product heating by slow conduction. It is unlikely that sufficient heat will get to the thermal centre and commercial sterility could be compromised. This is discussed in the chapter on process deviations.

Example Factors that affect starch gelatinization

When starch gelatinizes into a high-viscosity solution the starch granules first have to expand as water enters the granule and interacts with the amylopectin and amylase. There are several factors that affect this process.

- Time–temperature history
- Shearing (mixing) conditions
- Ingredients such as sugar and salt that affect the available water

Further complication with heat transfer rates is found with foods containing small particulates and semi-solid sauces. These exhibit both conduction and convection heat transfer simultaneously. Liquids heat by convection while the particulates will heat by conduction taking their heat from the hot liquid that surrounds them. This is referred to as mixed heating and is typical of the majority of in-pack processed foods. The rate of heating lies between that for pure conduction and pure convection.

10.3 HEAT DISTRIBUTION TESTING

The objective of heat distribution testing is to locate the position in the retort that delivers the slowest rate of heat transfer to the product. As described previously, temperature and heat are not the same; it is possible to have regions in the retort that are at the same temperature but the heat transfer rates may not be the same (Ramaswamy et al. 1983; Smout et al. 2000; Tung et al. 1984). This effect is mostly attributable to the presence of air within the retorting system although changes in media velocity across the surface can also cause some differences. With condensing steam retorts there is a venting period that serves to remove all air from the retort. This ensures effective steam condensation throughout the batch load. However, air is required in retorts for overpressure purposes but it is important that this does not interfere with the heat transfer mechanisms for the specific media used in that retort (see Table 10.4).

A TD test should be carried out prior to the HD test, as part of the standard procedures for evaluating a new retort and/or product/package combination. If the process authority doing the testing has sufficient probes, it is possible to carry out both TD and HD simultaneously. The probe locations for the TD testing are likely to be the same as those for HD because regions of restricted heat flow should lead to inconsistencies in temperature distribution. However, this should be critically evaluated when setting up the TD and HD testing parameters.

A HD test should be considered when using any of the retort media other than condensing steam. This includes when changes are made to packaged product loading conditions such as new or modified trays, baskets, loading configuration, as well as significant changes to retort temperature and overpressure.

Table 10.4 Methods used in different retorting systems to ensure good heat distribution.

Heat transfer media	Method to ensure good heat distribution
Water immersion	Circulation pump to provide flow of water across the surfaces of the packages.
Water droplet	Water distribution plates, with holes, to generate a shower of water that cascades down and through the baskets of packages.
Water spray	Circulation pump to provide powerful sprays of water that penetrate the baskets and contacts the package surfaces.
Steam-air	Circulation fan to force steam and air mixtures through the baskets of packages.

10.3.1 Conducting a HD test

There are several decisions to be made for HD that will dictate how the tests are conducted. These include the materials to fill the packages with, where to position the instrumented packages, how many packages to use, and what measurements are best to make for the analysis.

The first of these is the material to use in the packages. This should be consistent and not vary in its heat transfer rate to an extent that makes analysis impossible. Product-filled packages are not always ideal because of their natural variability. Non-product-based materials are more consistent and therefore suited for HD work. For example, bentonite clay suspensions of different concentration can be made to represent conduction (5%), convection (1%), or broken heating (3.5%) modes; however, HP tests should be carried out prior to their use to confirm the heating parameters (Robertson and Miller 1984).

- Conduction heating foods (meat, fish, and solid or high-viscosity foods). Polymers such as PTFE and polypropylene have similar thermal conductivities to the foods, as does a 5% bentonite suspension. Heat will be absorbed into the material at a similar but slow rate to the food. Blocks or cylinders can be machined easily and holes drilled for probes.
- Convection heating foods (liquids and semi-solids). 1% bentonite suspensions and silicone or mineral oils can be used. Heat will be absorbed at a fast rate similar to the food. The bentonite concentration or type of oil should be chosen so the f_h value is similar.

None of these non-product materials will break down thermally so can be used for an extended period of time. Data from the TD testing can be used to evaluate the stability of the materials, and determine when they need to be changed. However, the thermal and physical properties of foods will change as they break down in structure, making them unsuitable for repeat runs.

Note that broken heating foods present a challenge in that the type of broken heating has an impact on the rate of heat absorption into the material. IFTPS guidelines (IFTPS 2008) recommend not to attempt HD using simulant materials.

It may not be necessary for the entire retort load to be filled with a simulant material for the HD testing. This saves considerable cost and effort. If the TD studies identified regions of lower temperature it is possible to conduct the HD tests using instrumented packs filled with simulant materials or solid blocks of equivalent size to the packs. These can be located at the lower temperature regions, with the rest of the retort load or ballast comprising of product-filled packages of the type and size being evaluated. It is important that water-filled packages should not be used as ballast for these studies because they absorb heat at a much faster rate. This will change the quantity of steam that condenses with steam-air retorts and lower the water temperature with water cascade and spray retorts. Water-filled packages will have a reduced effect with water immersion retorts because of the much higher heat capacity of circulating water; however, they are still not to be recommended.

The positions in which to place instrumented packs will depend on the type of heating media. For water immersion, sprayed water, and steam-air, it is likely that the centre of the retort will create the greatest challenges for the circulation system. This is also the case when rotational programmes are used. For water cascade or raining water retorts operating in static mode, the lower part of the retort will have the least powerful water shower, as well as water with the lowest temperature (Campbell and Ramaswamy 1992; Smout et al. 1998). At least five instrumented packs (IFTPS 2008) should be clustered around the positions where lethality is likely to be at its lowest. These are general rules on least effective heat transfer location and it is necessary to consider specific design features of each retort before making the decision.

In a rotary retort, there will be differences in the angular velocity of packs on the rotational axis compared with those at the basket edges. The packs on the rotational axis will experience slower rotation speeds than those at the basket edges. Both internal and external heat transfer coefficients will be lower as a result of reduced centrifugal forces on the rotational axis. Consequently, these packs should result in lower heating factor values (f_h) and receive lower lethalities than those at the edges. It is sensible to position one instrumented pack at the basket edge for comparison purposes but to cluster remaining packs towards the basket centres.

Estimating the number of instrumented packs to use is always a challenging question. One can never have too many; however, the cost of the temperature sensors is a major consideration, as is the time required to set up high numbers of loggers and load them into the retort baskets. The IFTPS guidelines (IFTPS 2008) recommend a minimum of five instrumented packs located in separate areas of each basket. This could result in a total of 25 packs for a 5-basket retort, which will incur significant cost. The approach of carefully considering the likely location of slowest heating will reduce the number required; allowing five around that region together with further packs in other baskets for comparison. There is no definite answer to the number of packs and where they are located. It is the responsibility of the thermal process authority to make an educated decision based on knowledge of the retort. Duplicate HD tests are advisable. These can be done without taking the baskets apart so are not as time-consuming as duplicate HP tests where fresh product packs have to be made up.

As with TD and HP testing, it is essential to measure all parameters that influence the heating medium performance. This includes the circulating water flow rates or fan speed with steam-air systems. Increased flow of the heating medium over the pack surface increases the external heat transfer coefficient.

Data analysis is best done using heating factors (f_h) as the measure of the pack heating rates. The f_h value is a complex term that includes all of the resistances to heat transfer discussed in Section 10.2.1.3 above. The use of simulant materials inside packs or in blocks will reduce the variability that exists with actual food materials. As discussed above, the material should be chosen to display simple heating behaviour, that is, not broken heating such as a change from convection to conduction. Analysis of heat distribution for broken heating packs is complex and is not suitable for evaluating the performance of a retort. IFTPS guidelines (IFTPS 2008) suggest the heat transfer conditions are uniform when the %CV for measured f_h values is ≤5%, within and across replicate studies. Subsequent HP probes can then be located anywhere in the retort. If the f_h %CV is >5%, it should be confirmed that the sensors used were consistent with the accuracy requirements for the HD testing. If they are confirmed as accurate it is necessary to consider (i) if this variability in f_h is expected with this retort type or (ii) why the retort is not performing as it should. The latter must be resolved before HP testing takes place.

IFTPS guidelines suggest HD conditions 'do not necessarily need to be collected at the highest allowed overpressure permitted for the product/package being evaluated. This is in contrast to Heat Penetration studies where all worst-case retort conditions and parameters must be used'. Instead, the recommendation is that TD and HD are done using representative conditions of retort overpressure and temperature for the product/package being studied. This makes a complex test slightly easier to manage; however, it avoids the issue surrounding the effect of excessive air on external heat transfer coefficients.

All types of retorts have limits at which they should not be operated. For example, with water immersion retorts in rotary mode the centrifugal forces at rotation speeds above 25 rpm can cause the packs at the axis of rotation to be starved of water. Similar issues will arise with water cascade and water spray, possibly these retorts will experience problems of water starvation at lower rotation speeds. For steam-air retorts, there will be an overpressure level in which the balance of steam and air moves towards excessive air and insufficient steam to condense and deliver heat. It is important to understand the limitations of any retort to deliver heat to the product. HD tests have the advantage of highlighting operational heat transfer limits for retorts; TD may not have sufficient sensitivity to do this.

The recent addition of HD testing for overpressure retorts adds complexity to the work required for establishing thermal processes. The reasons for these tests are difficult to comprehend without understanding heat transfer mechanisms from a medium containing air to the colder surface of a food pack. This chapter presents the background to why heat differs from temperature as it is transferred from steam or water into a packaged food. The logic behind HD testing is sound. It must be achieved by careful consideration of the food product and the heating media's performance. A thermal process authority now needs to have an understanding of heat transfer as well as the thermal processing background required to carry out the testing procedures that are required to generate data to establish safe thermal processes.

References

Adams, H.W. and Hardt-English, P.K. (1990). Determining temperature distribution in cascading water retorts. *Food Technology* 44 (12): 110–112.

Bee, G.R. and Park, D.K. (1978). Heat-penetration measurement for thermalprocess design. *Food Technology* 32 (6): 56–58.

Campbell, S. and Ramaswamy, H.S. (1992). Distribution of heat transfer rate and lethality in a single basket water cascade retort. *Journal of Food Process Engineering* 15: 31–48.

Campden BRI (1977). *Guidelines to the Establishment of Scheduled Heat Processes for Low-Acid Foods. Technical Manual No. 3.* Chipping Campden UK: Campden BRI.

Campden BRI (1997b). *Guidelines for Establishing Heat Distribution in Batch Overpressure Retort Systems. Guideline No. 17.* Chipping Campden UK: Campden BRI.

DoH (1994). *Guidelines for the Safe Production of Heat Preserved Foods.* London: HMSO (out of print).

IFTPS (1992). *Temperature Distribution Protocol for Processing in Steam Still Retorts, Excluding Crateless Retorts.* Fairfax VA: IFTPS.

IFTPS (2008). Temperature and heat transfer distribution guidelines for processing in batch agitating retorts. Document Number VG.007.V1, www.iftps.org.

IFTPS (2014) Guidelines for conducting thermal processing studies. http://iftps.org/wp-content/uploads/2017/12/Retort-Processing-Guidelines-02-13-14.pdf

McKenna, A.B. and Tucker, G.S. (1991). Computer modelling for the control of particulate sterilization under dynamic flow conditions. *Food Control* 2 (4): 224–233.

NFPA (1985). *Guidelines for Thermal Process Development for Foods Packaged in Flexible Containers*. Washington DC: NFPA.

Park, D.J., Cabes, L.J. Jr., and Collins, K.M. (1990). Determining temperature distribution in rotary, full-immersion, hot-water sterilizers. *Food Technology* 44 (12): 113–118.

Ramaswamy, H.S., Tung, M.A., and Stark, R. (1983). A method to measure surface heat transfer from steam/air mixtures in batch retorts. *Journal of Food Science* 48: 900–904.

Robertson, G.L. and Miller, S.L. (1984). Uncertainties associated with the estimation of Fo values in cans which heat by conduction. *Journal of Food Technology* 19: 623–630.

Smout, C., Van Loey, A., and Hendrickx, M. (1998). Heat distribution in industrial-scale water cascading (rotary) retort. *Journal of Food Science* 63 (5): 882–886.

Smout, C., Van Loey, A.M.L., and Hendrickx, M.E.G. (2000). Non-uniformity of lethality in retort processes based on heat distribution and heat penetration data. *Journal of Food Engineering* 45 (2): 103–110.

Tung, M.A., Ramaswamy, H.S., Smith, T., and Stark, R. (1984). Surface heat transfer coefficients for steam/air mixtures in two pilot scale retorts. *Journal of Food Science* 49: 939–943.

Tung, M.A., Britt, I.J., and Ramaswamy, H.S. (1990). Food sterilization in steam/air retorts. *Food Technology* 44 (12): 105–109.

11 Process Validation: Heat Penetration and Process Calculations

The aim of an HP study is to determine the heating and cooling behaviour of a specific product, in order to establish a safe thermal process regime and to provide the data to analyse process deviations. Design of the study must ensure that all of the critical factors are considered to deliver the thermal process to the product slowest heating point. The numbers of instrumented sample containers and replicate retort runs is subject to much discussion (Bee and Park 1978; Campden BRI 1977; IFTPS 1995; NFPA 1985), with the final decision linked to the measured variability between samples and between runs.

Modern data logging systems can provide the facility for taking multiple temperature measurements, therefore large quantities of data can be taken more easily than was the case when the guidelines were first written in 1977 (Campden BRI 1977). These recommendations were to use three samples in three replicate runs, providing a total of nine measurements. This 3×3 system has served the industry for many years. The more common situation now is to take up to ten samples in two replicate runs, providing that the variability between runs is within acceptable limits (IFTPS 1995). However, there can be limitations on the number of probes that can be inserted through a packaging gland or through the central shaft of a rotating system, and in these situations at least two replicate runs should be completed.

The HP study should be carried out prior to commencing production of a new product, a change in the process or the package. Changes to any of the criteria that may change the time–temperature response at the product slowest heating point will require a new HP study to be conducted. The conditions determined in the study are referred to as the scheduled heat process and must be followed for every production batch, with appropriate

Essentials of Thermal Processing, Second Edition. Gary Tucker and Susan Featherstone.

records taken to confirm that this was followed. No further temperature measurement within containers is required in production, although some companies do measure temperatures in single containers at defined frequencies. However, the conditions used in single container testing do not represent the worst case, and it would be expected that the instrumented container shows a process value higher than that measured from the HP study. Such data are intended to show due diligence and are at best a comfort factor.

11.1 SETTING THE TARGET PROCESS VALUE

The first stage in a process validation study is to establish the processing targets based on a specific microorganism or group of microorganisms. Typical target organisms are considered in Chapters 2–6, which depends on the characteristics of the product and the pack storage conditions. Irrespective of the type of food, the target condition is to achieve commercial sterility for the product, which depends on the types and numbers of organisms present, both before and after the process, and on the intended storage conditions (DoH 1994).

Thus, a process value (F) is calculated from the decimal reduction time (D_T) and the number of log reductions of surviving spores, according to Equation (11.1):

$$F = D_T \cdot \log\left(\frac{N_0}{N}\right) \tag{11.1}$$

where N_0 is the initial number of organisms, N is the final number of organisms after a specific time–temperature history, and D_T is the decimal reduction time at a fixed temperature (T) to reduce the number of organisms by a factor of 10.

For example, for low acid, ambient stable products, spores of mesophilic *Clostridium botulinum* strains are the target. One of the guideline documents used by the UK food industry uses a starting number of 100 spores per unit mass or volume (N_0) and a final number of 10^{-12} spores (DoH 1994). This starting number (100) represents a high level of contamination from *C. botulinum* spores, and would only be found in practice from poor quality raw materials or poor factory procedures. The final number (10^{-12}) equates to a probability of 1 processed container in 10^{12} containing a viable spore. The F-value for this process is given the specific terminology F_0. A $D_{121.1}$ of 0.21 minutes is used in the

UK and the calculation is rounded up to the nearest integer, in this case to $F_0$3. This is the minimum process that should be given to low acid foods.

In commercial sterilization processes, however, it is common to operate at substantially increased safety margins, using F_0 values between 6 and 12 end of heating. These equate to log reductions between 18 and 36 (Equation 11.1). Pasteurization processes, on the other hand, are usually operated to only 5 or 6 log reductions of the target organism, because of the presence of a preservation hurdle that lessens the chance of surviving organisms germinating from their spore form and/or the vegetative cells growing. There is also an argument sometimes used that the target organisms for most pasteurization processes are less lethal than *C. botulinum*. Caution is needed here because some chilled food processes target *Listeria*, *Salmonella*, *Escherichia coli*, or psychrotrophic *C. botulinum* spores. All of these are pathogenic with high fatality rates. Further details on pasteurization treatments can be found in Campden BRI (1992, 2006) and are discussed in Chapters 5–7.

Having calculated the target process value according to Equation (11.1), the thermal process achieved in the food containers is measured using temperature sensors or log reduction methods. These methods are discussed in the subsequent sections. The data are then used to calculate the achieved process value for the conditions evaluated, and if required, further calculations are carried out at worst-case conditions. These enable the CCPs to be set for regular monitoring of the process.

As mentioned above, the current thought process used by most food companies is to validate the microbiological process safety under 'worst-case conditions' so that, by default, the process is safe under normal production conditions. Therefore, one of the key activities involved with establishing a thermal process is how to select the worst-case conditions likely to be experienced during normal production. This approach avoids the need to probe up containers during each production run, which can result in misleading information.

11.2 SELECTING THE CONDITIONS FOR THE HP STUDY

It has been stated above that the process validation study should be conducted using worst-case conditions. Therefore, by inference it should not be possible for a normal production batch to heat more slowly than the combination of factors evaluated as worst case. To determine the worst-case conditions, it is necessary to consider the product, process, and package separately, and to list all the factors that affect the heat transfer rate. The following lists suggest some of the factors that should be considered in a process validation study, although the lists are not intended to be exhaustive. These factors have been extracted from publications such as Campden BRI (1977, 1997a, 2008), Bee and Park (1978), NFPA (1985), and IFTPS (1995).

11.2.1 Product

- Formulation changes, including weight variation in ingredients that could lead to increased viscosity, e.g. high starch levels.
- Fill weight and in particular the % overfill of the key components, e.g. solids content.
- Initial temperature and the effects of delays in getting instrumented containers into the retort.
- Consistency or viscosity of the liquid components, both before and after processing.
- Size, shape, and weight of solid components, both before and after processing, to determine the critical particle.
- Potential for matting and clumping of the solids, e.g. sliced mushrooms tend to stick together forming large clumps.
- Preparation methods, e.g. blanching that can affect particle density.
- Rehydration of dried components, e.g. dried rice that will heat more slowly than the hydrated form and also absorb water from the liquid.
- Heating mode for selection of probe positions in the container, whether the product heats by convection, conduction, mixed or broken heating.

11.2.2 Container or package

- Type of material, including metal cans, glass jars, plastic pouches, and semi-rigid containers.
- Nesting possibilities with low profile containers, e.g. sardine cans.
- Vacuum and headspace that affect heat transfer through the top surface, and residual gases with flexible containers.
- Orientation the containers are loaded into the retort, e.g. symmetry of rotation.

11.2.3 Retort or processing system

- Mode of heating and cooling, and how it affects where the retort cold point is located, e.g. steam, steam/air, water immersion, raining water.
- Venting schedule, if steam.
- Overpressure profile and how this affects the volume of gas inside a flexible pack.
- Retort come-up time, which should be as short as possible to minimize the quantity of heat absorbed by the product during this phase.
- Racking and dividing systems that enable the containers to be supported but reduce the voidage for media to pass through.
- Rotation conditions.

The combination of conditions to arrive at the worst-case conditions should be determined by a competent individual, sometimes referred to as a thermal process authority (specifically a US FDA term). It is not advisable to 'stress' each condition at its limit of variation; otherwise, the containers used for process measurements will heat so slowly that all production will be grossly overcooked. A balance between theoretical worst case and what is likely to happen in practice has to be obtained. This is where the skill of the individual has its value.

A HP test is usually sub-divided into two further stages when conducting the tests:

(i) first to locate the product cold point in the container; and
(ii) second to establish the process conditions that will lead to the scheduled process.

11.3 LOCATING THE PRODUCT COLD POINT

Within each food container, there will be a point or region that heats up more slowly than the rest. This is referred to as the 'slowest heating point' or 'thermal centre' and should be located using thermocouples or some other sensing method positioned at different places in a food container. For foods that heat mainly by conduction, the slowest heating point is at the container geometric centre. Also, if the process utilizes rotation or agitation, the slowest heating point is forced to the container geometric centre. However, for foods that permit movement and heat by convection, this point is between the geometric centre and approximately

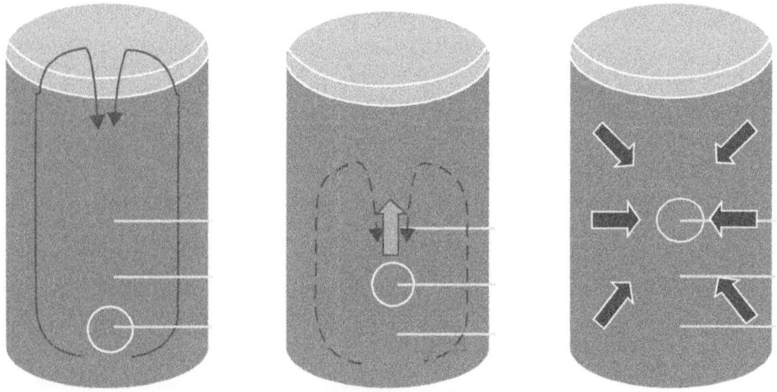

Fig. 11.1 Changes in the cold point location for a freely-moving liquid such as water, a mixed liquid such as a soup, and for a conduction heating product such as corned beef.

one-tenth up from the base (in a static process). During the retort process, the food viscosity decreases in response to increasing temperature, and as a result the slowest heating point moves downwards from the container geometric centre. Figure 11.1 illustrates how the cold point changes with viscosity. The critical time in the process is when the lethal effect on the target microorganisms is at its most significant, which is towards the end of the hold phase.

Convection and conduction heating products exhibit a single mode of heating, also known as simple heating, which results in a single coldest position within the container. This is not the case with broken heating products that change mode from convection to conduction or vice-versa. The cold point will move during the process and this requires the process authority to understand the changes that take place. Data to provide this information is obtained during the TD testing in which a single instrumented container should be placed in the retort load. This does not have to be in the retort cold spot, which may not be known at this stage of testing, but provides the time–temperature data that will inform about the changes in heating mode. It is the final mode of heating that is relevant to the process lethality. Temperature probes should be placed in locations relevant to this so that the product slowest heating point can be determined.

11.4 PROCESS ESTABLISHMENT METHODS

The theory behind microbiological reductions by heat is that the kinetics are approximately first-order. Thus, an F-value calculated using Equation (11.1) will be the same as that calculated from the time–temperature integration (Equation 11.2). Chapter 4 describes first-order kinetics in greater detail:

$$P \text{ or } F = \int_0^t 10^{\frac{T-T_{\text{ref}}}{z}} \cdot \text{dt} = D_T. \log\left(\frac{N_0}{N}\right) \tag{11.2}$$

where T is the product temperature, °C, T_{ref} is the reference temperature for the D_T value, °C, t is the process time, minutes, and the kinetic factor, z, is the temperature change required to effect a 10-fold change if the D_T value, °C.

This integration is usually done within the data logger software, to allow a thermal process to be operated until the target F- or P-value is reached. This method is referred to as the General method. Other methods that use the measured times and temperatures within predictive models are also acceptable and used widely. The most common are the Ball method and its numerous derivatives, and numerical methods using finite differences, for example CTemp (Campden BRI) and NumeriCAL (JBT FoodTech). These methods give flexibility to evaluate the process for different input variables such as product initial temperature, process temperature, and process time. Details of how each method works are not given here because of the complexities with using the methods. References are given at the end of this chapter to help those requiring information on the various methods.

In order that a process value can be calculated, it will need to be based on measurements of TD in the retort and HP in the product. TD is almost impossible to measure without temperature sensors; however, heat penetration usually uses temperature sensors but can also use log reduction methods. The next sections provide outline information on temperature measurement systems and log reduction methods, with references for further information should this be required.

11.4.1 Log reduction methods for HP testing

There are two main techniques that are classified as log reduction methods, and these use:

(i) microorganisms of known heat resistance, usually in spore form, in which the initial number (pre-process) and final number (post-process) are counted and used to calculate an F-value; and

(ii) biochemical systems, in which the initial concentration (pre-process) and final concentration (post-process) are measured and used to calculate an F-value.

Fig. 11.2 Picture of alginate spore beads and amylase TTI tubes.

Figure 11.2 shows an example of both types of technique. The alginate spore beads can be prepared using a high proportion of food material, typically more than 90%, which ensures the beads behave in a similar manner to the food. The silicone tubes used for enclosing the biochemical solutions are chosen so the density and thermal properties also match those of the food.

Irrespective of which technique is used, it is critical that the z-value is close to that for the target microbial species, preferably within 2–3 °C. If the z-value is not close, the processing temperature should be close to the reference temperature; otherwise, significant errors can arise between values estimated using log reduction and with probes (from Equation 11.2). Also, the decimal reduction time should allow sufficient log reductions to be measured in order that the process can be calculated. A test that results in no surviving spores or no biochemical activity does not allow the process F-value to be calculated and raises doubt as to where or when the total kill occurred. Some further detail is now given of each method.

11.4.1.1 Microbiological spore methods

These are often referred to as direct methods, but they in fact rely on measuring the achieved log reductions for a process using a non-pathogenic microorganism and converting this to a process value for the target pathogen using Equation (11.1). No factory tests for sterilization processes would use *C. botulinum* spores because of: (i) the hazards associated with handling these spores in a factory environment; and (ii) the low numbers of spores that

Table 11.1 Possible marker organisms for use with microbiological log reduction methods for low acid foods.

Spore-forming organism	T_{ref} (°C)	D_T (minutes)	z (°C)
C. botulinum	121.1	0.1–0.3	8–11
C. sporogenes	121.1	0.8–1.5	9–11
B. stearothermophilus	121.1	4–5	8–12
B. subtilis	121.1	0.1–1.0	10

would survive the process. The latter point is easily illustrated from the theory of an $F_0 3$ process that results in at least 12-log reductions in *C. botulinum* spores. With almost all commercial processes operating to safety margins many times larger than $F_0 3$, surviving spores will not be present and will only occur when the process goes wrong.

Hence, a non-pathogenic organism with a high $D_{121.1}$ value is used, for example spores of *Clostridium sporogenes* or *Bacillus stearothermophilus* (Table 11.1). An alternative is to use a gas-producing organism and estimate the severity of the process by the number of blown cans.

A microbiological method can be conducted using organisms distributed evenly throughout a food product (inoculation) or encapsulated in gel (alginate beads). The organisms are usually in their spore form because of the greater flexibility this provides in their use (Brown et al. 1984).

An inoculation test measures the average thermal process in a container (see Worked example 11.1). If the product is liquid it is relatively easy to introduce the spores but for solid products, it is necessary to first mix the spores into one of the ingredients to ensure that they are dispersed evenly throughout the container. Dispersion of the inoculum into the food dilutes the initial number of spores, usually to 10^3–10^5 per container, which reduces the *F*-value range that can be measured.

An encapsulation test allows the organisms to be placed at precise locations within a container or within a food particulate, by encapsulating spores in an alginate bead (see Worked example 11.2). The beads can be made with a high percentage of the food material so the heating rate is similar to the food. This method is used for continuous processes where the food contains particulates that require evaluation at their centres. This is considered separately in Chapter 7. The number of spores per bead is usually greater than for an inoculation test and can be of the order of 10^{6-8} per bead.

Worked example 11.1

The following example for an inoculation test uses 1 g of spores of *C. sporogenes* supplied by a microbiology laboratory with a measured $D_{121.1}$ value of 1.2 min and at a concentration of 5×10^7 per g. The can of food for process measurement has a mass of 400 g. The question is 'what is the maximum process value that can be measured with these spores?'

Equation (11.1) is used in the calculation:

$$F = D_T \cdot \log\left(\frac{N_0}{N}\right)$$

The initial numbers of spores in each can (N_0) is diluted to $5\times10^7/400$ or 125,000 spores per g. If we assume that the limit for spore numbers at the end of the process (N) is unity (1), then the equation becomes:

$$F = 1.2\log\left(\frac{125\,000}{1}\right) = 6.12$$

Hence, the maximum F-value that can be measured with these spores is 6.12 minute.

Worked example 11.2

The following example for an encapsulation test also uses spores of *C. sporogenes* with a measured $D_{121.1}$ value of 1.2 minute and at a concentration of 5×10^7 per g. The alginate bead weighs 0.2 g and is inserted at the centre of a 10 mm cube of carrot. The question is the same, which is 'what is the maximum process value that can be measured with these spores?' The initial numbers of spores in each bead (N_0) is $0.2\times5\times10^7$ or 1×10^7 spores in the bead. The same limit for spore numbers at the end of the process (N) applies (1), then the equation becomes:

$$F = 1.2\log\left(\frac{10\,000\,000}{1}\right) = 8.4$$

Hence, the maximum F-value that can be measured with these spores is 8.4 minute.

11.4.1.2 Biochemical systems

These can be enzymes, such as amylase or peroxidase, which denature (an unwinding of the structure) during heating with similar kinetics to that of the target microbial species. Most enzyme systems are limited to pasteurization because they are designed by microorganisms to operate at moderate temperatures. One exception is work by Tucker et al. (2007), in which amylase from the hyperthermophilic organism *Pyrococcus furiosus* is being used. This organism grows in saltwater pools in volcanic regions, such as Vulcano Island in Italy, and the amylase it secretes is naturally designed to be thermostable.

Calculation of pasteurization or P-value uses the initial and final enzyme activities from Equation (11.3), which also assumes the first-order thermal death time model applies. This is the same equation for calculating integrated process values when using microbiological spore methods, except that the number of organisms is replaced by the amylase activity:

$$P = D_T \cdot \log\left(\frac{A_{\text{initial}}}{A_{\text{final}}}\right) \tag{11.3}$$

where A_{final} is the final activity after a specific time–temperature history, A_{initial} is the initial activity and D_T is the decimal reduction time to achieve a 1-log reduction in amylase activity, minutes.

Biochemical systems are often referred to as time–temperature integrators (TTIs), although strictly speaking the term TTI is also applicable to microbiological systems. Encapsulating the TTI is an essential step that prevents it coming into contact with the food or processing environment. One TTI encapsulation method (Tucker et al. 2002) uses silicone tubing of 2.5 mm bore with the liquid TTI sealed in by silicone end plugs (Fig. 11.3). The amylase solution is recovered from the tube using a hypodermic syringe.

Amylase solutions are one of the few biochemical systems with appropriate kinetics (Tucker 2000), which requires: (i) the D-value to be large enough for residual activity to remain after the process; and (ii) the z-value to be in the 8–11 °C range typically measured for microorganism death. Table 11.2 illustrates some of the key attributes of a specific amylase TTI system.

Table 11.2 Key attributes of the *Bacillus amyloliquefaciens* α-Amylase TTI.

Category	Description
Operating principle	Reduction in amylase activity in response to time and temperature
Measurement method	Amylase assay to measure absorbance rate, using a spectrophotometer
Active temperature range	60–100 °C
Kinetic factor, or *z*-value	9.7 ± 0.3 °C
Decimal reduction time	$D_{80.7}$ = 18.7 min
Process value	'pasteurization value'
Sample size	0.02 mL

11.5 PROCESS CALCULATION METHODS

Several calculation methods are available for setting the scheduled process conditions. The most commonly used methods will be described in outline here, with the references provided for further information.

11.5.1 General method

The reference calculation method is known as the General method (Bigelow et al. 1920) and it provides a process value (*F*-value) that is sometimes sufficient for validation purposes. It converts measured times and temperatures to *F*-values from accumulated lethal rates. A lethal rate is a relative term that compares the microorganism killing effect at a measured temperature to one minute at a reference temperature (also discussed in Chapter 4). Equation (11.2) presents the lethal rate equation as part of the integrated *F*-value calculation, but the lethal rate (*L*) itself is as given in Equation (11.4):

$$L = 10\frac{T - T_{\text{ref}}}{z} \tag{11.4}$$

Integration of lethal rates over the measured times and temperatures is usually done using the trapezoidal method, with the calculation routines embedded within a data logger software. The General method allows the user to follow the *F*-value calculation during the process, usually in real time (with probes), but for loggers this is done when they are removed from the processing system. Time to achieve a target F-value is determined from the data and the process conditions can be established. While this is a simple method it does have some major limitations:

- F-values are relevant to the TD and HP test conditions on the day of testing and cannot be transferred to another set of conditions.
- Changes in critical variables cannot be evaluated without a new HP test, for example initial product temperature, retort temperature, come-up time, cooling profile.
- Deviations to the established process conditions cannot be evaluated by calculation and need to experimentally simulated.

The General Method was the first method used for the design and evaluation of thermal processes. The main limitation associated with the General Method is the need for time/temperature data in order to evaluate a particular thermal process; no means are provided to move between different processing conditions (e.g. different initial product or processing temperature). Despite these limitations, attempts have been made to apply the General Method to controlling canning retorts although with limited flexibility to adapt to changing retort conditions.

11.5.2 Ball method

In order to overcome the limitations of the General Method, formula methods that use theoretical or empirical equations for the description of the temperature evolution inside the container have been developed. One of the first formula methods was proposed by Ball (1923, 1927) and in spite of its limitations it is still widely used for the evaluation of thermal processes. Other formula methods worth referring to due to their use by the canning industry include Stumbo (1965), Gillespy (1951, 1953), and Hayakawa (1970, 1971).

Apart from log reduction and General method calculations, all process calculation methods use heating (f_h) and lag (j) factors. Heating factors originated in the canning industry (Ball 1923, 1927) as a measure of the product heating rate used for calculating process times for canned foods. By definition, the heating factor is the time taken for the difference between environment and product temperature to reduce by 90%. They are important terms in thermal process data analysis, because they provide information on the rate of heating for a container of food. Figure 11.3 shows an example of a logarithmic heating curve used to calculate f_h and j factors.

The Ball method is widely used within the USA and countries exporting thermally processed foods to the USA. Its use in Europe is less widespread. However, the Ball method offers options to the General method that enables some analysis of what-if scenarios to be calculated. The Ball method has three parts to it and can be described in its most simple form as:

(i) an equation to the straight heating line in Fig. 11.3, which is effectively the gradient at the end of heating,
(ii) this is connected by an experimentally determined complex hyperbolic function to
(iii) an equation to the straight-line cooling line.

The above description does not do the Ball method justice but does describe the constituent parts. The Ball method was derived for steam retorts because these were the main type used around the time Ball carried out his experiments (Ball 1923, 1927). Hence the complex connecting routine between heating and cooling is most accurate for steam retorts and adjustments to this are required for other retort types, for example water immersion and water spray or cascade retorts. It is important to note that the Ball method calculates an F-value at the end of cooling. Routines can be used to back-calculate an end of heating F-value but these begin to lose their accuracy. If an end of heating F-value is required, then it is easier to use only the first equation to the

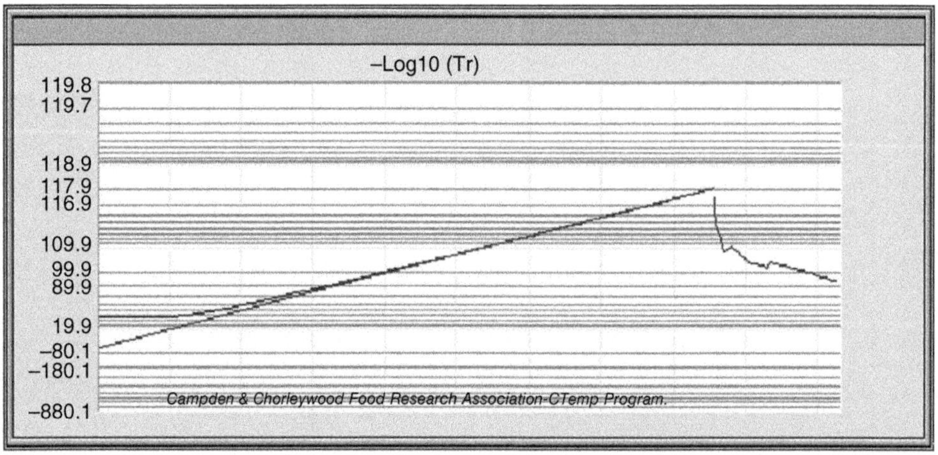

Fig. 11.3 Example of a logarithmic heating curve for canned meat used to calculate f_h of 50.4 minutes and j of 1.45.

heating part of the logarithmic heating curve, and express this as in Equation (11.5):

$$\frac{RT - T}{RT - IT} = j \cdot e^{-2.303t/fh} \tag{11.5}$$

where j is the heating lag factor, defined as a measure of the thermal lag before the can temperature responds to the changing environment temperature (dimensionless), f_h is the heating factor, defined as the temperature response parameter derived from the logarithmic heating curve (minutes), RT is the retort temperature during the hold period, °C, IT is the product initial temperature at steam on, °C, and T is the product temperature after heating time (t), °C.

Most Ball method calculations are now computerized versions of the tables and nomograms published by Ball (1923, 1927). Several attempts have been made since Ball to improve on the method, particularly for cooling lethality calculation.

11.5.3 Numerical methods

One major limitation of formula methods is the assumption that the external heating temperature must be constant during the heating and cooling periods. Variable heating medium temperatures, e.g. deviations in the retort temperature during the holding phase, cannot be modelled directly using formula methods.

In 1969 finite difference solutions of Fourier's conduction equation were shown to be a valuable alternative for the design and evaluation of thermal processes (Teixeira et al. 1969). Finite difference solutions of the conduction equation provide an accurate and flexible way of modelling the temperature evolution during thermal processing of packaged foods. Unlike formula methods, finite difference solutions of the conduction equation allow an easy incorporation of the experimental (time variable) retort temperature profile, thus allowing the evaluation of process deviations consisting of departures from the scheduled retort temperature.

The use of more advanced mathematical models to evaluate and predict process times and temperatures have increased as the computing power available on desk-top PCs has increased. Examples of numerical calculations are CTemp and NumeriCAL software, both of which utilize finite differences. These models deal with the physics of heat transfer and use finite differences to solve Fourier's partial differential equations. Therefore, they adopt a more scientific approach that allows variable retort profiles as the boundary conditions. This means that come-up and cool-down profiles can be defined as well as deviations in processes caused by retort temperature fluctuations (Tucker et al. 1996).

CTemp was the program written by Campden BRI that converted the finite difference models into a format that was relevant to analysis of heat penetration data. It has received widespread acknowledgement and is used globally by food companies establishing thermal processes. Campden BRI uses CTemp as one of the tools to establish process times based on heat penetration data. The models within CTemp originated as pure conduction heating back in 1989 and remained this way until collaboration with university of Leuven in Belgium enabled an approximation to convection to be included. This development took the approaches to modelling convection originated by Noronha et al. (1995) and incorporated them into a much-changed version of CTemp. The outputs from this improved model had a far greater impact than the original conduction heating finite difference models.

The accuracy of finite difference predictions for core temperature will diminish as the degree of convection increases. The convection approximation of Noronha et al. (1995) used the fact that the heating factor (f_h) was fixed in value within a solid geometry but the lag factor (j) varied from the centre to the edge. By adjusting j to suit the needs of the actual food material it was possible to obtain an improved match of predicted time–temperature date to experimental data.

The finite difference solutions of the partial differential conduction equations required an input value for the thermal diffusivity of the material. This value dictated the heat transfer rate through the product. Thermal diffusivity values can be looked up in tables or calculated from centre temperature data taken under conditions of constant retort temperature. It is a composite term that has similarities to the heat transfer coefficient (U) in Equation (11.5). Hence, it was referred to as the apparent thermal diffusivity to indicate that it was not a true representation of the thermal properties of the product, but also included the effects of heat transfer at outside and inside surfaces of the container as well as through the container wall.

The apparent thermal diffusivity (α) can be derived from the f_h value using the following relationships, Equations (11.6) and (11.7). These were derived from the first term approximations to the analytical solutions for a finite cylinder and for a rectangular parallelepiped:

For a finite cylinder:

$$\alpha = \frac{0.398}{60 \cdot f_h}\left(\frac{4}{d^2} + \frac{1.708}{l^2}\right)^{-1} \tag{11.6}$$

For a rectangular parallelepiped:

$$\alpha = \frac{0.233}{60 \cdot f_h}\left(\frac{1}{a^2} + \frac{1}{b^2} + \frac{1}{c^2}\right)^{-1} \tag{11.7}$$

Heating factor data to input into CTemp were calculated from heat penetration data, in the same way that was developed for the Ball formula method. Minimizing the influence of the retort come-up time was an essential component of these experiments.

In order to speed up the computations, the researchers based their finite difference models on a spherical geometry (Tucker et al. 1996). This was a further approximation that had a sound theoretical basis through similarities in the mathematics of analytical solutions to conduction heat transfer. CTemp was written to calculate can centre temperatures based on good quality heat penetration data that allowed the user to estimate the input parameters (f_h and j). Hence the model predicts temperatures from the position of the heat penetration probe. This will be at the point of slowest heating so that worst-case sterilization values are predicted for that container.

This was a semi-empirical method based on the numerical solution of the heat conduction equation for a sphere (Tucker et al. 1996). The model allowed the incorporation of the empirical heat penetration parameters f_h and j in the conduction model. The incorporation of the heat penetration parameter f_h in the conduction model for a sphere was performed by calculating an apparent thermal diffusivity using Equation (11.8):

$$\alpha = \frac{0.233}{60 \cdot f_h} \cdot R^2 \tag{11.8}$$

In order to incorporate the j value in the model Equation (11.9) was used. It shows how the theoretical j value varied across the radial position of a conduction heating sphere, initially at homogeneous temperature but immersed in a medium of constant temperature. In canning terms, this is represented by the can being heated by steam injected into the retort.

$$j(r)=\begin{cases} 2.0, r=0. \\ 0.63662\dfrac{R}{r}\sin\left(\dfrac{\pi \cdot r}{R}\right), 0<r\leq R \end{cases} \tag{11.9}$$

Using Equation (11.9) it was possible to determine a position inside the sphere that had a given *j* value. The model allowed simulation of the temperature evolution of any material showing j values from 0 at the surface to 2 at the core. Equation (11.9) was based on a step change in the ambient temperature. The classical method of determining heat penetration data by plotting the deficit of differences between process and product temperature in a semi-log plot is not appropriate as *j* values calculated using this procedure will be dependent on the duration and shape of the retort come-up time. Determination of the heat penetration parameters (f_h and *j*) for use with CTemp required an optimization routine (Saguy 1983) that minimized the sum of squares of the differences between experimental and predicted temperatures. The heat penetration parameters determined from the semi-log graph were used as initial estimates for the optimization procedure.

The convection approximation within CTemp extended the applicability of the numerical solutions to non-conductive foods whose heat penetration curves (in semi-log coordinates) showed a linear behaviour. This included those foods that exhibited convection or broken heating behaviour. Other finite difference programs, such as NumeriCAL, adopt similar methods to approximate for convection heating.

These predictive modelling approaches not only help with deciding the fate for batches of product that have undergone a process deviation, but the task of process establishment is made more straightforward. The models can be used to evaluate changes to initial product temperatures, shortened come-up times, or different retort temperatures. They offer considerable flexibility to make the job of process establishment easier. Applications for numerical models to process deviation analysis are discussed in Chapter 13.

11.5.4 Continuous flow with particulates

Continuous thermal processing of foods containing discrete particulates is not as common as with in-pack processing. There are various reasons for this. Equipment for processing and filling has to be specific for the type of food flowing within the heat exchange system, and this limits the range of foods that can be manufactured (see Chapter 8). Continuous systems are suited to long run times in which production efficiencies can be achieved. Changing products incurs waste at the interface because of the large amount of food inside the processing and filling systems.

When it comes to process calculations there are numerous assumptions for the flow and heat transfer conditions that need to be made. These differ substantially from in-pack processed foods because of the uncertainties with how long and what temperature is experienced by the liquid and particulate portions. Safety margins used in the calculations are greater, not solely down to the assumptions, but also the high processing temperatures compared with the reference temperatures (T_{ref}) used in the lethality calculations. To gain operational efficiency it is typical to process at 135–140 °C for a sterilization process. Lethal rates build up fast when the particulate centre temperatures are above the T_{ref} of 121.1 °C. Less than one second is required to achieve $F_0 3$ at 140 °C. The margin for processing error at such high temperatures is incredibly small.

One further complication is that the complexity and cost of validating the process achieved by a particulate are considerably more than with in-pack processes. This is because temperature sensors cannot be used in a flowing situation therefore other techniques must be used. There are self-contained temperature logging systems available, but they are currently too large and dense to fit into a flowing food particulate. The smallest commercial systems are based on i-buttons (www.ibuttonlink.com), which are stainless steel units approximately 16 mm diameter and either 3.1 or 5.9 mm thickness. Temperatures up to 140 °C can be measured with the most extreme versions. Smaller remote temperature sensors have been available for many years, for example to measure temperatures inside the human body, but these cannot withstand high temperatures.

The alternatives to temperature sensors are log reduction techniques such as microorganisms embedded in alginate beads or biochemical time–temperature integrators. These are made to represent the food's thermal and physical properties. They are introduced into the heat exchangers, travel with the food, and are recovered after cooling. These are difficult to set up and limited to a few specialist organizations who offer them. Section 11.4.1 describes the relevant features of both systems.

Prediction of temperatures and lethal rates for flowing particulates allows the processing conditions to be evaluated for specific food products. This is helpful in determining the operating conditions prior to testing the system with several tonnes of food product. This requires a flexible model that uses the measured liquid temperatures at key points in the process. Stoforos (1992) provided an overview of the challenges faced with validating aseptic processes for particulates. One example of a model for calculating temperatures inside food particulates was developed by McKenna and Tucker (1991). This was based on the finite difference models developed for the canning sector (Tucker 1991; Tucker and Holdsworth 1991) in which external heat transfer coefficients were applied. The program was similar to CTemp but designed for predicting temperatures at the centre of food particulates, given the name PCTemp. The core of the program was the 1D finite difference models that calculated internal temperatures at numerous positions within the spherical particulate geometry.

There were essentially three aspects to the PCTemp program:

a. Calculation of the heat transfer coefficient at the particulate surface. This was based on published correlations for Reynolds (flow term), Nusselt (heat transfer term), and Prandtl (physical properties) numbers. Reynolds and Prandtl numbers were only relevant if a slip velocity existed between the liquid and particulate. In this case, it required values for the liquid rheological properties and the thermal properties of both liquid and particulate at elevated temperatures. None of these data is easy to obtain and requires approximations to estimate realistic values. The default condition was zero slip velocity, which was the worst-case scenario and typically used by industry. The Ranz-Marshall correlation was the default value (Ranz and Marshall Jr 1952), which has a numerical value of 2.0 with zero slip velocity.
b. Calculation of the particulate surface temperature change with time. This required the calculated heat transfer coefficient from the chosen correlation, the default being Ranz-Marshall, coupled with the liquid time–temperature history. The latter is usually estimated based on fixed probe positions at the heat exchanger inlet, holding tube inlet and outlet, and cooler outlet.
c. Internal particulate temperatures were calculated from the finite difference models using the calculated thermal diffusivity and particulate dimensions. The latter were converted to those for an equivalent sphere using equations from Ball and Olson, as with the approach described earlier for CTemp. Temperatures were converted to lethalities so that the time required to achieve a target sterilization or pasteurization value was obtained.

It is common practice to adopt the worst-case assumptions when making these calculations, as is done for in-pack processing. However, for continuous processes these are more severe. These are that:

a. The particulate is stationary with respect to the surrounding carrier liquid (the food product), giving a zero slip velocity, as described above. This negates the need to calculate Reynolds and Prandtl numbers. If the Ranz-Marshall correlation is used the values for surface heat transfer coefficient (h_o) range from 27 to 140 W.m^{-2}.K^{-1} depending on values for thermal conductivity and particulate dimensions. Other correlations give similarly low values. If the slip velocity is increased marginally to 0.01 m.s^{-1} the h_o increases to over 700 W.m^{-2}.K^{-1}. It is likely that some movement of liquid across the particulate surface takes place during flow; hence, the assumption of zero slip velocity is conservative.
b. The particulate residence time in the holding tube is double that of the mean for the food product. This is based on Newtonian flow for the carrier liquid, as shown in Chapter 7. This requires Reynolds numbers in the laminar region of <4000 (for FDA purposes) or <2100 otherwise. This is a reasonable assumption for liquid foods that must be sufficiently viscous to carry particulates in continuous flow systems. Data from Worked Example 7.1 for tomato ketchup demonstrates low Reynolds numbers even with medium viscosity foods. Most continuously processed foods with discrete particulates are higher in viscosity therefore the Reynolds numbers will be even lower.
c. All lethality is accrued in the holding tube. This is despite the high temperatures within the particulates in the latter stages of heating and early stages of cooling. These will contribute towards a lethal effect but cannot be taken into account in the calculations.

It would be fair to state that these assumptions err considerably on the safe side. For example, it is not possible for the first two assumptions (a and b) to both be true in the same flow situation. This because the presence of a velocity profile within the liquid infers that a slip velocity must be present across the particulate surfaces. The increase in h_o from 27 to 140 to over 700 W.m^{-2}.K^{-1} illustrates the impact of a very low slip velocity (0.01 m.s^{-1}). In addition, temperature measurements with an in-pack product containing particulates will accrue lethality throughout heating and holding based on a probe located in the centre of the largest particulate. It is unusual to remove lethal rates during the retort come-up time, which for convection heating foods can make a contribution to total lethality.

The descriptions above on the process validation challenges for flowing particulate systems are intended as an introduction to the subject. It is every bit as complex as in-pack processing using retorts though not as widely commercialized. Readers are advised to consult the numerous articles and books dedicated to this subject.

References

Ball, C.O. (1923). *Thermal Process Time for Canned Food.* Washington DC: Bulletin of the National Research Council Vol. 7, Part 1, Number 37.

Ball, C.O. (1927). Theory and practice in processing. *The Canner* 64 (5): 27.

Bee, G.R. and Park, D.K. (1978). Heat-penetration measurement for thermal-process design. *Food Technology* 32 (6): 56–58.

Bigelow, W.D., Bohart, G.S., Richardson, A.L., and Ball, C.O. (1920). Heat Penetration in processing canned foods. *National Canners Association Bulletin* 16-L: 128, Washington DC.

Brown, K.L., Ayres, C.A., Gaze, J.E., and Newman, M.E. (1984). Thermal destruction of bacterial spores immobilsed in food/alginate particles. *Food Microbiology* 1: 187–198.

Campden BRI (1977). *Guidelines to the Establishment of Scheduled Heat Processes for Low-Acid foods. Technical Manual No. 3.* Chipping Campden UK: Campden BRI.

Campden BRI (1992). *Pasteurisation Heat Treatments. Technical Manual No. 27.* Chipping Campden UK: Campden BRI.

Campden BRI (1997a). *Guidelines for Performing Heat Penetration Trials for Establishing Thermal Processes in Batch Retort Systems. Guideline No. 16.* Chipping Campden UK: Campden BRI.

Campden BRI (2006). *Pasteurisation: A Food Industry Practical Guide. Guideline No. 51*, 2ee. Chipping Campden UK: Campden BRI.

Campden BRI (2008). *Heat Processing of Packaged Foods – Guidelines for Establishing the Thermal Process. Guideline No. 56.* Chipping Campden UK: Campden BRI.

DoH (1994). *Guidelines for the Safe Production of Heat Preserved Foods.* London: HMSO (out of print).

Gillespy, T.G. (1951). Estimation of sterilising values of processes as applied to canned foods. I. Packs heating by conduction. *Journal of the Science of Food and Agriculture* 2: 107–125.

Gillespy, T.G. (1953). Estimation of sterilising values of processes as applied to canned foods. II. Packs heating by conduction: complex processing conditions and the value of coming-up time of retort. *Journal of the Science of Food and Agriculture* 4: 553–565.

Hayakawa, K. (1970). Experimental formulas for accurate estimation of transient temperature of food and their application to thermal process evaluation. *Food Technology* 24: 1407–1418.

Hayakawa, K. (1971). Estimating food temperatures during various processing or handling treatments. *Journal of Food Science* 36 (3): 378–385.

IFTPS (1992). *Temperature Distribution Protocol for Processing in Steam Still Retorts, Excluding Crateless Retorts*. Fairfax VA: IFTPS.

IFTPS (1995). *Protocol for Carrying Out Heat Penetration Studies*. Fairfax VA: IFTPS.

McKenna, A.B. and Tucker, G.S. (1991). Computer modelling for the control of particulate sterilization under dynamic flow conditions. *Food Control* 2: 224–233.

NFPA (1985). *Guidelines for Thermal Process Development for Foods Packaged in Flexible Containers*. Washington DC: NFPA.

Noronha, J., Hendrickx, M., Van Loey, A., and Tobback, E. (1995). New semi-empirical approach to handle time-variable boundary conditions during sterilisation of non- conductive heating foods. *Journal of Food Engineering* 24: 249–268.

Ranz, W.E. and Marshall, W.K. Jr. (1952). Evaporation from drops. *Chemical Engineering Progress* 48: 141.

Saguy, I. (1983). Optimization of dynamic systems utilizing the maximum principle. In: *Computer-Aided Techniques in Food Technology* (ed. I. Saguy), 263–320. New York: Marcel Dekker.

Stoforos, N.G. (1992). An overview of aseptic processing of particulate foods. *Developments in Food Science* 29: 665–677.

Stumbo, C.R. (1965). *Thermobacteriology in Food Processing*, 1ee. New York: Academic Press, Inc.

Teixeira, A.A., Dixon, J.R., Zahradnik, J.W., and Zinsmeister, G.E. (1969). Computer optimization of nutrient retention in the thermal processing of conduction-heated foods. *Food Technology* 23: 845–850.

Tucker, G.S. (1991). Development and use of numerical techniques for improved thermal process calculations and control. *Food Control* 2: 15–19.

Tucker, G.S. and Holdsworth, D. (1991). Mathematical modelling of sterilisation and cooking processes for heat preserved foods. Application of a new heat transfer model. *Transactions of the IChemE* 69 (C1): 5–12.

Tucker, G.S. (2000). Estimation of pasteurisation values using an enzymic time-temperature integrator. *Food Australia* 52 (4): 131–136.

Tucker, G.S., Brown, H.M., Fryer, P.J. et al. (2007). A sterilisation time-temperature integrator based on amylase from the hyperthermophilic organism Pyrococcus furiosus. *Innovative Food Science and Emerging Technologies* 8: 63–72.

Tucker, G.S., Lambourne, T., Adams, J.B., and Lach, A. (2002). Application of biochemical time-temperature integrators to estimate pasteurisation values in continuous food processes. *Innovative Food Science & Emerging Technologies* 3: 165–174.

Tucker, G.S., Noronha, J.F., and Heydon, C.J. (1996). Experimental validation of mathematical procedures for the valuation of thermal processes and process deviations during the sterilization of canned foods. *Transactions of the Institution of Chemical Engineers, Food & Bioproducts Processing* 74 (C): 140–148.

12 Cooling Water Treatment

The importance of water disinfection cannot be underestimated because it is arguably the single most significant public health measure of the last century (see examples in Chapters 2 and 4). Almost all types of microorganisms can be found in water, and water is most often the vehicle of transmission for these organisms. Treatment of water to make it safe for human consumption is therefore very important, and today there are standards set for potable water.

All water that comes into contact with foods should be safe and comply with microbiological standards for drinking water. Water that may come into contact, intentionally or unintentionally, and needs to meet microbiological requirements such that the safety of the food for consumers is not compromised.

Intentional food contact applications include the use of water:

- As an ingredient.
- For intentional food contact applications, e.g. washing, fluming, blanching, brining, soaking.
- For intentional food contact surface applications, e.g. cleaning and sanitation of surfaces/equipment.

Unintentional applications are those for which cross-contamination of the water with food or food contact surfaces cannot be fully and consistently excluded, e.g. cleaning and sanitation of non-food- contact surfaces of processing equipment and production areas and can cooling water (FAO and WHO 2019).

The microbiology of the water used to cool the cans or other food containers is important (Warne 1988; Featherstone 2015). Contaminated cooling water can result in contamination of processed products in the containers. The cans (or other containers)

Essentials of Thermal Processing, Second Edition. Gary Tucker and Susan Featherstone.

are exposed to the water when still hot, and the seaming compound soft. As the metal cools a vacuum begins to form, which can result in minute amounts of water actually leaking into the containers. If the cooling water is contaminated, the containers are exposed to microorganisms at a time when the container is vulnerable. Bacteria or other microorganisms could enter the container, resulting in spoilage. If microorganisms such as *Escherichia coli*, other enteric bacteria, or mesophilic spore-forming bacteria are present in the water, it could make the product unfit for consumption. With contaminated can cooling water, leaker spoilage can often be found in cans, even with good quality double seams.

There are many ways that water can be rendered safe for food contact use and safe for use in the cannery. They are either chemical or physical treatments:

Chemical Treatments:

(i) Chlorine;
(ii) Chlorine dioxide;
(iii) Bromine;
(iv) Ozone;

Physical Treatments:

(i) Ultraviolet light;
(ii) Membrane filtration;

12.1 CHLORINE

Chlorine is used extensively as a disinfectant and oxidising agent in clean water applications. It is widely used as a disinfectant in swimming pools and cooling water recirculation systems. Typical applications include:

- Potable water distribution systems;
- Cooling tower recirculation water;
- Swimming pool water;
- Food and beverage processing.

When chlorine is dispersed in water, it hydrolyses very rapidly to form a variety of molecular and ionic species, which exist in a variety of forms, such as elemental chlorine (Cl_2), hypochlorous acid (HOCl), hypochlorite ion (OCl^-), and substituted amines or chloramines ($NH_{(3-X)}Cl_X$). Of these, the reactively dominant species for most aqueous chlorination reactions is HOCl. The hypochlorous acid is a weak acid that dissociates to form H^+ (hydrogen) and OCl^- (hypochlorite) ions, the concentrations of which are dependent on the pH of the solution.

Different forms of chlorine used in water treatment are Cl_2 (chlorine gas), $Ca(OCl)_2$ (calcium hypochlorite), and (NaOCl) sodium hypochlorite. The form of chlorine in aqueous solution influences the antimicrobial properties of the sanitizing solution. When elemental chlorine or hypochlorites are added to water, they undergo the following reactions:

- $Cl_2(g) + H_2O \leftrightarrow HOCl + H^+ + Cl^-$
- $NaOCl + H_2O \leftrightarrow HOCl + NaOH$
- $Ca(OCl)_2 + H_2O \leftrightarrow HOCl + Ca(OH)_2$

The hypochlorous acid (HOCl), being a weak acid, partially dissociates relative to the solution pH:

- $HOCl \leftrightarrow + H^+ + OCl^-$

At pH 7.5, the concentrations of HOCl and OCl^- (hypochlorous ion) are approximately equal. Above pH 7.5 the OCl^- will predominate and below pH 7.5 HOCl will be in a higher concentration. The relative concentration of HOCl is therefore dependent on solution pH.

The relative concentration of HOCl to OCl^- will also vary with temperature. As the temperature increases, the relative concentration of HOCl decreases.

The antimicrobial efficacy and sporicidal effectiveness of chlorine solutions decrease with increase in pH. HOCl is approximately 80–100% more effective than OCl^- in killing bacteria. It is clear that the activity of chlorine solutions is greatly affected by pH, with the greater activity at lower pH values.

12.1.1 Chlorine demand and residual chlorine

When one or more forms of chlorine are added to water, the amount of chlorine remaining in the water is less than the amount originally added. This is because the water has a certain 'chlorine demand'. The chlorine demand is the chlorine dose minus the chlorine residual (chlorine in the solution after a given contact period). The residual chlorine may consist of two forms, combined chlorine, and free chlorine. The term 'total chlorine' is used to refer to the sum of these two forms of chlorine. HOCl, OCl^- and Cl_2 are collectively referred to as free chlorine.

$$\text{Total chlorine} = \text{combined chlorine} + \text{free chlorine}$$

Ammonia (NH_3) or organic nitrogen species (urea, proteins, etc.) that may be present in water will react with the free chlorine to produce monochloramines and later dichloramines and trichloramines. These chlorinated ammonia compounds, as well as other chlorinated organic amines, represent the combined chlorine species. The disinfection ability of combined chlorine is about 1% that of free chlorine. Combined chlorine species may also produce odour and taste problems in drinking water, and have been categorized as suspected carcinogens.

12.1.2 Using chlorine

At the normal pH of cooling water, 'free' available chlorine is a more effective bactericide than 'combined' residual chlorine. It is usual to dose cooling water so that free available chlorine remains detectable after a contact time of 20 minutes. Excessive chlorination of cannery cooling waters is wasteful and it also should be avoided because chlorine is corrosive to the metals used in construction of thermal processing vessels.

As the lethal effect of chlorination increases at low pH (where hypochlorous acid predominates), the acidity of the cooling water must be ensured, but there are practical constraints as to how low the pH can be, given that normal cannery cooling water is in the pH range of 6.5–8.5.

Temperature must also be considered because, at high temperatures, chlorine solubility is reduced, with the result that chlorine gas is driven off.

High levels of organic matter will increase chlorine demand and, like inorganic impurities, they also protect bacterial contaminants.

Under good manufacturing conditions, it is sufficient to maintain residual free available chlorine levels of 2–4 ppm after a 20-minute contact time in order to be confident of holding total aerobic counts at less than 100 organisms/ml of cooling water.

Free available chlorine should be still detectable in the cooling water at the completion of the cooling cycle. This is almost certain to be one of the CCPs in a cannery. For quality control purposes (and for the HACCP), records of free available chlorine levels should be maintained to provide confirmation that cooling water chlorination procedures were adequate.

12.2 CHLORINE DIOXIDE

Chlorine dioxide (ClO_2) is a synthetic, green-yellowish gas with a chlorine-like, irritating odour. Chlorine dioxide is a neutral chlorine compound. Chlorine dioxide (ClO_2) and chlorine are both powerful oxidising agents but differ in their chemical structure and mode of action. In diluted, watery solutions, chlorine dioxide is a free radical. At high concentrations, it reacts strongly with reducing agents. ClO_2 has 2.5 times the oxidising capacity of Cl_2, and so ClO_2 is an effective biocide. However, it is less effective than chlorine at equal concentrations when the organic content of the water is high.

Chlorine dioxide does not hydrolyse in aqueous solutions and is therefore not subject to dissociation in the manner of hypochlorous or hypobromous acids. The ClO_2 is the active biocidal compound.

Chlorine dioxide is an unstable gas that dissociates into chlorine gas (Cl_2), oxygen gas (O_2), and heat. The end-products of chlorine dioxide reactions are chloride (Cl^-), chlorite (ClO^-), and chlorate (ClO_3^-). One of the most important qualities of chlorine dioxide is its high-water solubility, especially in cold water. Chlorine dioxide does not hydrolyse when it enters water; it remains a dissolved gas in solution. It is therefore not subject to dissociation in the manner of hypochlorous or hypobromous acid. The ClO_2 is the active biocidal compound.

There are many advantages of chlorine dioxide over other methods of water disinfection. The bactericidal efficiency is relatively unaffected by pH values between 4 and 10. Chlorine dioxide is superior to chlorine in the destruction of spores, bacteria, viruses, and other pathogen organisms on an equal residual base. Also, the required contact time for ClO_2 is lower. There are no similar corrosion issues as are associated with high free chlorine concentrations, although sodium chloride is generated as part of the chemical reaction when chlorine dioxide is made and can affect the water quality. Chlorine dioxide does not react with ammonia or amine compounds and destroys phenols.

Disadvantages of chlorine dioxide are it is explosive under pressure, difficult to transport and is usually manufactured on site. Chlorine dioxide is usually produced as a watery solution or gas. It is produced in acidic solutions of sodium chlorite ($NaClO_2$), or sodium chlorate ($NaClO_3$). For large installations sodium chlorite, chlorine gas (Cl_2), sodium hydrogen chlorite ($NaHClO_2$), and sulphuric or hydrogen acid are used to produce chlorine dioxide on site.

To produce chlorine dioxide gas, hydrochloric acid (HCl) or chlorine is brought together with sodium chlorite. There are two main methods of chlorine dioxide generation; the 2-chemical and the 3-chemical methods, named because of the reacting species involved (Table 12.1).

- $5NaClO_2 + 4HCl \rightarrow 4ClO_2 + 5NaCl + 2H_2O$
- $HCl + NaOCl + 2NaClO_2 \rightarrow 2ClO_2 + 2NaCl + NaOH$

Table 12.1 Comparing the 2- and 3-chemical method systems for the formation of chlorine dioxide.

	2-chemical method	**3-chemical method**
Precursor Chemical	Sodium Chlorite Hydrochloric Acid	Sodium Chlorite Hydrochloric Acid Hypochlorous Acid
Salt formation	On a molar basis, 25% more salt is formed than chlorine dioxide	On a molar basis, the same amount of salt is formed as chlorine dioxide
Reaction Yield	Relatively low (around 70%) meaning that relatively more reactant remains in solution	Relatively high (around 95%) meaning that relatively less reactant remains in solution
Corrosive Effect	Slightly more salt produced, and more product left in solution, so a slightly higher corrosive effect is expected	Slightly less salt produced, and less product left in solution, so a slightly lower corrosive effect is expected
Chemical Demand	Slightly more precursor chemical is required (9 mol to form 4 mol ClO_2)	Slightly less precursor chemical is required (8 mol to form 4 mol ClO_2)
Chemical Cost	As slightly more chemical is required, chemical/running cost should be slightly higher	As slightly less chemical is required, chemical/running cost should be slightly lower
Space Used	A relatively small generator is required with two precursor storage tanks and mixing pumps	A relatively large generator is required with three precursor storage tanks and mixing pumps
Capital Cost	Relatively low	Relatively high
Safety	Both reactions are stable and the generators safe to work with. The largest health/safety risk comes from the chemicals used and produced	
Operation	Both systems can easily be set up to output the required concentration (2000 ppm) of ClO_2 and both can meet the demand for dosing water as required (generally between 0.2 and 0.8 ppm). Flow rate is dependent on only generator size not system.	

For most applications, a short contact time is needed (about 20–30 seconds). The 3-chemical method is generally preferable as it results in less corrosion potential (pH and salt generation). Dosing is usually done, not at the cooling water dam, but rather on route to the cooker, thereby allowing a much smaller dose (as less is lost due to turbulence and sunlight). Furthermore, a smaller dose means less corrosion risk.

12.3 BROMINE

Bromide is a common constituent of natural waters at concentrations in the range of fractions of a part per million. Bromine also exhibits biocidal properties in water and is used as a water disinfectant, especially in cannery cooling waters. The bromine hydrolyses in water to give hypobromous acid (HOBr).

Bromine is usually added to water in the form of sodium bromide (NaBr). This dissociates in solution to form Na^+ and Br^- ions. Whenever chlorine or hypochlorite is added to water containing Br^-, there is a rapid formation of HOBr according to the reaction:

$$Br^- + HOCl \leftrightarrow HOBr + Cl^-$$

The resulting HOBr is also an electrophilic agent, but one that tends to react more rapidly than HOCl. It is therefore more bacteriocidal than HOCl. Under alkaline conditions, hypobromous acid dissociates to form the hypobromite ion, just as hypochlorous acid forms the hypochlorite ion. However, the dissociation to OBr^- and H^+ occurs at higher pH values. Bromine is therefore still effective (as a biocide) at much higher pH values than chlorine.

Much of the HOBr that reacts will be reduced to the bromide ion, to be reoxidized to HOBr by residual aqueous chlorine in the water. The overall result is an enhanced reactivity exhibited by aqueous chlorine in the presence of the bromide ion.

Bromination of cooling water is gaining popularity within Europe as the preferred biocide to chlorination.

12.4 OZONE

Ozone is a gas at ambient temperature, partially soluble in water and, like most gases, increases in solubility as the temperature of the water decreases. It is formed when oxygen molecules (O_2) are ruptured (i.e. by lightning, or high energy ultraviolet rays) producing oxygen fragments, which unite with other oxygen fragments to produce ozone (O_3). Ozone has the unique property of autodecomposition, producing numerous free radical species, the most prominent being the hydroxyl free radical ($OH^•$). As the pH of the solutions containing molecular ozone increases, the rate of decomposition of molecular ozone to produce hydroxyl free radicals also increases, such that at pH 10, ozone decomposes instantaneously.

Although the resultant hydroxyl free radicals are powerful oxidants, their half-life is so short (milliseconds) that no significant concentration can occur. This means that the hydroxyl free radicals do not have a significant effect on the microbes. Thus, the presence of molecular ozone is necessary to ensure microbial sanitation. It therefore has virtually no residual effect.

The half-life of molecular ozone is relatively long (~12 hours), but in aqueous solution it depends almost entirely on the amount of 'ozone demanding material' in the water. The dirtier the water, the higher the content of ozone-demanding materials and the shorter the half-life of the ozone (and vice versa). Ozone is therefore often not optimal for use in can cooling water recirculation systems where the water may contain a lot of organic waste and other debris.

Ozone has certain advantages over other biocidal solutions, for example:

- It has been shown to be a more powerful disinfectant than chlorine for the deactivation of a large number of microorganisms.
- It has been used safely and effectively in water treatment for many years.
- It is generally recognized as safe for treatment of bottled water because it does not remain in the water for very long and can thus be considered a processing aid rather than an additive with no safety concerns about residual ozone.
- It leaves no toxic residue. Excess ozone will rapidly disappear through its spontaneous decomposition in solution.
- Low concentrations (0.1–4 ppm) are required to effect practical disinfection (amounts vary according to the amount of organic matter present).

The disadvantages of ozone include the following:

- In terms of water treatment, ozone is more expensive than chlorine or the alternatives. The lower production costs of chlorine (relative to ozone), and its ability to maintain a residual concentration in distribution or recirculation systems, favours the use of chlorine over ozone.

- Ozone reacts with rubber and various synthetic rubber compounds. Teflon seals should be used in water systems where ozone is used as a disinfectant.

12.5 ULTRAVIOLET LIGHT

Ultraviolet (UV) light has been successfully used for water disinfection in various food-producing plants. This includes its use as a pasteurization treatment for clear liquids such as bottled waters. UV light acts on microorganisms by disrupting the cell's genetic material (DNA). Shortwave UV light irreversibly damages the nucleic acids, preventing reproduction and causing deactivation.

Unlike chemical methods, UV disinfection does not alter the taste, smell, pH, or chemical composition of the water. UV treatment is regarded as a viable alternative to the use of chlorine in clean water applications. A major disadvantage of UV light is that it can only be used to treat clear water, and any organic matter or other debris will shield the microorganisms from the UV rays. Furthermore, the penetration of UV rays into the water is very restricted.

UV light is absorbed by ordinary glass, and the water has to flow through expensive quartz tubes during the irradiation stage.

The lifespan of UV lights, which are also expensive, is reasonably short. Because of these disadvantages, UV light is not often used in the treatment of cannery cooling waters.

12.6 MEMBRANE FILTRATION

Membrane filtration is comparable, in capital cost and size, to a UV disinfection system, and effectively removes bacteria and larger microorganisms, although viruses will pass through the filters. The main problems with membrane filtration systems are:

- They can only be used for clean water applications, as organic material and other debris will rapidly block the filters.
- The filters need to be replaced frequently due to blockage and 'grow-through', which increases the running cost considerably. Grow-through can occur when there is a build-up of microorganisms on the dirty side and the organisms literally grow through the membrane pores onto the clean side, causing downstream contamination.

Because of the above drawbacks, membrane filtration is unsuitable for use in can cooling water recirculation systems. It tends to find applications for removal of microorganisms in water-like food products, for example water, apple juice, milk, and beer.

Reverse osmosis (RO) is a water purification process that uses a partially permeable membrane to remove ions, unwanted molecules, and larger particles from drinking water. The result is that the solute is retained on the pressurized side of the membrane and the pure solvent is allowed to pass to the other side. Reverse osmosis is not used to disinfect water, but can be used to take out some of the compounds in it that may make it difficult to disinfect or make it corrosive.

Reverse osmosis differs from filtration in that the mechanism of fluid flow is by osmosis across a membrane. The main removal mechanism in membrane filtration is straining, or size exclusion, where the pores are 0.01 μm or larger. Reverse osmosis instead involves solvent diffusion across a membrane that is either non-porous or uses nanofiltration with pores 0.001 μm in size.

References

Featherstone, S. (2015). *A Complete Course in Canning and Related Processes*, 14e, vol. 1. Oxford, UK: Woodhead Publishing/Elsevier.

Warne, D. (1988). *Manual on Fish Canning. FAO Fisheries Technical Paper – T285*. Rome: FAO.

FAO and WHO (2019). Safety and quality of water used in food production and processing – meeting report. Microbiological Risk Assessment Series no. 33. Rome.

13 Handling Processing Deviations

13.1 WHAT CONSTITUTES A PROCESS DEVIATION

By definition, a process deviation occurs whenever the actual process is less than (or different to) the scheduled process or when any critical factor does not comply with that specified in the processing schedule.

The scheduled process is the document that states the process times and temperatures, which ensures the product is processed adequately. In addition to the retort conditions, it also includes the critical control points (CCPs) for the product and packaging that are deemed essential for product safety. Sometimes there may be additional factors that ensure the product quality is acceptable.

All thermal processing factories will experience process deviations from time to time, and the long-term success of that factory depends on how these are managed. Good manufacturing practices for thermal processing procedures are critical in minimizing deviations, as is the adherence to the HACCP plan. As retort control becomes more automated, the reliance on manual decisions is reduced with the result that the incidences of process deviations are also reduced.

Chapters 10 and 11 considered the methods for establishing safe processing conditions. These are usually calculated from process measurements taken under worst-case conditions and include such factors as retort come-up time, hold temperature and time, initial product temperature, fill weight, pH, and a_w. This is not an exhaustive list and there are further details in Chapters 11 and 16. A deviation can occur when any of these factors go out of control.

Whenever there has been a deviation from the scheduled process for any thermally processed low acid or acid food, the processor is obliged to set that product aside from other production

Essentials of Thermal Processing, Second Edition. Gary Tucker and Susan Featherstone.

and immediately evaluate the process with respect to that product potentially being a public health threat.

Such evaluation of the process given must be done by a competent thermal processing authority according to recognized procedures, using all the available critical data for that particular batch.

Complete product and processing records must be kept for all aspects of any deviation and the corrective actions taken. If it is determined that the product is not a public health threat, then it may be distributed as normal, but if this cannot be guaranteed, then it must be destroyed under controlled conditions.

13.2 WHAT CAN GO WRONG

There are many reasons why deviations can occur. Below is a list of some of the general reasons with examples. This list is not all inclusive and there are other reasons why deviations could occur. Failure or error in one or more of the basic processing parameters:

- low initial temperature – i.e. below the recommended minimum;
- low, but uniform, process medium temperature;
- shorter process time than the scheduled process states;
- incorrect scheduled process given to that product/can size.

Change in heating medium temperature during processing:

- momentary drops in temperature (due to loss in steam, etc.);
- long drops in heating temperature;
- excessive heating medium temperature fluctuation.

A failure in the retort operation:

- no venting cycle;
- improper venting cycle (too short);
- partially opened valves, low steam pressure;
- excessive water or air inside retort;
- wrong container packing or position.

Equipment malfunction or failure:

- inaccurate temperature monitoring devices;
- recorder chart malfunctioning;
- no ink in recorder chart;
- missing information records.

Product-related errors:

- product out of specification with regard to critical factor (e.g. pH);
- product style variation;
- inadequate headspace in agitation is used;
- wrong ingredients/net mass.

A thermal processing schedule is a document that defines all the factors that must be controlled, so that food products are processed safely. With so many factors needing to be controlled it is inevitable that, on occasion, one or more factors will fall outside of the acceptable limits. Table 13.1 gives examples of some of the more common process deviations experienced by canneries. The consequences of some of these mechanical errors can be wide ranging, for example, air pressure failure will prevent valves from operating correctly as well as affecting overpressure control. Hence Table 13.1 is purely for illustration purposes and simplifies what can be a complex situation.

Despite the possibilities for there being many causes of a process deviation, one of the most common types is a failure in the steam system caused by boiler problems. This often results in a temporary drop in retort temperature while the boiler is down, as shown in Fig. 13.1. Often the downtime is only a few minutes because a boiler is a critical part of most food production sites and demands immediate attention. In Fig. 13.1, the deviation is clear and lasted a few minutes before the retort temperature was restored to the set-point value.

Table 13.1 Examples of the more common process deviations experienced by canneries.

Mechanical error	Possible impact
Steam supply or boiler failure	Drop in retort temperature causing inadequate lethality.
Instrumentation or air failure	Incorrect valve operation preventing the correct application of scheduled process. Poor pressure control causing packs to change shape during the process.
Automatic loader failure	Product stuck at different stages in the retort cycle, causing many issues such as low product initial temperature or inadequate cooling.
Incorrect process (time/temperature) given to product	Inadequate lethality or excessive lethality causing quality issues.
Incorrect product pH or a_w	Inadequate lethality or excessive lethality causing quality issues.
Rotation or agitation stops	Reduced heat transfer resulting in inadequate lethality.

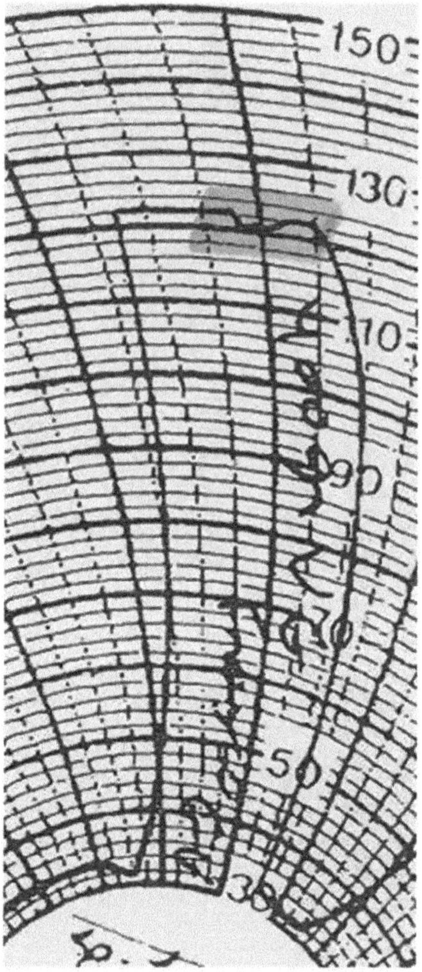

Fig. 13.1 Chart recording showing a process deviation caused by boiler failure during the process hold phase.

13.3 ACTIONS REQUIRED

In the unfortunate event of a process deviation occurring, there must be a procedure in place so that actions can be taken. This involves a number of steps, as described below:

- Put the product on hold pending investigation by the thermal processing authority (TPA) or approved persons, or extend the hold phase, or reprocess immediately if appropriate for the product.
- Collect all the production records to provide a full description of the deviation for the TPA.
- Segregate the lot of containers that have undergone the process deviation and ensure they do not leave the factory.
- Contact the TPA so that a prompt analysis can be made.

The TPA term is one used by the US FDA and USDA and all companies exporting thermally processed foods to the USA. A processing authority is a person or organization having expert knowledge of thermal processing requirements for foods packed in hermetically sealed containers and having adequate facilities to make these process determinations. Within European establishments, it is more common for the Technical Manager or Quality Manager to act in this capacity.

Putting product on hold requires that the containers are clearly labelled and isolated from normal production containers where possible. A marked area should be available in all factories that produce thermally processed foods. A batch should be shrink wrapped to keep the containers together and avoid odd containers getting knocked off and finding their way into areas where normal production is kept.

Until the deviation is analysed it is not possible to decide on the fate of the batch. If the containers have left the factory and are in the distribution chain, it will be a costly exercise to recover them for reprocessing or destruction. The worst scenario is for the containers to have reached the retailer shelves, which inevitably means that some will be with the consumers. Recalls at this stage also involve substantial adverse publicity but are extremely rare. The few that have happened over the long history with canned foods have tended to be limited to isolated food poisoning outbreaks. Brand damage is so severe that a company will struggle to repair its image – as the following example proves (Box 13.1):

Box 13.1 Brand damage

In July 2007, two people in Texas and two in Indiana (USA) were made seriously ill and hospitalized with botulism poisoning suspected as originating from eating Castleberry's Hot Dog Chili Sauce. The precise cause of the incidents is not known but is suspected to have arisen from a series of process deviations due to improper functioning of retorts.

Although Castleberry recalled everything made on the one manufacturing line, the only products linked to illness were the chili sauces. The recalled products were sold under a multitude of brand names. Castleberry's Food Company is an establishment owned by Bumble Bee Foods LLC. Damage to the integrity of the Bumble Bee brand was also severe in addition to that for the brand names manufactured at Castleberry's Food Company.

13.3.1 TPA actions

The TPA will require the documentation that describes what went wrong in the process deviation, in addition to documentation for the scheduled process. This includes the temperature distribution and heat penetration data taken during the process validation trials. It will be necessary to establish the process lethality delivered by the deviant process and assess the adequacy for commercial sterilization.

Decision-making can be helped by process calculations of lethality generated during the process deviation (Fig. 13.2). There are several calculation methods that can be used, although the modern finite difference methods such as CTemp and NumeriCAL offer the greatest flexibility. Formula methods, such as Ball, are limited for most deviations because the method requires the lowest retort temperature to be used in the calculations. With reference to Table 13.1, the following are examples of process deviations that can benefit from the use of a calculation method:

- *Steam supply or boiler failure*. The effects can be calculated based on the retort recorder data as input to the calculation method, providing the process failures do not change the product heating rates.
- *Instrumentation or air failure*. For some valve responses, such as the steam valve, it might be possible to calculate the product response based on the retort recorder data.
- *Automatic loader failure*. The effects can be calculated if the product IT or uniformity of temperature prior to processing were the only changes. However, some products will change their heating rate because of viscosity increases caused by water absorption into dry components such as rice, beans, and pasta. Calculation methods should not be used.
- *Incorrect process (time/temperature) given to product*. It might be possible to use one of the calculation methods if the process changes are not too different from the scheduled version.
- *Incorrect product pH or a_w*. These affect the types of microorganisms that can grow in the product. Calculation methods need to evaluate different microorganism death kinetics.
- *Rotation or agitation stops*. These can be difficult to calculate and tend to require an experimental simulation of the deviation process.

Worked example 13.1 illustrates how the Ball method results in a highly conservative assessment because of the low retort temperature reached during the deviation. Had the deviation occurred later in the process time, when the product was above 100 °C, the energy contained within the product would prevent such a large drop in retort temperature. The Ball method might be possible to apply. Worked example 13.2 uses the CTemp program to calculate the effect of the deviation by using the deviation time-temperature profile as the input to the calculations (see Worked examples 13.1 and 13.2).

Worked example 13.1 Process deviation caused by boiler failure – using the Ball method

Product is a pouch of egg fried rice processed in a steam-air retort for 35minute hold at 121 °C to achieve an $F_0$6 end of heating. The process uses slow rotation at 2 rpm to tumble the rice and enhance heat transfer.

Heating factor (f_h) 19.2 minute, Lag factor (j) 1.51

The process deviation was caused by boiler failure 10 minute into the hold period. Retort temperature (RT) dropped from 121 to 80 °C in 2 minute, remained at 80 °C for 5 minute while the boiler was down, and was brought back up to 121 °C in 3 minute once the boiler was repaired. This was a similar deviation to the one illustrated by the chart in Fig. 13.1.

Deviation analysis using the Ball method requires a constant RT and so the process is re-calculated with a RT of 80 °C, which gives:

- $F_0$0 after 35 minute hold at 80 °C;
- $F_0$6 after 77 317 minute hold at 80 °C.

The calculated $F_0$0 after a 35 minute hold at 80 °C would result in the batch being thrown away. Note that the Ball method cannot be used to calculate a processing time at the low RT of 80 °C.

Worked example 13.2 Process deviation caused by boiler failure – using the CTemp method

The same product, process, and deviation are used in this example but the analysis uses the CTemp program to calculate the effect of the boiler breakdown. Figure 13.2 shows an Excel graph taken from the CTemp predictions showing the drop in RT and its effect on the rice pouch temperature. With no extension to the 35 minute hold, the end of heating lethality is below $F_0 1$. Even with cooling lethality included in the analysis, the total F_0 is less than 3.0.

The decision from the deviation analysis using CTemp is clear and the batch of product should be thrown away.

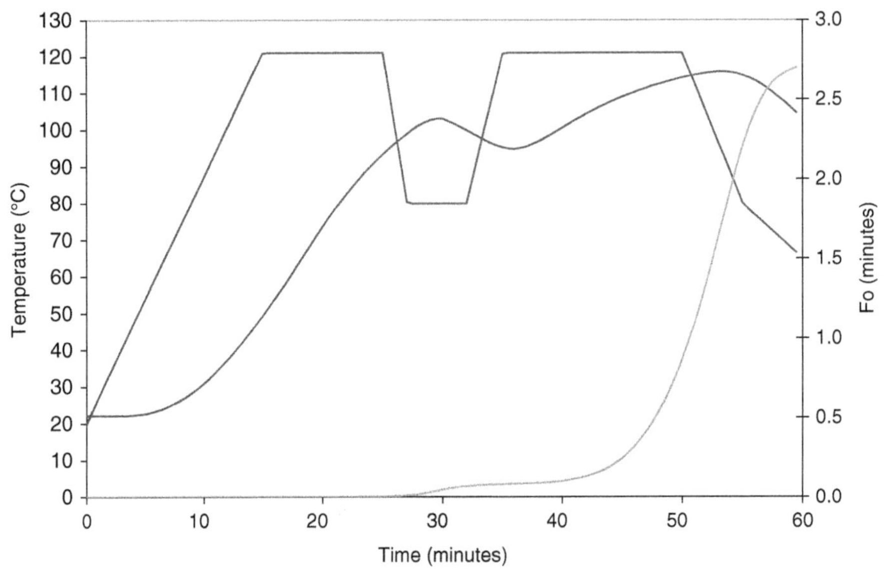

Fig. 13.2 Time-temperature and time-lethality graph for the process deviation, showing F_0 end of heating <1.0 and F_0 end of cooling <3.0.

Many companies apply simple rules for hold time extension when deviations such as a boiler breakdown occur. If steam supply is lost for a number of minutes it is common to add at least that number of minutes onto the hold time. Worked example 13.3 shows how this rule can be used in the rice pouch deviation.

13.3.2 Process deviation analysis for broken heating products

The worked examples above are for a simple heating product; that is a product that only shows one rate of heating or heating factor (f_h). However, there are numerous products that contain complex thickening agents, which change their viscosity so

Worked example 13.3 Process deviation caused by boiler failure – using the CTemp method but with a 10-minute hold extension

The same product, process, and deviation are used in this example and the analysis uses the CTemp program to calculate the effect of the 10 minute boiler breakdown, but the hold time was extended by 10 minute.

A 10-minute extension to the 35 minute hold time increased the end of heating F_0 to 4.9 and the end of cooling F_0 to 8.8. While F_0 4.9 end of heating is less than the target of F_0 6.0, it is still above the minimum botulinum cook of $F_0$3. Hence this batch is safe and can be distributed.

drastically that the heating rates change during processing from conduction to convection and vice-versa. Petfood chunks in gravy or jelly are examples, as are soups and sauces made using starches in their powder form. Figure 13.3 illustrates a semi-log plot (logarithmic heating curve) for a chunks-in-jelly pet food that shows two break points.

Within a process calculation model such as Ball or CTemp, the calculations use the time to reach each break point as the determining factor that fixes when the thermal properties change in the calculations. However, the time to each break point in Fig. 13.3 is not only determined by the time spent by the product in the retort, but also it is a function of the product temperature and shearing conditions within the product. For example, starch gelatinization

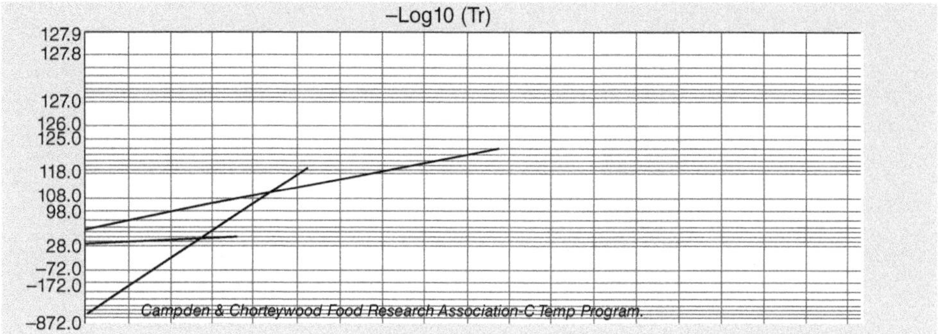

Fig. 13.3 Semi-log plot (logarithmic heating curve) for a chunks-in-jelly pet food showing two break points.

is a function of time, temperature, and shear rate. A different set of conditions, as with a process deviation, can cause the break points to occur at different times. This makes calculation of the lethality during process deviations almost impossible if the process involved structure changes that are influenced by both container agitation and the integrated time-temperature relationship for the product.

Examples of foods that display broken heating points that depend on time, temperature, and shear are those in which heat and mass transfer occur simultaneously within the product. Pulses that rehydrate in a sauce are classic examples, such as beans in tomato sauce, in which starches leach out of the pulse and water diffuses in. Many of these foods are processed in reel & spiral cooker-coolers using high rates of intermittent can rotation. If the reel stops rotating (a process deviation), agitation of the product also stops and heat and mass transfer takes place at a different rate. This cannot be predicted using the types of models used for thermal process analysis. Only a full computational fluid dynamics (CFD) solution can get close to calculating the temperature rise under these conditions. Even so, there will be a high level of uncertainty about the assumptions made for product properties such rheological behaviour and hence the predicted results will be uncertain. For these reasons, it is not recommended to use CFD for process deviation analysis.

If one of these complex products experiences a process deviation, it is not safe to attempt process calculations unless the product has gone through its last break-in heating rate. After this last break point, there will not be any more changes in thermal properties and it is acceptable to calculate product temperatures whether a deviation has occurred. However, even though the thermal properties will be constant, calculations should not move too far from the original conditions, and severe deviations are best not modelled.

The alternative to process calculations is to experimentally recreate the deviation conditions in a retort simulator. With batch retorts, this is relatively straightforward but with continuous retorts of the hydrostatic and reel and spiral types, this requires specialized pilot equipment. Figure 13.4 shows a pilot reel and spiral simulator and Fig. 13.5 a pilot hydrostat simulator.

The approaches to resolving the fate of a batch of a 'broken heating' product that has suffered a process deviation can be summarized in the following two points:

Fig. 13.4 Pilot scale reel and spiral simulator.

Fig. 13.5 Pilot scale hydrostat simulator. Courtesy of Stork Food and Dairy Systems BV.

- If the retort deviation occurs before the final break-in heating rates there is sufficient uncertainty about the way that product will heat throughout the process. A calculation method such as Ball, CTemp, or NumeriCAL will not be able to predict centre temperatures accurately therefore the batch should be thrown way. However, if the deviation is considered minor and the

value of the batch substantial, the full deviant process can be experimentally recreated in a process simulator.
- If the retort deviation occurs after the last break-in product heating rate, the deviant process can be calculated using one of methods mentioned above.

The calculation examples in this chapter must only be undertaken by a competent person who understands the calculation model used, and more importantly, of its limitations. If there is any doubt whether the calculation will provide an accurate representation of the change in lethality experienced by the deviant process the batch must be thrown away or reprocessed. The next section considers when reprocessing is appropriate.

13.3.3 Reprocessing

Reprocessing is not ideal, but it is sometimes an option when the batch size and/or the value of the batch is substantial. There are a few factors to consider when deciding to reprocess a batch, these include:

- Has the initial temperature changed?
- Have the heating characteristics changed (products with starch, coagulated protein, etc.)?
- Has the cold spot changed?
- Can the product quality withstand a recook?

There are numerous examples of products which heat slower upon reprocessing or heat slower depending on storage time between the original process and the reprocess. Products of concern are those containing starches, binders, extenders, and the like. To optimize product quality, condensed soups or sauces are often formulated with thickeners that activate late in the thermal process or during cooling. Some products experience thickening of the sauce component during storage. If changes during storage have not been considered, a reprocess using the original cook time may be inadequate. In one case, a product with an original process 110-minute cook time, when reprocessed, required a 180-minute cook time.

Timeliness of reprocessing may also be important depending on the severity of the original processing deviation. If the product is severely under-processed, there may be a concern for the failure to destroy *Staphylococcus aureus*. If this organism survives in the product prior to reprocessing, and sufficient time permits growth and production of enterotoxin, reprocessing will not deactivate the toxin and may even intensify the heat-stable enterotoxin.

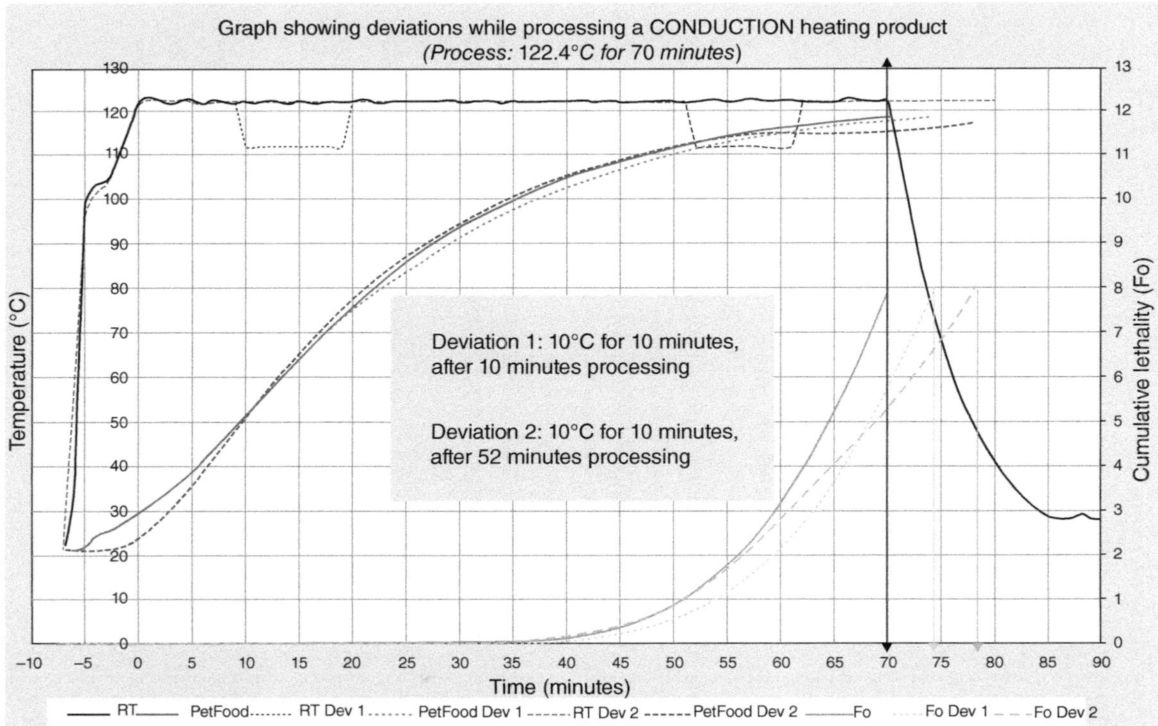

Fig. 13.6 Graph showing two scenarios of the same deviation, but at different times in the process.

Reworking or blending into new formulation is another method of reprocessing. Whenever performed, the scheduled process must cover the level of rework material and procedure for including rework. Some original scheduled processes may allow for a certain level of rework (e.g. 2%). If rework is an included option in the processing schedule it is essential to experimentally evaluate the potential for rework to affect the scheduled process.

Worked example 13.4

In an example using a conduction heating-type product, two scenarios of the same deviation, but at different times in the process (see Fig. 13.6). In the first scenario, Deviation 1, a deviation occurred where the processing temperature dropped by 10 °C, 10 minutes into the 70-minute process. In the second scenario, Deviation 2, the temperature also dropped by 10 °C for 10 minutes, but after 52 minutes of the process time had elapsed. The calculated time that the process was required to be extended is significantly different, depending on where in the process (in the heating of the product) the deviation occurred: 4 minutes for Deviation 1 and 8 minutes for Deviation 2.

14 Packaging Options for Heat-Preserved Foods

Food preservation is about delaying the onset of natural degradation and biodeterioration. Biodeterioration may be as a result of the metabolic processes of one or many microorganisms or it can be caused by insect, rodent, or bird damage. Besides biodeterioration, there is also natural degradation that occurs in foods caused by intrinsic enzymes; those enzymes present in the product that cause degradation or decay after death. For example, loss of food quality by intrinsic enzymes is an important topic, as it can cause quality deterioration and render food unacceptable. Food, whether slaughtered or picked, is therefore an active organic system that can react favourably or unfavourably with its packaging. Understanding packaging options and the way they work together with the product is therefore critical to the success of any food packaging system (Tucker 2008).

The functions of packaging are to contain, protect, preserve, portion, inform, promote, and make portable. The packaging options for thermally processed foods include metal (mainly tinplate, but some aluminium) and glass and specific plastics (laminates and composites) (Fig. 14.1).

The primary packaging is the packaging that is in direct contact with the food. The protection and preserving functions are arguably the most important functions of primary food packaging, as it must keep the food in good condition until it is sold and consumed. If adequately packaged, the shelf life of food may be extended, which allows the natural life of the food to be prolonged. This enables consumers to have a choice in terms of the food available, and so food resources can be more equitably distributed. Correct packaging prevents any wastage (such as leakage, damage, or deterioration) that may occur during transportation and distribution. Good packaging and presentation encourage consumers to buy products (Fig. 14.2).

Essentials of Thermal Processing, Second Edition. Gary Tucker and Susan Featherstone.

Fig. 14.1 Selection of packaging types for thermally processed foods.

Fig. 14.2 Display of packaged thermally processed foods in South Africa.

Primary packaging for heat-preserved food requires the following attributes:

- Able to maintain a hermetic seal over the shelf life;
- Able to withstand the process temperature and pressure differentials;
- Provide a physical barrier that can afford the food the required microbiological, gas, and light barriers;
- Able to withstand the physical stresses of transport, storage, and distribution;
- Must not react adversely with the food.

14.1 METAL CONTAINERS

The most common form of packaging for thermally processed products is the metal can. The strength and non-breakability of metal make it an ideal material for food packaging. Metal is also an excellent light and gas barrier, is quick and easy to seal, can withstand the temperature extremes of food processing, and is 100% recyclable. Cans are made from tinplate, tin-free steel, or aluminium.

14.1.1 Tin plate

A base plate of low-carbon, mild steel, onto which a layer of tin is electrolytically deposited, is called a tinplate. The tin coating imparts a degree of corrosion resistance to the steel base, which would otherwise readily rust. The thickness of the tinplate varies, depending on the size of the cans that are to be manufactured and their intended application. It is usually between 0.15 and 0.25 mm thick. With the worldwide trend of minimising packaging and waste, the thickness of tinplate has progressively reduced. Nowadays, many cans are made from a thinner, stronger DR (double reduced) tinplate, where the steel sheet is cold rolled twice prior to being tin-coated, compared with the SR (single reduced) tinplate that was the norm. Tin coating mass varies, according to end-use and whether or not lacquers are to be applied, and ranges from around 1.4–14.0 $g\,m^{-2}$.

Tinplate on which the tin coating mass is the same on each surface is known as an equally coated plate (E), whereas a plate with different tin coating masses on each surface is referred to as a differentially coated plate (D). When specifying tin coating masses, it is customary to quote for each surface (inside/outside for differentially coated plate). E1 means there is 2.8 g of tin per m^2 of the plate, while D3/1 means that the tinplate is differentially coated and has 8.4 g of tin per m^2 of the plate on one side, and 2.8 g of tin per m^2 on the other surface. The convention is that the first figure after D denotes the internal surface.

14.1.2 Tin free steel (TFS or ECCS)

The same mild carbon base plate, as is used for manufacturing tinplate, can be electro-chromium coated (a surface chromium/chromium oxide treatment) to produce tin-free steel. Plain TFS cannot be readily welded, and it lacks the corrosion resistance of conventional tinplate (since there is no sacrificial protection of the steel by an outer layer of tin), but it provides an excellent key surface onto which can be applied protective lacquers. TFS is mainly used for two-piece draw-redraw (DRD) cans and sanitary ends (i.e. non-easy-opening ends).

14.1.3 Aluminium

Aluminium alloys are widely used as food contact materials. The alloys contain about 96% aluminium and elements like magnesium, silicon, iron, manganese, copper, and zinc, which are all present in very specific amounts to improve properties like formability, ductility, strength, etc. When exposed to air, aluminium forms aluminium oxide, a colourless resistant film that protects the container contents. At low and high pH, however, (<4.5 and >8.5) aluminium oxide becomes soluble and corrosion can occur.

Aluminium is also commonly used for the manufacture of drawn and wall-ironed (DWI) beverage cans. Aluminium alloys are used to make beverage can ends, closures for food cans and some DRD food can bodies. It is easy to fabricate, lightweight, relatively inert, and like steel is 100% recyclable. However, it lacks the strength of steel.

14.1.4 Protective coatings (lacquers)

Under certain circumstances, cans may be plain (unlacquered), but many food products are reactive and corrosive, and so they can may require an inner protective coating. The primary function of internal lacquers is to protect the can from the contents and also the contents from the can. Lacquers can extend the shelf life of a product by reducing the rate of dissolution of tin by corrosive products, prevention or masking sulphide staining, reducing the bleaching action of tin on colour, and reducing or preventing adhesion of the product to the can.

Internal food can lacquers should have the following characteristics:

- Non-toxic;
- Not adversely affect the flavour or colour of the food;
- Provide an effective barrier between food and container;
- Easy to apply on tinplate;
- Not peel off or blister during canned food sterilisation and storage;
- Have mechanical resistance to can manufacturing operations;
- Economical.

The internal lacquers have particular properties developed to meet the various product requirements. They are based on a wide variety of resin types. Some of the resins occur naturally but most are produced by chemical synthesis.

The main types of lacquers used on the interior of most food cans and ends are discussed below.

14.1.4.1 Vinyl lacquers

These are very flexible, have good adhesion properties and are free from odour and taste. Although they are air-drying, vinyl lacquers are normally baked to accelerate drying. A disadvantage of vinyl lacquers is that they decompose at about 170–185 °C when they are in contact with steel. They are therefore applied over an epoxy base coat.

Vinyl lacquers are not suitable for products that are processed in steam or water above 100 °C because they take up water, which is known as 'blushing'.

14.1.4.2 Organosol lacquers

These lacquers are dispersions of high-molecular weight PVC resins in a hydrocarbon diluent. They also contain a plasticiser.

The advantages of organosol lacquers are that they are very flexible and are high-temperature process resistant, unlike the vinyls. They also have a high-solid content and are therefore cost-effective, yielding good coverage.

Organosols can be pigmented with aluminium but they cannot be mixed with zinc. They are often used on full aperture tinplate easy-opening ends and drawn cans.

14.1.4.3 Epoxy-phenolic lacquer

These are the most commonly used lacquers. They are epoxy resins based on epichlorohydrin and bisphenol A (see Box 14.1). They are combined with phenolic resins, which are produced by the reaction of a phenol with formaldehyde. This combination gives a gold lacquer with good flexibility, adhesion, and chemical resistance. Epoxy-phenolics are cured by heat polymerisation (cross-linking) at temperatures of about 200 °C.

Various additives may be incorporated into the lacquer to give it special properties. For instance, aluminium pigments or zinc oxide or zinc carbonate may be added to mask sulphide staining of the tinplate. Meat release agents may also be incorporated to help solid meat packs slide easily from the can on opening.

A disadvantage of epoxy phenolic lacquers is that they include bisphenol A, so minute traces of this compound and its derivatives may migrate to the food contents although, with strict manufacturing control, this can be contained to well within safe limits.

14.1.4.4 Polyester lacquer

Isophthalic acid (IPA) and terephthalic acid (TPA) are the main carboxylic acids used in polyester coatings, but a wide variety of polyester resins can be synthesised by condensation reactions between a polyvalent acid and polyalcohols or epoxides. Polyester resins are easy to handle during the production process and adhere well to the metal surface. However, they are usually not stable under acidic conditions and have poor corrosion resistance, therefore, they cannot be used for acidic food types. Polyethylene terephthalate (PET) coatings are laminated onto the inside and sometimes also the outside surface of non-welded cans.

There is a lot of interest in polyester coatings, as the first-generation BPA-replacement coatings are mostly based on polyester or acrylic compounds. The second-generation solutions under development include spray coatings based on polyethylene and polypropylene.

14.1.4.5 Acrylic lacquers

Ethyl acrylate is the most commonly used monomer to synthesise acrylic coatings. Acrylic resins display corrosion and sulphide stain resistance, but can be brittle. They have a clean appearance when pigmented with titanium dioxide, but may change the organoleptic properties of food. Because of these properties, they

Box 14.1 Migration Compounds

Packaged foods can become contaminated with traces of chemicals that migrate from the packaging material into the food during processing and or subsequent storage. These chemicals can be contaminants in the basic material compounds from which the packaging is made or can be elements of the packaging material itself. All efforts are made to ensure that any transfer of any substances is kept to as low a level as possible, and that no transfer of any substances that may be harmful occurs.

Most studies investigating chemical migration from food cans focused on BPA, BADGE, and their derivatives. The total migrates from cans may also contain oligomers, catalysts, reaction accelerators, epoxidised edible oils, amino resins, acrylic resins, various esters, waxes, lubricants, and metals. Furthermore, non-intentionally added substances (NIAS), such as impurities, reaction by-products, and degradation products, generally constitute a part of the migrate. Exposure estimates for these, often complex mixtures are much more difficult or even impossible to calculate, because many NIAS are unknown or unidentified substances.

Bisphenol A (BPA) is an organic monomer used to make many different plastic materials and resins. Minute traces of BPA can be found in a wide range of food products, because it is used in the production of polycarbonate and epoxy-phenolic resins, which are in turn used for food packaging (e.g. the lacquers of some food cans, linings of glass jar lids, bottle tops, and water bottles). It is also used in the manufacture of baby bottles, compact discs, white dental fillings, nail polish, contact lenses, false teeth, adhesives, water pipes, and flooring.

Bisphenol A diglycidyl ether (BADGE) is a chemical derivative of BPA used to make epoxy resins, which are used in many applications including food can lacquers. Over time, traces of bisphenol A can migrate from the packaging into food.

BPA and some of its derivatives mimic oestrogen and are endocrine disruptors. Endocrine Disrupting Chemicals (EDCs) have the capacity to mimic hormones and to interfere with the hormonal messaging systems that regulate our everyday body functions. It is suspected that these compounds can cause reproductive damage and birth defects and that their ingestion may lead to prostate and breast cancer (Diamanti-Kandarakis et al. 2009).

In 2002, the safety of epoxy resin can coatings was confirmed by an analysis of the European Commission's Scientific Committee on Food (SCF). The SCF estimated total dietary intake of BPA from all food contact sources, including epoxy resin coatings, to be in the range of 0.00048–0.0016 mm/kg body weight per day, which is below the Tolerable Daily Intake set by the SCF of 0.05 mm/kg body weight per day (EFSA 2007).

The potential human exposure to BPA from can coatings is therefore minimal and poses no known risk to human health. Can coatings have been and continue to be recognised as safe by the US Food and Drug Administration, the UK Food Standards Agency, the EU Scientific Committee on Food, and other government bodies worldwide. Many studies on BPA and its derivatives are ongoing.

The Plastic Materials and Articles in Contact with Food (England) (No. 2) Regulations 2006 permit the use of BPA in the manufacture of plastic materials and articles intended to come into contact with food, provided that no more than 0.6 mg/kg migrates into the food.

Plastic types; (i) (PET), (ii) (high-density polyethylene), (iii) (low-density polyethylene), (iv) (polypropylene) and (v) (polystyrene) do not use BPA during polymerisation or package forming, and thus cannot leach the substance into food or beverages (www.Bishpenol-A.org).

BPA has been banned in France since 2015. According to research done by EFSA (European Food Safety Authority) and the FDA (Food and Drug Administration) in the USA on the migration of BPA, they see BPA as a low-risk chemical. EFSA reduced the migration limit of BPA from packaging from 0.6 to 0.05 mg/kg as of 8 September 2018. (EU 2018/213).

are commonly used as external coatings. Acrylics and their blends are currently under investigation as replacements for BPA-based epoxy coatings

14.1.4.6 Side stripe lacquers to cover the weld

After welding of the lacquered 3-piece can body, the side seam is repaired by coating along its length. The best protection of the welded area is provided by a powder stripe. The powder is electrostatically charged to achieve deposition on the welded area and then fused when cured. Liquid side strip lacquers may also be used; either epoxy phenolic or organosol.

14.1.5 Internally plain (unlacquered) cans

In some instances, the reaction between the tinplate of the can and the product results in a beneficial reaction (the colour of some fruit and vegetables remains brighter and appears more 'natural', as in Fig. 14.3), making the product more desirable. For these types of products, internally plain cans, with a higher tin-coating mass are used. The overall shelf life of these products is generally determined by the rate of tin pickup from the can. The statutory maximum level of tin for tin pickup differs from country to country in the range 150–250 ppm.

Tin pickup is usually slow; approximately, 20 ppm in the first month and 2–5 ppm per month thereafter can be used as a normal guideline, although it will vary for different products (see Box 14.2). The rate of tin pickup is influenced by many factors include the following:

- Amount of oxygen present (this is related to the headspace size and vacuum level);
- The acidity of the product (lower pH usually results in more tin dissolution);

Fig. 14.3 Fruit cans do not use an internal lacquer and detinning takes place over time (newly packed can on left, internally detinned can on right).

Box 14.2 Tin in food (UK, Food Standards Agency 2002)

Canned foods make the biggest contribution to dietary intake of tin. Although most foods contain very low concentrations of tin, usually below 10 mg/kg, the concentration in canned foods packed in internally plain cans will increase with time as a result of the gradual dissolution into the food of the tin coating.

High concentrations of tin in food irritate the digestive tract and may cause stomach upsets in sensitive people at tin concentrations above 200 mg/kg, with an increased risk of effects at concentrations above 250 mg/kg. Such high levels are seldom encountered. These effects, the symptoms of which include fever, headache, nausea, vomiting, diarrhoea, abdominal cramps, and bloating, are short-term, with recovery expected soon after exposure. There are no long-term effects associated with occasional intakes from tin in the diet, even at a higher level.

Source: Food Standards Agency, Food Survey Information Sheets 29/02 (August 2002). Tin in Canned Fruit and Vegetables.

- Types of acids present (different acids influence the reaction differently);
- Presence of pigments;
- Levels of nitrates and sulphides.

14.1.6 External covering

Tinplate containers and ends may be supplied to a specific tin coating mass and with or without an external lacquer, depending on the end-use and the environment that the can is going to have to withstand (e.g. hot humid conditions will require a top-end can specification).

14.2 CAN CONSTRUCTION AND HANDLING

Metal cans are available in many shapes and sizes to suit all types of products. They can be made from two pieces or three pieces of metal.

Three-piece cans are manufactured from a rectangular piece of tinplate (known as a body blank), which is formed into a cylindrical shape and then joined along a vertical seam by welding. To this cylinder two ends are added, one by the can maker and the other is seamed on, after filling, by the canner (Fig. 14.4). These are referred to as the maker's end and the canner's end, respectively.

Two-piece cans are made either by a draw and re-draw (DRD) process using aluminium, TFS, or tinplate or by a draw and wall-ironing (DWI) process using tinplate. DRD cans are made from circular blanks of pre-lacquered plate, which are first drawn into shallow cups and then re-drawn, once or twice depending on the can's final dimensions, causing an elongation of the wall and a simultaneous reduction of diameter. DWI cans are made from a similar process; however, the walls are stretched further by a wall-ironing process carried out by a narrowing series of three tool rings (Fig. 14.5).

The advantages of two-piece cans over three-piece cans is that they have no side seam, and only one double seam, thus reducing the risks of leakage arising from imperfect seam formation. Also, excellent lacquer coverage is achieved as the DWI can is internally spray lacquered after fabrication, compared with roller coated as a flat sheet before can manufacture.

Fig. 14.4 The different stages of three-piece can manufacture.

Fig. 14.5 The different stages of 2-piece DWI can manufacture.

14.2.1 Product specification

The formulation of the product must be considered when choosing a can specification because the inclusion of certain ingredients or intrinsic compounds in the food can cause a change in product/container compatibility, which can affect the (product) container corrosion and required shelf life.

The presence of nitrate in certain foods will result in accelerated corrosion and should be controlled to as low a level as possible. Pack nitrate levels should ideally be less than 10 mg/kg. If it is unavoidably higher, an appropriate can specification that will afford the required amount of protection must be chosen. With acid products ($pH < 5.0$), packed into internally plain tinplate containers, the nitrate content of the product, as packed, should be less than 5 ppm, determined as nitrate.

Sulphur dioxide/sodium metabisulphite ($Na_2S_2O_5$) is commonly used as a preservative with some fruit products. Excess sulphur dioxide may play a role in many corrosion reactions inside tinplate cans resulting in the formation of tin sulphide, inhibition of tin dissolution, the promotion of iron corrosion, and the formation of hydrogen sulphide taints in the product. Sulphur dioxide should not be present in internally plain bodied cans and should not exceed 15 ppm total SO_2 in internally lacquered cans.

Some trace metals can also be corrosion accelerators and must be controlled if possible and at least known. Copper levels should not exceed 1 ppm in internally plain bodied cans.

14.2.2 Storage and handling of empty unused cans and ends

Cans and ends should be handled and stored with care to prevent damage that could compromise the final integrity. Storage areas should be free from dust, insects, dampness, condensation problems, and excessive heat (heat could dry out the compound in the ends and accelerate corrosion).

If there is a significant difference in day and night temperature (5–7 °C difference), the storage areas should be closed to control the humidity and the packaging should be wrapped. Depending on the humidity, condensation could occur on the packaging, causing corrosion, especially on cut edges of the can and end, so-called warehouse sweating.

At all times during storage and conveying, care must be taken to minimise the impact and physical stresses on the can. Damaged flanges can result in defective seams and abuse can result in corrosion and/or spoilage.

Loose easy-opening ends or cans fitted with easy-opening ends must be treated even more carefully, as even slight damage to the tab or score may result in subsequent failure or leakage.

14.2.3 Cleaning of empty unused cans

Although there is very little microbiological build-up on empty unused cans and ends, they may become contaminated with a variety of foreign objects and therefore good manufacturing practice dictates to clean all cans before filling. The most efficient way is by inversion and the use of air jets and/or potable water jets. If water is used, care must be taken not to leave wet cans for extended periods as corrosion may start. Sterilisation of the unused cans and ends is unnecessary as the canning process will eliminate the small microbial load that may be present.

14.2.4 Double seam formation and inspection procedures

The seam joining the can end and the body is known as the double seam. It is the formation of this hermetic seal that is critical if the container is to function correctly. Errors in or inadequately formed 'double seaming' can lead to leakage and subsequent canned food spoilage.

The double seam is formed by interlocking the can body flange and the can end curl during two rolling actions. The first action roll curls the edge of the can end up and under the flange of the can body and folds the metal into five thicknesses while embedding the flange into the compound. During this operation, the circumference about the edge of the can end is reduced causing the 'extra' metal to wrinkle. The second action roll flattens and tightens the seam so that the hermetic seal is formed. This action causes the wrinkles (formed in the first operation) to be ironed out while the compound is forced into any gaps between the metal surfaces (Fig. 14.6).

As product safety depends on the existence and maintenance of the hermetic seal, double seam formation must be checked regularly during production; after all jams under the sealing machine, after adjustment to the machine, and after machine start-up following a long delay in production. Good manufacturing practice guidelines indicate that visual inspection of double seams should be at least every 30 minutes. A lot of information as to the quality of a double seam can be obtained by a visual and tactile examination of the rolled seam. Skilled operators can often see if the double seam is out of specification.

Fig. 14.6 A photograph of a cross section through a double seam.

Full double seam tears down procedures should be followed on samples for each seaming head at least every four hours. Can manufacturers and can seaming machine suppliers usually supply directions for seam formation and standards against which double seams are evaluated. The critical parameters of the double seam that must be controlled are as follows (Fig 14.7):

- Seam tightness (percentage tightness rating/wrinkle grade and free space)
- Actual overlap
- Percentage body hook butting (%BHB)

Any deviations from the standard require immediate corrective action (Codex Alimentarius 1993).

14.2.5 Washing of filled cans

Containers should be washed with hot water after seaming to remove product residues from the outside of the can. This will reduce the risk of external corrosion during processing and prevent a build-up of product contamination in the sterilisation system.

For some fatty foods, detergents may be required in the can wash water. These must be specifically chosen so that they will not cause corrosion of the exterior of the can.

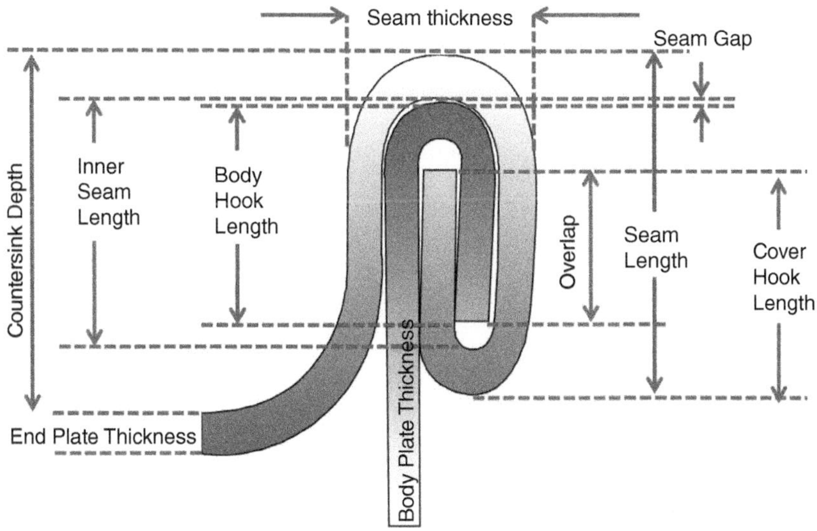

Fig. 14.7 Diagram of a double seam showing the various dimensions.

Cans with easy-open ends, which have an oil-based score protection system, should not be washed with water-containing detergents, because this will reduce the effectiveness of the protective oil.

Cans with aluminium easy-open ends should not be washed with water having high alkalinity, because residues can lead to aluminium corrosion at sites of metal exposure on the tab and around the scoreline.

The wash water should not be too cold, because this can reduce the initial temperature of the cans. This is especially important with starch-based hot-filled products (e.g. sauces), as in extreme situations, cans being held in the cold wash water may change the pack heating characteristics.

14.2.6 Processing of cans

Processing conditions for all containers, and particularly those with easy-open ends, should be controlled to ensure that permanent deformation of the ends (peaking) does not occur during processing. The internal can pressure developed during processing is mostly counter-balanced by the external steam pressure. The differential or excess pressure inside the can should be insufficient to cause permanent deformation of the ends if the cans have been filled to achieve an acceptable final vacuum level and headspace and if the strength of the end is sufficient. At the end of processing, as cooling starts, if the retort pressure is suddenly relieved, the excess pressure within the can may cause peaking of the ends.

Generally, pressure cooling is required when cans of 73 mm, or greater, in diameter are processed at temperatures of 121 °C (250 °F) and above and cans of 153 mm in diameter are processed at temperatures of 110 °C (230 °F) and above.

Pressure cooling should always be used with easy-opening ends.

14.2.7 Cooling of cans

'Stackburn' occurs when cans are packed, palletised, and warehoused too hot, causing retention of heat in a stack of cans for long periods. This can result in microbial spoilage from thermophilic organisms, deterioration of product quality, or accelerated corrosion.

Conversely, over-cooling will result in the cans failing to dry adequately, leading to external rusting and increased risk of post-process microbial contamination.

14.2.7.1 Corrosion prevention

The presence of air during the processing and cooling cycle may also exacerbate corrosion and must be excluded by adequate venting or exhausting at high enough temperatures (>85 °C).

Incorrect quality cannery water can give rise to corrosion at sites of exposed metal during any stage of the processing cycle, in both steam and water, or after processing while the containers are still wet. In order to minimise corrosion and to reduce scale, a typical processing or cooling water should have the following attributes:

- pH 7.0–8.0;
- Chloride + sulphate content ≤ 100 ppm;
- Free halogen < 5 ppm;
- Hardness (as $CaCO_3$) 30–150 ppm.

Soft water can result in a highly corrosive environment and should be treated with an appropriate water treatment system to minimise risks of corrosion.

The easy-open end must be thoroughly dried immediately after cooling to prevent corrosion of the score and tab, independent of cooling water quality, and treatment. Where cans are left in crates after cooling, it is preferable to limit water retention by sterilising the cans with the easy-open end down.

Corrosion may occur on wet containers irrespective of water quality and treatment, and so cans must be dried as soon after processing as possible. Drying equipment should ensure that:

- As much water as possible is removed from the can and end, especially in the case of an easy-open end, in particular from under the score and tab, preferably by use of twists and an air knife;
- The residual heat within the container (should be 35–40 °C) can be used to complete the drying of moisture remaining under the curled edges of the tab. If the residual heat is inadequate, a heated air stage should be used.

When cans are labelled and packed immediately after processing, there should be a residence time of at least two minutes during and after drying, but before packing, to allow for complete evaporation of moisture.

The main risks of inadequate drying are rusting of tinplate cans and ends, and stress corrosion cracking of the score of aluminium ends.

Wet cans should never be manually handled because there is the possibility of infecting the cans with microorganisms, through the cooling double seams before the hermetic seal stabilises.

14.2.8 Secondary packaging

All labels and secondary packaging that come into contact with the cans must be suitably non-corrosive. Label paper, paper board, and adhesives must comply with certain levels for pH, chloride, and sulphate measured on a water extract. These levels are generally accepted as pH 5.5 to 7.5, 0.05% as NaCl, and 0.25% as Na_2SO_4, respectively.

Water-based glues should not be applied in excess or at low temperatures, as drying is inhibited and corrosion or mould can result.

14.3 GLASS

Glass is one of the oldest packaging materials used for food. It is almost completely chemically inert and therefore is suitable for packing many products and all foods. The glass used for food packaging is approximately 70–74% silica (SiO_2) and 15% sodium carbonate Na_2CO_3 (added to silica to lower the melting point) by weight. It is known as soda-lime glass. It contains traces of aluminium (Al_2O_3) and magnesium (MgO) for better durability.

Box 14.3 Properties of glass

Glass is a super-cooled liquid (i.e. a liquid that solidifies without any significant structural changes). Although generally considered fragile, newly formed glass fibre has the strength to support twice the weight of equivalent steel fibre. Unfortunately, in practice, glass only has about one-thousandth of its theoretical strength because it is a liquid and so any defect will spread unhindered throughout its mass.

Glass is a poor conductor of heat and sudden temperature changes (Δ60–65 °C) may cause breakage due to heat shock. Sudden heating, which puts the surface of the glass under compression, is less dangerous than sudden cooling, which puts the surface in tension (Rees and Bettison 1991).

Source: Based on Rees and Bettison 1991.

Various compounds can also be added to give the glass some colour (to block out light from transmitting through the glass), for example FeO and Cr_2O_3 will produce a green tint, and cobalt (II) oxide-aluminium oxide ($CoAl_2O_4$) will result in blue glass (see Box 14.3).

The principles of processing in glass are the same as for cans, but certain modifications are necessary to improve the thermal properties of glass, which make it less vulnerable to rapid changes in temperature of more than 60 °C. The advantages of glass are many and include the following:

- Forms a 100% gas barrier;
- Impervious to flavours and odours;
- Imparts no migrating chemicals to the food products;
- Is resistant to any chemical attack from food;
- Can seal efficiently;
- Relatively strong;
- Allows the product to be viewed.

The main disadvantage of glass is that it can be broken and the broken pieces can be very sharp and constitute physical hazards. Other disadvantages are that clear glass is not a light barrier, so product deterioration can take place due to light (e.g. Vitamin B_3 and colour loss), although this can be overcome by using tinted glass. Also, it is relatively heavy and is not suitable for products that are to be frozen.

14.3.1 Glass manufacture

Glass containers for food are usually made from clear or flint glass (Bansal and Doremus 1986) (see Box 14.4). The raw materials for a batch of glass are typical:

• Silica sand	45%
• Soda ash	13%
• Limestone	10%
• Minor ingredients (e.g. feldspar, calumite, selenium)	2%
• Cullet	10%

These are melted in a furnace at 1500 °C. The molten glass is fed into container-forming machines. A gob of glass is loaded into a parison mould, which forms the blank shape of the container. It is then transferred to a blow mould to be blown into the final shape. Once formed, the hot containers (650 °C) enter an annealing oven for controlled cooling. If they are allowed to cool too quickly, they will develop internal stresses and will be more susceptible to impact damage. Treating the surface before (hot end – with a vapour of organic titanium or an inorganic tin compound) and after (cold end – e.g. with oleic acid), the annealing ovens increases the strength and lubricity of the containers.

A glass container for thermal processing must be able to form a hermetic seal with the cap. The most important feature on the glass container for this to be possible is the sealing surface, which must be of exactly the right dimensions and be smooth.

Box 14.4 Recycled glass

Broken or waste glass is known as cullet. Glass is infinitely recyclable and does not affect the quality or integrity of the new product. The benefit of using cullet is that there are no fusion losses in the melting process, i.e. one ton of cullet will generate the ability to remake one ton of glass. Before recycled glass is melted, non-glass contaminants must be removed (e.g. metal caps, lids, and neck rings, lead collars from wine and champagne bottles, stones, and dirt).

The different colour glass must not be recycled together or they will affect the tint and properties of the new glass containers.

14.3.2 Closures for sealing glass food containers

The suppliers of caps/closures will have the correct specific information for each product type, size, and processing combination. Closures for glass containers are made with either lacquered tinplate, tin-free steel, or aluminium, into which has been placed a flowed-in plastisol lining compound (or a rubber ring with a pry-off cap) that acts as a sealant between the glass rim surface (called the 'finish') and the cap. The type of sealing compound, glass neck finish, and cap design are specific for different applications (Fig. 14.7).

For bottles, the main types are:

- Crown caps;
- Roll-on-pilfer-proof (ROPP) caps;
- Snap-on caps;
- Corks.

For jars to be thermally processed, the main types of caps are:

- Twist-on-twist-off (TO) caps;
- Push-on twist-off caps (PT).

All lids and caps should neither affect the product nor be affected by it, and they should seal the container for its expected shelf life. This is evaluated by testing trial containers with the product to be packaged to make sure that there is no interaction between the pack and the product. Expert advice should also be sought from the packaging suppliers when selecting the type of closure to be used.

14.3.3 Sealing mechanisms

Like cans, glass must be hermetically sealed to prevent product contamination after sealing and thermal processing. The closure is held in place by the vacuum in the container and/or the friction between the glass finish and the cap. The sealing surface of the glass may be across the top of the finish as with twist caps, around the side of the finish as with pry-off caps, or around both the top and side seals as with push-on twist-off (PT) caps. It is important that the glass sealing surface is free of defects and protected from damage, as otherwise there is an unacceptable risk that the container will leak and draw in contaminants.

It is recognised as good manufacturing practice to ensure that the diameter across the finish of the jar is less than that of the diameter across the body of the container. This prevents the closure from suffering undue damage through striking the closures on adjacent containers as they move along conveyors. Fortunately, with most containers that have lost their hermetic seals prior to processing, the caps will fall off during retorting and thus alert operators to pack failure.

In addition to obvious loss of vacuum, other faults to be aware of when using glass include the following:

- *Cocked caps*: usually caused by misalignment of lug type closures while passing under the sealing machine, so that the lug sits on top of the thread rather than underneath it. Cocked caps are readily visible as part of the top of the closure is raised.
- *Crushed lugs*: occur when the sealing machine forces the lug of a twist cap down over the thread, rather than engaging it correctly, while winding the closure down onto the finish.
- *Stripped caps*: result when the cap is over-tightened so that the lugs strip and splay-out over the thread of the finish.
- *Tilted caps*: occur when pry-off and PT caps do not sit down uniformly on the finish.

14.3.4 Inspection procedures

The frequency of inspecting for the adequacy of seals with glass containers should be sufficient to ensure the consistent formation of hermetic seals and defined opening torque. As a guide, this means that intervals between non-destructive testing should be no more than 30 minutes, while destructive testing should take place at least every 4 hours per sealer. In addition to this, a visual inspection should follow every occasion that the capper jams. The results of all closure examinations should be recorded (see Box 14.5).

Box 14.5 Tamper evidence

The vacuum inside the container, the hermetic seal, and the properties of tinplate and tin-free steel allows a feature to be designed into the cap where an audible pop and a visible deflection of a button in the cap occurs when the closure is first opened after processing. This is an excellent feature and can indicate if the container has been tampered with.

14.3.5 Packing and processing

14.3.5.1 Inspection and preparation of containers

All incoming glass containers must be inspected for cracks, chips, and small bubbles in the glass, as these will weaken the container and result in breakage during processing. New jars and bottles should be rinsed with clean water. Care is needed to make sure that the containers are not heated too quickly, as the heat shock may cause them to break.

14.3.5.2 Filling

If very high-fill temperatures are used, the glass containers may have to be pre-warmed to prevent thermal shock. Fill temperatures of at least 65 °C are recommended to ensure that there is an adequate vacuum generated in the container and that the product is sufficiently de-aerated.

The initial fill temperature of the product should always be higher than the temperature of the process water at the start of the process.

Particulate foods may need to be blanched to remove entrapped air that could cause excessive pressure inside the container during processing. Similarly, some products (e.g. pastes) may need to be de-aerated.

Clean-filling conditions must be used to prevent contamination of the sealing surface, as this could affect the integrity of the seal. Product on the outside of the jar could affect the grip on the outside of the container and therefore affect the capper performance.

The fill level should be consistent because different size headspaces can result in varying amounts of internal pressure build-up inside the container during processing. Generally, a headspace of at least 6% (with a maximum of 10%) of the brim-full capacity of the container is recommended.

14.3.5.3 Capping

The choice of sealing compound in the cap is very important, because different compounds are designed for different processing temperatures and cap application styles (e.g. hand-applied, steam flow closed, pressure processed, pasteurised, etc.). Suppliers of caps will have the correct specific information for each product type, size, and processing combination. The good performance of the capper is critical and must be controlled.

When using PT caps and sometimes with TO caps, steam is used to soften the sealing compound and to create a vacuum. The efficiency of the vacuum generation by steam is monitored by cold water vacuum checks, jars are filled to a nominal headspace of 10–12 mm with cold water, and the vacuum in the jar is measured. It should be in the region of 65 kPa (50 cm Hg).

14.3.5.4 Atmospheric processing

Pasteurised products can be processed in a steam tunnel or a boiling water bath. Cooling must be done by a water spray, with first warm water and then cooler water, to avoid heat shock.

14.3.5.5 Pressure processing

Food in glass containers is usually pressure processed in full water immersion retorts. Sophisticated water spray retorts can also be used, providing that the retort temperature and pressure can be strictly controlled. Generally, food in glass containers is not processed above 121 °C (NFPA 1984).

Overpressure is required to ensure that the cap does not lift during processing. Too much overpressure can result in the sealing compound is 'cut through'. This is a condition that can lead to seal failure. A general rule for twist-off caps is that 70–80 kPa pressure, more than what is required to reach process temperature, should be maintained with sterile compressed air. The amount of overpressure required for PT caps is greater because there is no physical holding of the cap onto the jar. The correct overpressure for each product type and process temperature combination should be established, but a general rule is that the maximum that should be used is 220 kPa (32 psi), and the absolute maximum is 245 kPa (35 psi). The fluctuation of more than 20 kPa (3 psi) is not permissible.

To minimise pressure fluctuations during processing glass jars with caps, first ensure there is at least 15 cm of water above the top layer of jars, and second ensure there is at least 15 cm between the top of the water and the retort shell.

To ensure effective temperature distribution during processing, the process water should be recirculated from the bottom of the retort to the top during processing. As for processing cans, the quality of the retort water must be continually monitored; it should be sterilised using a system such as chlorination or bromination.

14.3.5.6 Cooling

At the end of the complete process cycle, the jars must not be removed from the retort until the product temperature is colder than it was at filling. This ensures the internal pressure is less than before the process and so the cap surface will be depressed further.

As for cans, over-cooling can result in the jar failing to dry adequately, leading to rusting of the cap (if scratched) and entrapment of water beneath the cap sides, which could go mouldy.

14.4 PLASTICS, FLEXIBLES, AND LAMINATES

The term 'plastic' can refer to any of a group of synthetic or natural organic materials that may be shaped when soft and then hardened, including many types of resins, resinoids, polymers, and cellulose derivatives. The development of plastics in general has revolutionised food packaging. Plastics are very versatile and can be engineered to replace most other packaging materials. See Table 14.1 for melting points of different plastic materials used for food packaging. Generally, plastics are lighter than glass or metal, they can be transparent or opaque, they can be very thin and flexible, or highly rigid (Hutton 2003).

Plastic polymers are used to make many different packaging types. There are three main categories of plastic-based packaging used for thermally processed foods. According to Codex Alimentarius:

Table 14.1 Table of some plastics used for food packaging and their melting points.

Plastic	Melting point
High-density polyethylene (HDPE)	130 °C
Low-density polyethylene (LDPE)	110 °C
Polypropylene (PP)	160 °C
Polyester – polyethylene terephthalate (PET)	260 °C
Polyester – polyethylene naphthalene dicarboxylate (PEN)	270 °C
Ethylene-vinyl acetate (EVA)	90–120 °C
Ethylene-vinyl alcohol (EVOH)	280 °C
Ionomers e.g. Surlyn	180–215 °C
Polyamide (PA) e.g. Nylon 6	220 °C
PVC	100–260 °C
Cellophane	175–205 °C
Poly lactic acid (PLA)	150–160 °C

(i) A '**Flexible container**' means that the shape or contours of the filled, sealed container are affected by the enclosed product.

(ii) A '**Semi-rigid container**' means that the shape or contours of the filled, sealed container are not affected by the enclosed product under normal atmospheric temperature and pressure but can be deformed by external mechanical pressure of less than 0.7 kg/cm^2 (10 psig) (i.e. normal firm finger pressure).

(iii) A '**Rigid container**' means that the shape or contours of the filled and sealed container are neither affected by the enclosed product nor deformed by external mechanical pressure of up to 0.7 kg/cm^2 (10 psig) (i.e. normal firm finger pressure).

These sorts of containers can be made from various polymers or laminated or co-extruded materials (various plastic polymers, paper, and/or aluminum) and have been one of the major packaging growth areas in recent years both in technological and materials advances and in actual market share. Packaging films are used to construct packages for many different types of convenience foods. Some important material properties are necessary to meet the requirements of these demanding convenience markets and include the following:

- A balance between toughness and stiffness;
- High strength and puncture and tear resistance;
- High-temperature resistance (>130 °C is required for some reheat/cook-in-package applications);
- Barrier properties (e.g. moisture, gas, odour, light);
- Grease resistance;
- Sealability;
- Aesthetic properties and printability;
- Processability/machineability (balanced surface friction);
- Environmental friendliness (including recyclability).

14.4.1 Advantages of retortable plastics

There are several advantages proposed for plastic containers, which include the following:

- Transport and storage of flat, empty pouches are compact, thus cheaper than for cans.
- Filling lines can easily be changed to accommodate different sizes.

- Pouch dimensions allow rapid heat penetration and therefore faster sterilisation, resulting in energy savings during processing (note that rapid agitation in rigid cans can provide heat transfer benefits far greater than can be achieved by thin geometry pouches).
- Short-processing times result in better product quality (although the comment above is also relevant for rapid agitation with cans).
- The product is commercially sterile and therefore requires no specialised storage conditions (e.g. refrigeration).
- Pouches are easy to open, either by tearing or cutting.
- The pouch web material cannot corrode and therefore no rusting of the container can occur.

14.4.2 Disadvantages of retortable plastics

Each container type (e.g. metal, glass, or plastic) also has its disadvantages, which includes the following:

- The capital investment in the equipment for pouch manufacture and processing is large.
- Production/filling speed is generally slower than competitor packaging (realistically, about half that of cans).
- Heat processing is complex (temperature and overpressure control) and therefore the retort design is more critical and expensive.
- As the pouch is not rigid, some solid products lose their shape.
- Gentler handing of the packed pouch is required to ensure that the integrity is not compromised.
- The shelf life of products in pouches is usually less than that of cans and is dependent on the barrier properties of flexible materials.

14.4.3 Polymers used for retortable packaging

14.4.3.1 Polypropylene (PP)

PP is very popular for retortable containers, because it is a microwavable package and a good water vapour barrier. However, it is a relatively poor gas (e.g. oxygen and carbon dioxide) barrier. Where longer shelf life is required, PP is often used together with a barrier material such as EVOH (see section below). It is used for rigid plastic pots/bowls and bottles for various sterilised products, as well as part of the structure in some laminated films.

14.4.3.2 Polyethylene terephthalate (PET)

PET is the most commonly used polymer of a group called polyesters. It is a medium to a high-oxygen barrier and has good strength and chemical resistance. It is an important polymer used in retortable packaging. In order to make PET resistant to high temperature, so it is suitable for retorting, it is used in the crystalline form.

14.4.3.3 Ethyl-vinylalcohol (EVOH)

EVOH is an important barrier polymer in packaging. EVOH, like other barrier materials, has very poor mechanical strength, so is only used in multilayer applications. It has very high resistance to oxygen in its dry state, which is significantly reduced as the moisture content increases. When used together with PP in a multilayer application, the EVOH becomes wet during processing (moisture barrier properties of PP are poor at high temperature), so higher levels of EVOH have to be used, compared with when it is used in non-retortable packaging.

14.4.3.4 Polyvinylidene chloride (PVDC)

PVDC is an important barrier polymer. It is not affected by moisture in the same way as EVOH and is therefore a better barrier in retort applications.

14.4.3.5 Polyamide (PA)

Nylons belong to the group of PA polymers. Nylon-6 is often used in food packaging because of its overall superiority in the combination of cost, physical properties, and process adaptability. All PA has good barrier properties, although not as good as EVOH and PVDC.

14.4.3.6 Aluminium

Although not a polymer, aluminium foil is very important as a barrier material for retortable packaging. Its main advantage is that it is an excellent gas and moisture barrier, and this does not change at the high temperatures of retort processing. The main disadvantage is that it lacks physical strength and flexibility and can form pinholes and cracks. The opacity of aluminium foil can be regarded as an advantage or a disadvantage. Where the product requires a light barrier, it will be an advantage, but if the customer would like to see the product, it is a disadvantage.

14.4.3.7 Glass-coated barrier films

Films treated with silica oxide provide excellent barrier properties as well as transparency and microwavability. Many of the recent developments in retortable films contain transparent plastic materials coated with aluminium or silica oxides.

14.4.4 Types of packages used for thermally processed foods

14.4.4.1 Retort pouches

Arguably a successor to the tinplate can, the materials and manufacturing technology of retort pouches have had an enormous amount of work put into them to allow them to be able to compete in this highly sensitive food sector, where food safety and stability are paramount.

The standard food pouch is a pillow pouch, which is made by heat sealing on three sides, leaving an opening through which to fill the pouch, and closure is done by a final, simple heat-sealing operation (Fig. 14.8). Pouches are made from flexible web stock that has the

Fig. 14.8 Selection of glass food packages, showing a range of jar and cap designs.

Example 14.1 Retort pouches for tuna

An example of the construction of one of the current tuna retort pouches is as follows: Cast polypropylene as the inner food contact layer (it has a good heat seal surface, is flexible, strong and is food compatible). This is followed by an aluminium foil layer that provides a light, gas, and odour barrier, which determines the shelf life of the product and also gives the pouch stiffness and allows for the tear notch. The next layer is nylon, which provides abrasion resistance. Finally, the outside layer is polyester, which provides the high-temperature resistance, toughness, and printability of the material (Kerry 2012).

appropriate properties for the specific application. They are usually made from the appropriate, printed flexible web and filled in one operation on a form-filled seal machine. Pouches can also be premade, loaded into a magazine then filled and sealed by the packer.

The stand-up pouch (SUP) is a flexible pouch, with a bottom gusset that can stand erect without external support when filled, and has become a highly desirable packaging option. The standup feature of the standard pouch makes it able to compete ably with cans and cartons for ambient shelf-stable, dry and chilled products. SUP manufacture is fairly complex. Whether made from monolayer or multilayer materials (depending on use), the structure suffers from stresses of gravity and uneven weight distribution from the contents. The sidewalls must have engineered stiffness and strength and the seals and adhesives must be able to withstand temperature (low and/or high), physical and fatigue abuse.

To be able to differentiate the retort pouch and offer a clear window to see the product, alternate non-metal, barrier materials have been developed to replace the aluminium foil (e.g. Silica oxide on PET by plasma or vacuum depositing).

Depending on the final use of the package (i.e. required to withstand pasteurisation conditions <100 °C or sterilisation conditions 110–122 °C), different films and laminates are used. Regardless of the use, the package must provide a strong hermetic seal, which should normally be at least 3 mm wide and continuous. For heat sealing, the sealing surfaces should be a plane parallel to each other and the temperature of the jaw should be uniform across the entire sealing area. Since the integrity of the heat seal is critical to the safety of the product, it should be tested routinely. Typical testing protocols include the following:

- Seal strength tests normally used to determine the best combination of time temperature and seal pressure;
- Burst-pressure tests;
- Seal thickness tests;
- Dye penetration tests;
- Visual appraisal of seal quality (no creases, bubbles, etc.).

Whether considering flexible retort pouches, semi-rigid or rigid packs, all of these options offer the common attraction of providing a means to minimise the nutritional and sensory quality losses that often are associated with traditional thermal processing in rigid containers, while simultaneously providing the opportunity to display visually appealing products.

14.4.4.2 Plastic cans and pots

Foil-free rigid plastic containers are particularly desirable for the ready-meal market, because the shelf-stable product can be microwave heated in its packaging prior to consumption. Such bowls/trays are generally opaque and made from laminates to give the desired shelf life (usually 12–24 months). Clear pots and bottles are often used for fruit to display the cut style and fruit colour (Fig. 14.9 shows a seal integrity test on a fruit cup).

For convenience foods, the desirable attributes of these packs are as follows:

- Easy to open;
- Have a hermetic seal;
- Easy to fill, seal and process;
- Provide an adequate barrier for the required shelf life;
- Aesthetically pleasing.

14.4.4.3 Retortable composite carton

Retortable cartons designed for shelf-stable food products are available (www.tetrapak.com) (Fig. 14.10). They were designed as an alternative to the metal can for foods such as beans, vegetables, tomatoes, soups, and sauces. The main advantages are efficient shelf-space usage; square packages are easily aligned on the shelf for a greater brand impact and they are also easy to open. They are made from pre-printed aluminium foil, laminate-based board. The thermal process is very similar to that for cans and jars using overpressure.

Fig. 14.9 Retortable pouches.

Fig. 14.10 Seals of plastic cups being tested. Photograph courtesy of HG Molenaar.

Fig. 14.11 Retortable cartons.

14.4.4.4 Processing considerations – control of headspace

Rigid plastic pots are generally sealed with flexible-lidding material, and so the amount of headspace entrapped inside the package must be controlled, otherwise too much will result in excessive pressure created during processing at high temperatures. To minimise the effect of pressure build-up, these sorts of containers are processed with overpressure, typically 1 bar over the retort pressure for that temperature.

Foil or film laminates are usually used, bonded to the container by heat, induction, or ultrasound sealing (Box 14.6). The laminate structure and types of adhesives are specifically chosen for product type and process temperature. The sealing conditions are also very specific for product type and lidding system.

Box 14.6 Induction sealing

Induction sealing is a non-contact method of heating multilayer laminated foil-lidding film, to hermetically seal plastic, and glass containers.

Although direct application or sometimes 'cap-less' induction sealing can be used, most frequently a plastic 'over-cap' is supplied to the packer with a multilayer foil liner already inserted. The top layer is a paper pulp that is generally spot-glued to the cap. The next layer is an adhesive that is used to bond a layer of aluminium foil to the pulp. The bottom layer is a polymer film laminated to the foil. Once applied, the container passes under an induction coil, which emits an oscillating electromagnetic field that causes the conductive aluminium foil to heat. The heat melts the adhesive and releases the foil from the over-cap. The polymer film layer also heats and melts onto the container sealing surface. When cooled, the polymer creates a bond with the container, resulting in a hermetically sealed product. The total time taken is in the order of a few seconds.

Sealing can be done with either a hand-held unit or on a conveyor system. Care must be taken to ensure that the correct pulse, exposure time, and top pressure is used for the specific seal and lidding system, as it is possible to over or underheat the film.

References

Bansal, N.P. and Doremus, R.H. (1986). *Handbook of Glass Properties*. New York: Academic Press Inc.

Diamanti-Kandarakis, E., Bourguignon, J.P., Giudice, L.C. et al. (2009). Endocrine-disrupting chemicals: an endocrine society scientific statement. *Endocrine Reviews* 30 (4): 293–342. https://doi.org/10.1210/er.2009-0002.

Codex Alimentarius (1993). Recommended International Code of Hygienic Practice For Low and Acidified Low Acid Canned Foods, CAC/RCP 23-1979, Rev. 2 (1993) 1. FAO/WHO.

Commission Regulation (EU) 2018/213 (2018). On the use of bisphenol Ain varnishes and coatings intended to come into contact with food and amending Regulation (EU) No 10/2011 as regards the use of the substance in plastic food contact materials. 12 February 2018. Official Journal of the European Union. https://eur-lex.europa.eu/legal-content/EN/TXT/HTML/?uri=CELEX:32018R0213&from=EN.

EFSA (European Food Safety Authority) (2007). Opinion of the Scientific Panel on food additives, flavourings, processing aids and materials in contact with food (AFC) related to 2,2-BIS(4-HYDROXYPHENYL) PROPANE. Published 29 January 2007.

Food Standards Agency, Food Survey Information Sheets 29/02 (August 2002). Tin in Canned Fruit and Vegetables. www.food.gov.uk/science/surveillance.

Hutton, T. (2003). *Food Packaging: An Introduction. Key Topics in Food Science and Technology No.7.* Chipping Campden, UK: Campden BRI.

Kerry, J.P. (2012). *Advances in Meat, Poultry and Seafood Packaging*. Chapter 12, 333–362. Cambridge, UK: Woodhead Publishing.

NFPA (1984). *Processes for Low-acid Canned Foods in Glass Containers, Bulletin 30-L*, 5ee. Washington DC: National Food Processors Association.

Rees, J.A.G. and Bettison, J. (1991). *Processing and Packaging of Heat Preserved Foods*. London: Blackie and Son Ltd.

Tucker, G.S. (2008). *Food Biodeterioration and Preservation*. Oxford: Blackwell Publishing Ltd.

www.Bishpenol-A.org (n.d.)

15 Incubation Testing

Incubation tests are used for a variety of reasons that are explained later. They can provide information on the adequacy of a thermal process, particularly for spoilage microorganisms; however, their value for determining whether food safety organisms have survived a process is less certain and so incubation should not be used as the sole criterion of product safety. As described in several chapters in this book, the assurance of product safety is achieved through a properly developed and implemented HACCP program (Chapter 16).

15.1 PURPOSE OF INCUBATION TESTS

Incubation testing is intended to identify surviving microorganisms within a container of food that has undergone a thermal process. These organisms are likely to be present in very small numbers unless there is a gross processing or container fault, and so the first step in incubation is to grow the surviving microorganisms to reach numbers that can be detected. A variety of holding times and temperatures can be used to achieve this, and the choice depends on which group of microorganisms is of interest. Times and temperatures for incubation are discussed later in this chapter.

There are several reasons why a company would wish to undertake incubation tests on thermally processed foods. For example:

- If a process is thought to be of marginal severity, possibly as a result of a process deviation;
- If leaker spoilage is suspected, based on results of seam evaluation or seal testing, or from the poor microbial quality of cooling water;

Essentials of Thermal Processing, Second Edition. Gary Tucker and Susan Featherstone.

- If new package types or packaging materials are being evaluated, which may include different types of closure, and information on the seal integrity needs to be generated and assessed;
- If different raw materials are being used for the food product and there are concerns that heat-resistant spoilage organisms may be present;
- It is a legal requirement in some countries;
- Some customers require incubation testing as standard.

A company exporting thermally processed (canned) foods to the US will need to comply with FDA or USDA regulations. The USDA requires incubation testing to be part of a company's HACCP plans. It is common for companies to adopt a standard incubation approach, irrespective of whether the FDA or USDA regulations apply. Recommended sample numbers for incubation are one container for each batch retort load or one container for every 1000 containers with a continuous processing system such as a reel and spiral or hydrostat. Recommended incubation conditions are 35 ± 2.8 °C (this originates from 95 ± 5 °F) for at least 10 days. Normal production containers should be selected for incubation. It is not recommended to use the containers with heat penetration probe housing.

Spoilage of thermally processed foods occurs from time to time, and it is important to understand what caused the containers to spoil so that it can be avoided in future. If spoilage has occurred, results from an incubation test can identify the types of organism responsible and, in doing so, will help in determining the reasons why the spoilage occurred. The next section looks at the causes of spoilage in thermally processed foods.

15.2 CAUSES OF SPOILAGE

In the context of HACCP, microorganism contamination of a food product can occur in several ways. Although HACCP is intended for food safety assessment, the thought process helps determine the causes of spoilage. To study why spoilage occurred in a thermally processed food, the following classification for microbiological contamination is useful:

- Presence or *growth* of organisms already in the product. This is unlikely for a thermally processed food unless it bypasses the process altogether.

- *Introduction* of organisms after the thermal process. This is usually during the cooling operation through permanent or temporary holes in the packaging.
- *Survival* of organisms through the process. This requires either an under process or organisms of high heat resistance to be present.

Spoilage of a thermally processed food is usually caused in one of two ways; first by the growth of microorganisms following leakage or second by under processing. Other causes are incipient spoilage, gross spoilage, and spoilage from thermophilic organisms. Incipient spoilage occurs before the containers enter the retort process and usually results in food that has spoiled but with no surviving organisms to detect. Gross spoilage is caused by major damage to the container and usually results in mixed microflora. Thermophilic spoilage usually occurs with inadequate cooling but can also occur if raw materials become contaminated with heat-resistant microorganisms.

When a container of spoiled food has no viable microorganisms, spoilage may have occurred before the thermal-processing step (incipient) or the microorganisms causing spoilage may have died during storage. The latter is known as autosterilisation and makes the determination of the causes of spoilage difficult.

15.2.1 Leaker spoilage

The most common cause of spoilage, leaker spoilage, requires three factors to be present for this to occur. These are as follows:

(i) holes in the container that can be temporary or permanent;
(ii) bacteria or organisms present around the holes; and
(iii) water for their transport.

A pressure gradient between the outside and inside of the container is ideal to assist in the movement of organisms. The differential pressure between the headspace of a steam-closed metal can and the outside provides the perfect driving force to assist with drawing liquid from around the can seam into the can headspace. This driving force is less pronounced for flexible packages such as retortable pouches, although capillary action presents a means for bacteria to travel along with seal wrinkles.

Leaker spoilage is reported to be the most common type of spoilage, as can be seen from the data in Table 15.1 (Segner 1979). These data are for metal cans, but it is expected that this pattern of

Table 15.1 Causes of canned food spoilage.

Product category	Incipient	Gross	Leaker	Thermophilic	Insufficient process
Vegetables	9	17	255	84	107
Meats	21	3	130	13	31
Poultry	7	2	62	3	22
Soups	1	0	29	3	15
Pasta	1	1	23	11	13
Dairy based	0	0	82	5	8
Petfoods	16	2	154	55	58
Total % of all	55	25	735	174	254
spoiled lots	4	2	60	14	20

Source: Segner 1979.

Box 15.1 Example of leaker spoilage: *Staphylococcus aureus* in canned peas (1957, UK)

This was a landmark spoilage incident because it highlighted the need to avoid handling metal cans while they were still warm and wet. Details of the incident are given below:

- A10 cans of peas processed in vertical steam retorts cooled in a cooling canal with typically 1–2 ppm free chlorine.
- Eleven outbreaks, with infected cans largely edible and unblown.
- Poor seams identified with 10% failure in pressure tests and 2–3% 'blowers' from production.
- Cans are left to dry in baskets for varying times, then unloaded by hand while sometimes wet, and later packed into cartons.
- Nine of 17 handlers were *S. aureus* carriers, one lady with boils matched phage-type and she unloaded cans from retort basket at the time of the outbreak.
- Normal fitness for work restrictions was ignored.

spoilage is similar for other container types, and although published in 1979 these data are still relevant today. A viable mixed microflora of bacterial rods and cocci is indicative of leakage, although it is not unknown for such leakage to be from single bacteria types (see Boxes 15.1 and 15.2).

Box 15.2 Example of leaker spoilage: *Salmonella typhi* in canned corned beef (1964, Aberdeen, Scotland)

This incident highlights the importance of using good-quality water for cooling processed cans. Can seams are vulnerable to leaks when hot and the metal is expanded in the seam. Details of the incident are given below:

- 6 lb catering cans of corned beef produced in Argentina.
- More than 500 people were diagnosed with typhoid, although only three elderly or already very sick patients died.
- No gas production or product breakdown.
- A meat-slicing machine transferred the *Salmonella* onto other meats and the retail counters nearby, so spreading the infection.
- A chlorination plant had been introduced following three previous (less publicised) outbreaks in UK, but was out of action.
- Cans were cooled in unchlorinated river water, because the factory cooling wells had collapsed.
- The canning plant was downstream from town sewage outflow, with typhoid endemic in the local population.

15.2.2 Under processing

The second most common cause of spoilage is under processing (see Box 15.3), also referred to as an insufficient process, and this can be caused by several events, for example:

- retort operations that are faulty because of inaccurate or improperly functioning thermometers, gauges, or controls;
- excessive contamination of the product for which normally adequate processes are insufficient;
- changes in formulation or handling of the product that results in a more viscous product or tighter packing in the container, with consequent lengthening of the heat penetration time;
- accidental bypassing of the retort operation altogether.

Often with batches of containers that have undergone an insufficient process the level of heat applied to the food will knock out most of the microorganisms leaving a single more heat-resistant species. This is usually a spore-forming bacterium from the Clostridia or Bacilli groups.

Box15.3 Example of under processing: *Clostridium botulinum* in canned vichyssoise soup (1971, USA)

This incident illustrates the importance of controlling and monitoring viscosity of liquids and semi-solids. These rely on convection currents to enhance the heat transfer. Details of the incident are given below:

- A New York man died and his wife had become seriously ill due to botulism after eating a can of vichyssoise soup.
- Five cans of soup were found to be contaminated with the botulism toxin.
- The factory had introduced an improved 'thicker' formulation. They checked with US National Canners Association (NCA) for process times and temperatures, but failed to mention the change in soup viscosity.
- *Clostridium botulinum* spores survived the inadequate process and were able to germinate and grow during ambient storage of the soup.
- Vichyssoise is eaten cold so the botulism toxin could not be destroyed by heating the soup.
- HACCP systems for low acid-canned foods were introduced in 1973 in USA as a result of this incident.
- Soup viscosity should be a CCP.

15.2.3 Thermophilic spoilage

This is most commonly associated with inadequate cooling of processed containers in which spores of thermophilic organisms have time to germinate and grow. The time required at ideal growth temperatures for spore germination requires several hours with the food held at 45–60 °C. Hence, the major cause of thermophilic spoilage is when the cooling cycle of a batch retort is interrupted and the food containers remain in warm water while a retort fault is repaired. Thermophilic spoilage usually results in a single organism isolated from processed food.

One such fault that is increasing in occurrence is with automatic loading systems that on occasion break down and cause a bank of retorts to stop at varying places in their cycles. In this instance, cooling is not always considered of concern, because the food safety organisms (e.g. *C. botulinum*) will be removed by this stage, leaving only the very heat-resistant organisms that are not a safety hazard. Note that thermophilic organisms are not known to cause food poisoning. However, spoilage from these organisms represents considerable commercial damage to companies because of the costs of recalls, compensation, and damage to their reputation.

The other common cause of thermophilic spoilage is when raw materials are contaminated with heat-resistant microorganisms, either when raw materials suppliers are changed or from batch to batch microflora changes. Chapters 1 and 3 give examples of organisms of high-heat resistance, for example *Geobacillus stearothermophilus* and *Clostridium sporogenes*, together with process calculations that illustrate what process levels are required to ensure their adequate destruction. Log reductions of 3–5 are used to reduce the numbers of these spores to commercially acceptable levels.

Foods with a simple formulation can sometimes be given a process close to the minimum botulinum cook of $F_0 3$, for example beetroot and other vegetables in which texture can be damaged by higher process levels. This process level will only achieve around 1–2 log reductions with most thermophilic organisms and so contamination levels must be minimised. However, formulated foods such as soups and sauces usually contain many ingredients, with certain ones known for being contaminated with heat-resistant microorganisms. Starches, herbs, and spices have a reputation for containing high numbers of thermophilic spore-forming organisms (see Box 15.4). A process sufficient to achieve 3–5 log reductions

Box 15.4 Example of thermophilic spoilage: *Bacillus sporothermodurans* in retorted rice

Thermophilic spoilage incidents do not often make headline news and so information on specific incidents tends to remain confidential. This example was found in retorted pouches of rice. Some details of the incident are given below:

- Isolated cases of pink spots in rice with mushrooms identified.
- Rice pouches retorted to F_0 6–8 end of heating, and F_0 9–11 end of cooling.
- *B. sporothermodurans* were identified as the responsible organism and introduced from the mushrooms.
- Heat resistance showed it to display thermophilic tendencies, with $D_{121.1}$ of 2.8 minute. Only 5 log reductions were achieved during process.
- Initial spore loading per packs were around 10^5, so process was borderline for survival.
- The alternative source of mushrooms sourced, using canned mushrooms with low/zero initial loading.

of these would require F_0 12–20, depending on the organism and its level of heat resistance.

15.3 DESCRIPTIVE TERMS FOR CANNED FOOD SPOILAGE

Much of the published work on incubation and spoilage of thermally processed foods refers to canned foods. This is because there is a 150-year history in which considerable expertise has been developed over this time period as key spoilage incidents have occurred (see incidents highlighted in the above sections). Information on pouches and plastic trays is less available, primarily because of the reduced time period that these containers have been available. There are several descriptive terms for cans that are useful in understanding the nature of canned food spoilage. They describe incremental increases in gas pressure within the can as follows:

- *Flat*: A can with both ends concave, and they remain in this condition even when the can is brought down sharply on its end on a solid, flat surface.
- *Flipper*: A can that normally appears flat, but when brought down sharply on its end on a flat surface, one end flips out. When pressure is applied to this end, it flips in again and the can appears flat.
- *Springer*: A can with one end permanently bulged. When sufficient pressure is applied to this end, it will flip in, but the other end will flip out.
- *Soft swell*: A can bulged at both ends, but not so tightly that the ends cannot be pushed in somewhat with thumb pressure.
- *Hard swell*: A can bulged at both ends and so tightly that no indentation can be made with thumb pressure. A hard swell will generally 'buckle' before the can bursts. Bursting usually occurs at the double seam over the side seam lap, or in the middle of the side seam.

The terminology has evolved from the canning sector and unfortunately cannot be easily transferred to other packaging types such as semi-rigid containers and glass. The following tables use these descriptions.

15.4 METHODS FOR INCUBATION TESTING

15.4.1 Sample size

The number of containers examined bacteriologically must be large enough to give reliable results. The exact number of containers for incubation depends on the suspected number of failures, which presumes some prior knowledge of the pack integrity. Sample size can be calculated using statistical procedures (BSI 1999). Table 15.2 presents the required sample size in order to be 95% sure of finding a defective container (Campden BRI 2001). The sample size is dependent on the level of contamination (or defective containers) and not on the batch size. These sampling plans were published in BS 6001-1: 1999.

If no defective containers are found during the incubation test, it can be assumed there is a 95% chance that the batch is clear of defective containers. This is an unsatisfactory result because the absence of defective containers does not necessarily imply that the batch is sound. This absence of evidence scenario is one of the reasons put forward by some companies for minimising the incubation testing done as a routine practice.

If the incubation is carried out as part of a spoilage investigation and the containers for incubation are all known to be spoiled, then the required sample numbers are reduced. This assumes that each of the spoiled containers already contains high numbers of organisms. When the cause of spoilage is clear-cut, incubating 4–6 containers may be adequate, but in some cases, it may be necessary to incubate 10–50 containers before the cause of spoilage can be determined.

Table 15.2 Required sample size to be 95% sure of finding a defective container.

Detection level	Number of containers to test
1 in 100	300
1 in 500	1500
1 in 1000	3000
1 in 10000	30000

15.4.2 Temperatures and times for incubation

The exact temperatures and times for incubation testing will depend on the types of microorganisms that will grow. For example, if a low acid food such as beans in tomato sauce is suspected of being under processed, then incubation at 37 °C would be appropriate. Table 15.3 shows some recommended temperatures and times. The general rules for incubation temperatures are 25 °C for yeast and mould, 37 °C for mesophilic organisms, and 55 °C for thermophilic organisms.

Control of temperature in an incubator needs to be within ±1 °C; otherwise, there could be various types of microorganism growing in the food. Heat transfer from the incubator air to the food containers will be poor unless a circulation fan is used, and the containers are stored on a mesh-racking system. An incubator must have a permanent record of the air temperature.

As shown in Table 15.3, the different types of microorganisms require incubation at different times and temperatures. These conditions are guidelines only, and it is likely that specific laboratories will use local variations that are preferred. This is because microorganisms are notoriously hard to categorise and can evolve their growth characteristics to suit their environment.

15.4.2.1 Thermophilic organisms

Thermophilic organisms tend to grow between 45 and 65 °C although some, such as *Bacillus coagulans*, can grow slowly below 40 °C but more optimally at 45 °C. Therefore, it is essential to know what organism is expected to grow, so the conditions can

Table 15.3 Recommended temperatures and times for incubation testing.

Incubation temperature (°C)	Incubation time (days)	Food group and target organisms
25	14	Medium/high acid, targeting acid-tolerant bacteria, yeasts and moulds
30	14	Low acid foods, targeting leaker spoilage with mixed flora
37	10	Low acid foods, targeting under sterilisation and less optimally leakers
55	7	Low acid foods, targeting thermophiles

be tailored for that organism. Incubation times of 5–14 days are required but some organisms, such as *G. stearothermophilus*, can grow and die off within seven days. This is known as autosterilisation. Regular examination of the containers is essential to ensure the examination is done while the organisms are growing actively.

A further complication for thermophilic growth is the slow increase in food temperature from ambient (5–25 °C) to incubation temperature (45–55 °C). With high incubation temperatures, large food containers that heat up slowly, and poor heat transfer within incubators, it is necessary to ensure containers are separated in the incubator to maximise heating. If not, mesophilic organisms could start growing while temperatures in the containers are in the 25–45 °C range, which would confuse the situation.

Incubation testing for thermophilic organisms is essential if thermally processed foods are for export to tropical climates where ambient temperatures regularly exceed 45 °C. Process levels must be higher than with foods for sale in temperate climates, so that spores from thermophilic organisms are destroyed. F_0 values of more than 18–20 are typically used.

15.4.2.2 Mesophilic organisms

Spore-forming mesophilic bacteria prefer growing between 30 and 37 °C and so incubation times of up to 21 days are used. A lower incubation time of 10 days is recommended with the higher temperature of 37 °C. This tends to be the optimum growth temperature (37 °C) for many pathogenic bacteria, although a number of these (including *Salmonella*, *Escherichia coli,* and *Listeria*) do not produce spores.

Many non-spore-forming mesophilic bacteria are implicated in spoilage from containers that bypass the retort or through post-process leaker contamination. Incubation conditions of 14 days at 30 °C or 21 days at 25 °C are used, with the choice depending on the suspected organism. Mixed flora can be found in either scenario.

Yeast and mould spoilage can also be caused by containers bypassing the retort but is more commonly associated with post-process recontamination. Unless the mould is heat-resistant, such as *Neosartorya* spp. or *Byssochlamys* spp., incubation temperatures should be below 30 °C. A minimum incubation time of 14 days at 25 °C is recommended; however, the product pH and water activity will be instrumental in determining the precise conditions to optimise growth of the surviving yeast or mould.

15.4.3 Post-incubation inspection of containers

One of the objectives of incubating food containers is to encourage the surviving organisms to grow under the ideal incubation conditions and reach high numbers so they can be identified by microbiological examination. Information on the identification methods is given in reference sources such as Campden BRI Technical Manual No.18 (1987) and the FDA Bacteriological Analytical Manual (2001).

Further physical checks on the food after incubation involve pH measurement to identify if the pH has changed as a result of metabolic products of microorganism growth, and loss of vacuum caused by gas generation.

15.5 BIOTESTING

Biotesting is a specific type of incubation test that is designed to evaluate new container formats for which little data exist on the seam or seal quality (Campden BRI 1992). It involves the testing of high numbers of containers and also requires significant populations of microorganisms. As such, it is a test that can only be carried out by specialist laboratories.

A typical biotest involves filling containers with a nutrient material and applying a standard heat process to all of the containers. Sometimes a proportion of the containers will be deliberately damaged in a controlled manner in order to create a defect. These containers will be a positive control. Once the heat process is finished, the containers are cooled in water into which a gas-producing organism such as *Erwinia herbicola* has been added. All of the cooled containers are then incubated at 30 °C, or the appropriate temperature for the organism used and inspected daily.

References

BSI (1999) *Sampling Procedures for Inspection by Attributes. Sampling schemes indexed by acceptance quality limit (AQL) for lot-by-lot inspection.* British Standard, BS 6001–1. https://shop.bsigroup.com/ProductDetail?pid=000000000030219543.

Campden BRI (1987). *Examination of Suspect Spoiled Cans and Aseptically Filled Containers. Technical Manual No. 18.* Chipping Campden UK: Campden BRI.

Campden BRI (1992). *Safety Aspects of the Biotest Methodology. Technical Manual No. 36.* Chipping Campden UK: Campden BRI.

Campden BRI (2001). *Guidelines on Incubation Testing of Ambient Shelf Stable Heat Preserved Foods. Guideline No. 34.* Chipping Campden UK: Campden BRI.

Segner, W.P. (1979). Mesophilic aerobic spore-forming bacteria in the spoilage of low-acid canned foods. *Food Technology* 33 (1): 55–59.

US Food & Drug Administration (FDA) (2001). *An Examination of Canned Foods, Chapter 21a.* Center for Food Safety & Applied Nutrition, Bacteriological Analytical Manual Online.

16 Critical Factors in Thermal Processing

16.1 BACKGROUND

The HACCP (Hazard analysis critical control points) system, which is science-based and systematic, identifies specific hazards and measures for their control, to ensure the safety of food. HACCP is a tool to assess hazards and establish control systems that focus on prevention rather than relying mainly on end-product testing. (Codex Alimentarius 2003). Food processors should control food hazards through the use of systems such as HACCP.

This chapter does not attempt to cover all of the implementation of a HACCP system for thermal processing, rather it discusses various critical factors to assist in the identification of critical control points (CCPs). Food processors should be as follows:

- **identify** any steps in their operations that are critical to the safety of food;
- **implement** effective control procedures at those steps;
- **monitor** control procedures to ensure their continuing effectiveness;
- **review** control procedures periodically, and whenever the operations change.

Hygiene control systems should be applied throughout the food chain to control food hygiene throughout the shelf life of the product through proper product and process design. Control procedures may be simple, such as checking stock rotation or calibrating equipment. In some cases, a system based on expert advice, and involving documentation, may be appropriate.

16.2 KEY ASPECTS OF HYGIENE CONTROL SYSTEMS FOR FOOD PROCESSING (FROM CODEX ALIMENTARIUS)

Inadequate food temperature control is one of the most common causes of food-borne illness or food spoilage. Such controls include time and temperature of cooking, cooling, processing, and storage. Systems should be in place to ensure that temperature is controlled effectively where it is critical to the safety and suitability of food. Temperature control systems should take into account:

- the nature of the food, e.g. its water activity, pH and likely initial level, and types of microorganisms;
- the required shelf life of the product;
- the method of packaging and processing;
- the end-use of the product.

Such systems should also specify tolerable limits for time and temperature variations. Temperature-recording devices should be checked at regular intervals and tested for accuracy.

Examples of other steps that contribute to food hygiene and safety include:

- chilling;
- thermal processing;
- irradiation;
- drying;
- chemical preservation;
- vacuum or modified atmospheric packaging.

Each of these steps can be considered as processing systems (Chapter 8), but are discussed in detail as microbiological hurdles in Chapter 3.

Contamination of food can occur via biological, chemical, or physical sources. This book deals primarily with contamination by microbiological means; however, chemical and physical contamination must be understood so that it can be controlled within the thermal-processing environment.

Transfer of microorganisms to the food product, processing equipment, or to packaging needs consideration at various stages of the processing lines. Pathogens can be transferred from one food to another, either by direct contact or by food handlers, onto contact surfaces or through the air. Raw, unprocessed food should be effectively separated, either physically or by time, from processed food. Access to processing areas should be restricted or controlled. These are good manufacturing practices and are usually handled as one of the prerequisites in the HACCP plan of a company.

Systems should also be in place to prevent contamination of foods by foreign bodies, such as glass or metal shards from machinery, dust, harmful fumes, and unwanted chemicals. In manufacturing and processing, suitable detection or screening devices should be used where necessary.

No raw material or ingredient should be accepted by a food processor if it is known to contain parasites, undesirable microorganisms, pesticides, veterinary drugs or toxic, decomposed or extraneous substances that would not be reduced to an acceptable level by normal sorting and/or processing. Specifications for raw materials should be identified and applied and raw materials or ingredients should be inspected and sorted before processing.

One of the functions of the packaging of a heat-preserved food is to provide adequate protection for products to minimise re-contamination after processing. It is imperative that packaging materials are non-toxic and do not pose a threat to the safety and suitability of food under the specified conditions of processing, storage, and use. This includes changes to the packaging materials as a result of the applied heat process and should also consider the material breakdown that could arise from the flexing of containers during the process.

Water quality is important. Only potable water should be used in food handling and processing, with the following exceptions: for steam production, fire control, and other similar purposes not connected with food. Water recirculated for reuse should be treated and maintained in such a condition that no risk to the safety and suitability of food results from its use.

16.3 IDENTIFYING CRITICAL CONTROL POINTS IN THERMAL PROCESSING

An in-depth understanding of the product being processed is vital to be able to identify critical factors and set up critical control points and measurements. Critical factors of thermal processing are physical and chemical factors that can influence the thermal response of a product to the thermal process, the variation of which may influence the scheduled process, including container, product, retort, and processing conditions (ITFPS 2014; FDA Guidelines n.d.).

Critical factors that relate to the product characteristics are numerous and include the following:

16.3.1 Microbial load or bio-burden

The manner in which microorganisms are destroyed by wet heat will be considered in many of the chapters of this book. It is clear that the starting number of microorganisms (initial load) will have a direct influence on the final number of microorganisms given the same process. It follows that if a product or ingredient is unusually contaminated, the risk of surviving viable cells or spores will be higher. It is therefore imperative that good quality raw materials are used and where possible washed, and if so, in clean water.

16.3.2 pH of the product

The term 'pH' is used to designate the intensity or degree of acidity of the product. It is an intrinsic characteristic of the product, although factors such as ripeness in fruit or rigor mortis in meat or simply variety can show small but significant variations in the pH. The value of pH, the logarithm of the reciprocal of the hydrogen ion concentration in solution, is determined by measuring the difference in potential between two electrodes immersed in a sample solution. This is discussed in greater detail in Chapters 5 and 6.

As has been discussed, low acid foods have a pH of greater than 4.6 and acid or acidified foods have a pH of less than or equal to 4.6 (Table 16.1). The pH of the product immediately after processing is considered to be a significant pH.

The kinds of microorganisms that can grow and survive in heat-treated foods are strongly dependant on the pH of the food. While strict control of pH is critical for acid and acidified foods, especially around the range of 4.0–4.8, as pH is generally one of the two hurdles that keep these products safe and stable, it is also important for low acid foods. The thermal death time (TDT) of most microorganisms is closely linked to the pH of the medium or food. The efficacy of the thermal process is linked to its pH, and this should not be allowed to vary significantly. *More than 0.2 pH units are considered to be significant for most products.*

Table 16.1 Examples of common variation in the pH of some fresh fruit.

Fruit	Min pH	Max pH
Pears	3.6	4.7
Cherries	3.6	4.4
Tomatoes	3.9	4.7

16.3.3 Water activity (a_w)

Water activity is a measure of the amount of water that is available to support microbial growth (see Section 3.5). It is one of the hurdles to the growth and survival of microorganisms (Table 16.2). Changes in water activity can affect the efficacy of the thermal process and hence the risk of survival of spoilage microorganisms or microorganisms of public health significance. Lower water activity reduces the ability of microorganisms to take up nutrients and compounds for metabolism.

Salt and sugar are both examples of compounds that can be added to reduce water activity; salt in cured meats, sugar in jam.

Special attention needs to be given to products that use both salt and sugar as microbiological hurdles, when reformulating for healthier foods as taking out sugar or salt from some formulations could result in a less stable product that may require more severe heat treatment or other compensation.

16.3.4 Consistency

The consistency or viscosity of the product has a direct relationship with the rate of heat transfer into the product. In foods, heat transfer can be by two mechanisms; conduction or convection, and the consistency of the food determines which is dominant. Corned beef, for example, is a solid pack that heats slowly by conduction, whereas peas in brine heat rapidly by convection. Not all heating is either conduction or convection, as there are combinations of the two that are determined by-product characteristics. The viscosity of the sauce and the amount, size, and shape of solid particles all affect the heating patterns. These factors are considered in some detail in Chapter 11 on Measurement and Validation of Thermal Processes.

Table 16.2 Minimum water activity required for growth.

Organism	Minimum a_w
Normal bacteria	0.91
Normal yeasts	0.88
Normal moulds	0.80
Halophilic bacteria	0.75
Xerophilic fungi	0.65
Osmophilic yeasts	0.60

Source: Mossell and Ingram 1955.

The formulation is fundamental to the physical characteristics of the product. Any changes or deviations from what is standard will affect the rate of heating and hence could affect the safety of the thermal process. Besides the consistency of the product, which includes the viscosity, the solid ingredients are also critical to the characteristics of the product.

16.3.5 Presence, concentration, and types of preservatives

In some products, chemical preservatives are used in the product formulation, either for flavour or to enhance the efficacy of the thermal process or for both.

Cured meats have sodium and potassium salts such as chloride, nitrite, and nitrate, as well as sometimes polyphosphates and ascorbate that give the products their characteristic flavour and appearance, but also have a preserving action so that a sub $F_0 3$ thermal process is sufficient. The preserving action of salt is primarily due to the lowering of water activity.

Unfortunately, not all microorganisms are susceptible to the action of nitrite, but of those that are (e.g. anaerobic spore-formers such as *Clostridium botulinum*) nitrites are taken up into the cell and inhibit the ability to form energy-rich compounds and therefore inhibit growth and cell division. Nitrites are effective particularly at acidic pH, specifically less than pH 6.0 (Rahman 2007). Nitrates do not have any direct action, but are converted into nitrites, which inhibit growth.

Typically, cured luncheon meats contain 3.0–4.0% NaCl (salt on water phase) and 150 ppm (0.015%) nitrite and are given a thermal process equivalent to F_0 1.0–1.5. If this balance is changed, the safety of the process will have to be re-evaluated.

16.3.6 Rehydration

Heat transfer in dry material will be slower than in moist, so dehydrated ingredients must be thoroughly and consistently rehydrated. Dried ingredients cannot be substituted for fresh without the appropriate testing.

It is important to understand that the storage conditions, such as time, temperature, and humidity, of dry ingredients can affect the way that they rehydrate. Rehydration of solids must be done in a known controlled fashion. If the cooking process has to rehydrate ingredients, any changes to these procedures (e.g. higher temperature, shorter time) may cause inadequate hydration.

It is often found that dried materials have higher levels of microbial contamination than fresh, and therefore achieving complete rehydration before the thermal process killing step is critical.

16.3.7 Blanching

Blanching is used in the preparation of raw materials for many products. It is a short heat treatment, usually either in steam or in water. The benefits of blanching are many and include the following:

- wilting of plant tissue to facilitate packing into containers;
- expelling air from plant tissue. Removal of oxygen can improve the colour and extend the shelf life of the product, as well as changing the density;
- inactivating enzymes, to improve the colour, flavour, and nutritional value, as well as preventing enzymatic action before the containers enter the thermal process;
- preheating the product, which affects the vacuum in canned food, as well as reducing the required process time as a result of elevated initial product temperatures;
- reducing the initial microbial load on the product, specifically of vegetative cells.

It is worth noting that over blanching or under blanching can affect the container fill weight and the solids to liquids ratio. Both of these are critical factors.

Also of concern with blanching operations is that thermophilic bacteria can grow inside blanchers. To prevent this happening, blanchers should be operated in the 88–93 °C (190–200 °F) range. Maintaining product at or above 82 °C (190 °F) helps to control thermophilic growth. If blanchers must be operated at temperatures lower than 82 °C because of product considerations, then they should be regularly evaluated for thermophilic growth and build-up.

16.3.8 Size and style of in-going ingredients

Larger pieces of solids will take longer to heat at their centre, because the solids will heat by conduction. The thermal process is designed to deliver a process at the coldest spot in the container, which might be in the centre of a solid ingredient. The relationship between particulate size and process time requirements is illustrated by the data in Table 16.3.

Table 16.3 Recommended process times for various size shrimps in brine in 307 × 113 cans processed at 250 °F (121.1 °C) with a minimum initial temperature of 45 °F (7.2 °C).

Shrimp size	Minutes at 250 °F
Medium, small, tiny	14
Large	16
Extra large, Jumbo	18

Source: Data from NFPA (1982).

Table 16.4 Recommended process times for asparagus in brine in 211 × 304 cans processed at 250 °F (121 °C) with a minimum initial temperature of 70 °F (21.1 °C).

Asparagus style	Minutes at 250 °F
Cuts	13
Spears, tips down	19
Spears, tips up	21

Source: Data from NFPA (1982).

The style of the ingredient also can affect the rate of heat penetration. For example, French cut (diagonal) and cross-cut green beans do not have the same rate of heating. Similarly, sliced or diced carrots constitute very different packs in terms of their heating rates. Asparagus spears heat up differently if the tips are positioned up compared with the tips pointing down (Table 16.4).

As a general rule, the higher the mass of the product the longer it will take to heat, although the physics of conduction heat transfer dictates that the thickness is the critical dimension. Also, the higher the mass of the solid components of the product, the longer it will take to heat because of the predominance of conduction heat transfer.

It is not only the visible solid ingredients that can result in slower heat penetration, the same viscosity tomato sauce, but with higher percentage tomato solids/fibre can be expected to heat slower.

16.3.9 Container, packing, and filling considerations

16.3.9.1 Headspace

The headspace is the closed space containing air or gasses above the food inside a closed container. The size of headspace is critical when forced convection heating is used in agitated retort, especially with end-over-end or axial agitation. This is because the headspace acts as a bubble that travels through the product and mixes the regions of higher and lower temperatures. If the headspace volume is too small, inefficient or no mixing will take place and the rate of heating of the product will be markedly reduced.

Not all of the headspace actions relate to heat transfer. An excessive headspace will result in too much air inside a container and excessive oxygen (~21% of air is oxygen). This can result in oxidation of the product, especially at the headspace product interface and can cause unsightly browning or discolouration. Excessive oxygen can also cause corrosion of the interior of the container.

If the process is static, a large headspace can result in a layer of air that acts as an insulation layer and may reduce heat transfer. This becomes important for containers where the diameter is larger than the height, and so heat transfer occurs primarily through top and bottom surfaces.

A general rule is that the headspace should be about 5% of the container volume.

16.3.9.2 Container vacuum and exhausting of containers

The vacuum in a container serves as a visual indicator that no gas-forming microbiological spoilage has taken place. It also reduces the amount of air (and hence oxygen) inside the container, which reduces the potential for oxidative deterioration of the product and corrosion of the container.

Excessive air inside cans can result in large pressure build-up during processing, which can cause the ends to deform irreversibly (peak) or bulge (flippers). As mentioned above, too much air inside the container can result in a layer of air that can insulate and reduce heat transfer.

Air can be removed from the container and a vacuum-formed by several methods:

- *Hot filling*: Filling hot product into the container and sealing it while hot will result in a vacuum being formed as the product cools and contracts. The fill temperature is directly proportional to the final vacuum, but is also influenced by headspace. Figure 16.1 illustrates the relationship between fill temperature and the vacuum formed.
- *Exhaust box*: The product is filled into the container and then is passed through a steam or hot water exhaust box. Usually only for metal cans.

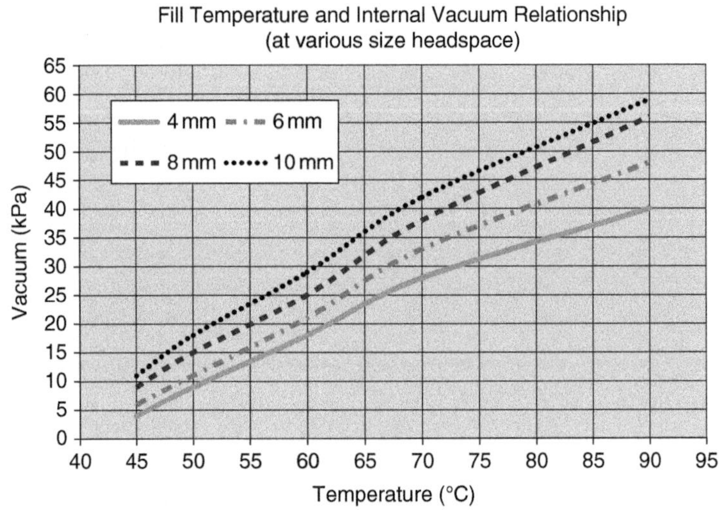

Fig. 16.1 Relationship between fill temperature and the vacuum formed, for various headspace volumes. For a colour version of this figure, please see the colour plate section.

- *Mechanically*: The containers are filled and passed into a chamber, where a vacuum is drawn and the container end is seamed on. Usually only for metal cans.
- *Steam injection*: Steam is injected into the headspace of the container immediately prior to seaming or fitting the lid. The steam displaces the air in the headspace and forms a vacuum when it condenses.

16.3.9.3 Container size and geometry

The shape and size of the container are very important to the position and the rate of heating of the cold spot in the product. The cold spot in larger containers will take longer to reach the desired temperature or lethality (Table 16.5). In addition, the shape of the container will affect the position of the cold spot. This can be especially problematic for flexible containers, where even slight variations in fill volume or headspace can affect the geometry of the container.

16.3.9.4 Initial temperature of product

Initial temperature is defined as the average temperature of the contents of the coldest container to be processed at the time that the thermal processing cycle begins. It is determined after thorough

Table 16.5 Recommended process times for baked beans in tomato sauce in various can sizes, processed at 250 °F (121.1 °C) with a minimum initial temperature of 70 °F (21.1 °C).

Can size (US)	Minutes at 250 °F
211 × 400	69
307 × 214	80
307 × 400	100
603 × 700	275

Source: Data from NFPA (1982).

Table 16.6 Recommended process times for cream style corn with various initial temperatures, processed at 240 and 250 °F (115.6 and 121.1 °C).

Initial temperature (°F)	Process time at 240 °F (minutes)	Process time at 250 °F (minutes)
140	70	52
160	67	49
180	62	45

Source: Data from NFPA (1982).

shaking or stirring of the filled and sealed container. The initial temperature is a critical control point because it has a direct influence on the time taken for the cold spot in a container to achieve the minimum required temperature or lethality (Table 16.6).

16.3.10 Process-related critical factors

16.3.10.1 Processing method

The type of retort (e.g. still, horizontal, vertical or crateless, or agitating) can affect the rate of heat transfer into the cans. If agitation is used, the method of agitation (end-over-end, axial, and intermittent) can significantly affect the rate of heating. Batch processes compared with continuous processes can also vary.

16.3.10.2 Processing medium

The heating medium (e.g. saturated steam, steam/air mixtures, water immersion, water spray, and water cascade) is critical to the rate of heat transfer from the media to the container surface. Air is one of the worst heat transfer media, despite its widespread use in refrigeration and ovens, and prolonged contact with container surfaces will result in poor heat transfer. It is for this reason that venting of retorts is so important to processing, and the agitation or circulation of heating media is critical to the even distribution of heat and also during cooling.

16.3.10.3 Type and characteristics of heat-processing system

The type of processing system and the way that it operates should be understood and must be reproducible. For example, the method of loading baskets (e.g. crateless, jumble packed, stacked with or without layer pads) must be fixed and consistent between batches. The orientation of the containers (vertical, horizontal or jumble packed) must also be consistent, because different orientations can affect the rate of heat transfer, especially if the pack's heat by convection.

For steam retorts, the vent schedule must be established to achieve air removal, and there should be a minimum come-up time specified before the hold time can commence.

The pressure inside the retort must be stable and must correspond to the required pressure for that given temperature and altitude (only for condensing steam retorts). A retort operating at a pressure above the saturated steam pressure could indicate inadequate venting. This is because a mixture of air and steam will have a higher pressure than saturated steam. This situation is further complicated by variation in atmospheric pressures at different altitudes, as shown in Table 16.7.

16.3.10.4 Processing temperature

Operating a retort at the correct processing temperature is vital. The official temperature measuring device for processes filed with the FDA is the mercury-in-glass (MIG) thermometer, which must be annually calibrated and regularly validated. Although this situation is currently changing and thermometers based on platinum resistance principles are now acceptable to the FDA, many companies still rely on the MIG as their master temperature indicator (MTI). The MTI, the temperature controller and the recording chart must be all set to read the same, as otherwise there is no reliable record of the process. Small deviations in process temperature can result in dramatic changes in the accumulated lethality of a process. This is illustrated with the data in Table 16.8 for canned pilchards in tomato sauce.

Table 16.7 Gauge pressure (psi) corresponding to specified process temperature at various altitudes, note that psi × 6.895 = kPa.

Temp (°F)	Sea level	Feet above sea level							Temp (°C)
		500	1000	2000	3000	4000	5000	6000	
200	–	–	–	–	–	–	–	–	93.3
205	–	–	–	–	–	–	0.5	0.9	96.1
210	–	–	–	0.4	0.9	1.4	1.8	2.3	98.9
212	0.0	0.2	0.5	1.0	1.5	2.0	2.4	2.9	100.0
215	0.9	1.1	1.4	1.9	2.4	2.9	3.3	3.8	101.7
220	2.5	2.7	3.0	3.4	3.9	4.4	4.9	5.3	104.4
225	4.2	4.5	4.7	5.2	5.7	6.2	6.6	7.1	107.2
230	6.1	6.3	6.6	7.1	7.6	8.0	8.5	9.0	110.0
235	8.1	8.3	8.6	9.1	9.6	10.0	10.5	11.0	112.8
240	10.3	10.5	10.8	11.3	11.7	12.2	12.7	13.1	115.6
242	11.2	11.4	11.7	12.2	12.7	13.1	13.6	14.1	116.7
245	12.6	12.9	13.1	13.6	14.1	14.6	15.0	15.5	118.3
248	14.1	14.3	14.6	15.1	15.6	16.0	16.5	17.0	120.0
250	15.1	15.4	15.6	16.1	16.6	17.1	17.5	18.0	121.1
252	16.2	16.4	16.7	17.2	17.7	18.1	18.6	19.1	122.2
255	17.8	18.1	18.3	18.8	19.3	19.8	20.2	20.7	123.9
260	20.7	21.0	21.2	21.7	22.2	22.7	23.1	23.6	126.7
265	23.8	24.0	24.3	24.8	25.3	25.8	26.3	26.8	129.4
270	27.3	27.5	27.8	28.3	28.8	29.3	29.8	30.3	132.2
275	30.9	31.2	31.5	32.0	32.5	33.0	33.5	34.0	123.0

Table 16.8 Accumulated process lethality for pilchards in tomato sauce in 73 × 110 mm cans, at various process temperatures with an initial temperature of 60 °C.

Process temperature (°C)	F_0 achieved at 75 minutes (minutes)	Process time to achieve $F_0$6 (minutes)
116	5.4	78
118	7.7	71
121	12.9	62

Small deviations from the product or processing specification, which may seem negligible or insignificant, may seriously affect the adequacy of the process for that product, therefore any changes or deviations should be evaluated for the effect on the adequacy of the process.

16.3.10.5 Processing time

The processing time is also critical to the safety of a thermal process. It is important, when using automatic timing devices linked

Table 16.9 Lethality increase (F_0) per minute of process time at various temperatures.

Can centre temperature (°C)	Increase in F_0 per minute (minutes)	Can centre temperature (°C)	Increase in F_0 per minute (minutes)
111	0.10	121	0.98
112	0.12	122	1.23
113	0.15	123	1.55
114	0.19	124	1.95
115	0.25	125	2.45
116	0.31	126	3.09
117	0.39	127	3.89
118	0.49	128	4.90
119	0.62	129	6.17
120	0.78	130	7.76

Table 16.10 Boiling temperature of water at different altitudes.

Altitude (metres)	Boiling point of water (°C)	Altitude (feet)	Boiling point of water (°F)
0 (sea level)	100	0 (sea level)	212
500	98.4	1640	209
1000	96.7	3280	206
1500	95.1	4921	203
2000	93.4	6561	200
2500	91.7	8202	197
3000	90.0	9842	194
3500	88.2	11482	191
4000	86.4	13123	188

to the retort controller, to ensure that if there is a temperature overshoot, and then an automatic correction, that the timer only starts once the temperature has stabilised at the process temperature. Table 16.9 presents data for the increase in F_0 per minute for container centre temperatures between 111 and 130 °C. This illustrates the importance of correct timing for process hold periods.

16.3.10.6 Processing at high altitude

As altitude increases and atmospheric pressure decreases, the boiling point of water decreases (Table 16.10). Using the processing time for canning food at sea level may result in spoilage if you live at altitudes of 300 metres (1000 feet) or more. Lower boiling temperatures are less effective for killing bacteria. Increasing the process time or pressure compensates for lower boiling temperatures.

When processing in atmospheric conditions e.g. steam or boiling water (i.e. pasteurising) to compensate for the lower boiling point of water, the cooking time must be increased. The temperature of the water cannot exceed its boiling point – unless if using a pressure cooker. For example, when boiling or poaching eggs, it will take longer to cook eggs at high altitudes, because water boils at a lower temperature, but the proteins require the same temperature to denature.

References

Codex Alimentarius (2003). *Recommended International Code of Practice General Principles of Food Hygiene*. CAC/RCP 1-1969, Rev. 4-2003.

FDA Guidelines for Low Acid Canned Food Manufacturers Part 2 – Processes/Procedures Manufacturers. (n.d.) http://www.fda.gov/Food/FoodSafety/Product-SpecificInformation/AcidifiedLow-AcidCannedFoods/GuidanceCompliance/default.htm.

IFTPS (2014). *Guidelines for Conducting Thermal Processing Studies*. www.iftps.org.

Mossell, D.D.A. and Ingram, M. (1955). The physiology of spoilage of foods. Journal of Applied Bacteriology 18: 232–268.

NFPA (1982). *Thermal Processes for Low Acid Foods in Metal Containers, Bulletin 26-L*, 12ee. June 1982. Washington DC: National Food Processors Association.

Rahman, M.S. (2007). *Handbook of Food Preservation*, 2e. Taylor & Francis Group, Boca Raton FL: CRC Press.

17 Environmental Aspects of Thermal Processing

Awareness of the importance of the environment reached a pinnacle in 2019. Campaigns were driven by the younger generation, led by individuals such as Greta Thunberg, as well as seasoned media presenters such as Sir David Attenborough. The focus was on climate change and plastic pollution that are damaging the environment. Climate change relates to the amount of greenhouse gases (GHG) that are being released into the atmosphere as the global population increases exponentially. It is damaging to nature through the increases it is causing to global average temperatures. This has many negative effects including the changing of climates for growing food globally, which is destabilising food supply systems. Implications for agricultural systems in most growing regions of the world are substantial. Plastic waste is the other issues that have hit the headlines recently. Many plastics are difficult to recycle and do not break down naturally for hundreds of years, if ever. The waste ends up in landfill sites and is discarded as rubbish where it builds up in the countryside or finds its way into the oceans where it is very damaging to sea life and often ends up back in the food chain.

One aspect of the environmental concerns is the growing need to satisfy customer demand for products from sustainable sources. This is leading businesses to consider the full environmental and social impacts of their activities. Many food producers and retailers are undertaking lifecycle assessments (LCA) that cover a range of environmental impacts for the lifecycle of their products. This requires a study of all stages in the production of a unit of food and includes production of raw materials, the manufacturing process, transport and distribution, retail, consumer use, and waste disposal. In a food context, this includes farming activities, packing, transport to market, retail, in-home preparation and consumption, and waste disposal at all stages of the lifecycle.

Essentials of Thermal Processing, Second Edition. Gary Tucker and Susan Featherstone.

Many governments are making pledges to cut GHG emissions by the year 2050 in order to address the pending climate change disasters. Some countries are setting ambitious targets of net-zero emissions. This is a difficult target to achieve without recourse to carbon offsetting initiatives that put the onus on other countries to supply certain high-emission goods and services. Irrespective of how this can be achieved, there is a positive role that thermal processing can have on reducing GHG emissions. This includes the recyclability of most packaging formats (e.g., glass and metal), avoiding the need for refrigeration of fresh produce, local products harvested and processed in season and using localised labour forces.

Assessing the GHG emissions arising from the manufacture of a food product was the subject of a UK initiative that resulted in a publication referred to as PAS 2050 (Publicly Available Specification for the assessment of the lifecycle greenhouse gas emissions of goods and services). This document, published by the British Standards Institution (BSI), was based on existing internationally accepted standards (ISO 14044:2006 and ISO 14040:2006).

Thermal processing, as a technology for preserving food, can be environmentally beneficial because the goods are stored ambient (avoiding the need for refrigeration), and agricultural products are processed close to where they are grown (avoiding transportation emissions and forcing crops to grow out of season). An example of a GHG assessment is given in this chapter for a bottled apple juice.

17.1 LIFECYCLE ASSESSMENT (LCA)

LCA was developed to study the environmental impacts arising from the production, use, and disposal of products or services. These do not necessarily have to be foodstuffs. LCA provides a mechanism for investigating and evaluating such impacts from the extraction of basic materials from Nature, from production through packing, distribution, retail, product use (e.g. home cooking and consumption of food), and end-of-life management (e.g. disposal of waste from peeling fruit). LCA considers the environmental impact of the 'product systems' in terms of the environmental consequences of flows (principally flows of substances) between such systems and the environment (Fig. 17.1).

In conducting an LCA on a specified product system, all inputs to the system are ideally traced back to primary resources. For example, the impacts of electricity generation are not only those

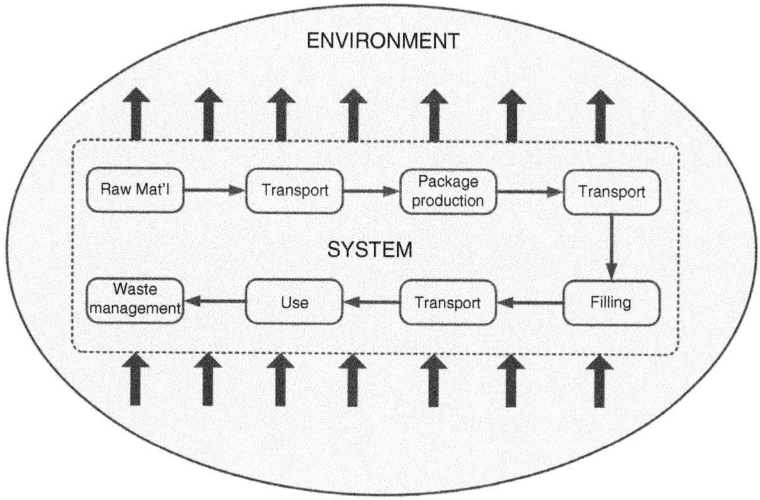

Fig. 17.1 Product systems and the environment. Courtesy of Chris Foster.

of power station operation and electricity transmission, but include extraction of primary fuels, such as coal, oil, and uranium and the burdens of refining. All outputs from the system are ideally followed forward through subsequent processing until they enter the environment. LCA considers impacts on all environmental media – air, water, and land.

The LCA process has become standardised. The ISO 14040 series contains the main standards applicable to LCA. ISO 14040 (2006) describes the four main elements of an LCA study as being:

(i) goal definition and scoping;
(ii) lifecycle inventory development (compiling the inventory of relevant inputs and outputs);
(iii) lifecycle impact assessment (LCIA) (evaluating the potential impacts associated with the inputs and outputs identified in the lifecycle inventory);
(iv) improvement analysis (interpreting the results in relation to the objectives).

17.1.1 Impact categories

LCA assesses a number of environmental impact categories. The main categories that are relevant to the agri-food sector, and specifically thermal processing, are as follows.

17.1.1.1 Global warming potential (GWP)

A global warming potential assessment is sometimes referred to as a carbon footprint. Lifecycle emissions of greenhouse gases are assessed and expressed in units of carbon dioxide equivalents (CO_2e), calculated by multiplying the mass of a given greenhouse gas by its global warming potential. This is the method described in PAS 2050 (Tucker et al. 2008).

Thermal processes are in general low in GHG emissions because foods are heated and cooled efficiently, are stored in ambient conditions (non-refrigerated), and the raw materials can be processed close to the growing regions. The relative GHG efficiency of a thermal process will be highlighted in the bottled apple juice example.

17.1.1.2 Pesticide use/ecotoxicity

Impact assessment of ecotoxicity in LCA is complicated by the fact that not all toxic chemicals have the same mechanism of action (whereas all greenhouse gases do). There is also a large (and to some extent evolving) number of substances that can contribute to the impact category, ranging from metals, through persistent organic pollutants to pesticides. The most commonly used methods relate relevant releases to a reference substance (e.g. 1,4-dichlorobenzene) by means of calculated toxic equivalency potentials, and one of these may be used in this instance.

Thermal processing is no different from other means of food preservation in that the crops and raw materials are grown using standard agricultural practices, which include pesticide use.

17.1.1.3 Abiotic resource use

This captures the extent to which the activities in the product's lifecycle contribute to the depletion of non-living resources – essentially fossil and mineral resources. Several assessment methods are available, since the 'environmental impact' of mineral or fossil resource use is somewhat difficult to define. The widely used CML (Centre for Environmental Sciences, Leiden University) method relates all abiotic resource use to the extraction of a single substance (antimony) on the basis of the relative abundance of different resources. The eco-indicator, 99 method takes a different approach, assessing the significance of resource consumption now in terms of anticipated additional energy demand extraction of the same resource at a designated point in the future. The latter method considers fossil energy use and mineral use separately – something of an advantage given the high profile of fossil fuel consumption.

Thermal processing can be an attractive preservation method in terms of its minimal depletion of abiotic resources for certain packaging materials, when compared with other technologies. For example, metal and glass packaging both require high quantities of energy for their extraction (and so are relatively high in their GWP), but have a minimal effect on the abiotic resources of the planet. After oxygen and silicon, aluminium is the third most common element and iron is the fourth. Both metals can be recycled infinitely without degradation to the metal quality. For example, 54% of steel is recycled in the EU and since 150 years ago, 75% of all primary aluminium is still in use. Glass is extracted from silica sand using limestone and soda ash, which are readily available raw materials. In addition, recycled glass (cullet) can be added in the extraction process up to levels of 81% for green glass, and lightweight of bottles is now commonplace.

17.1.1.4 Acidification potential

Acidification impact assessment quantifies acid-gas releases from the system and/or the subsequent damage they cause. Both mid-point methods (quantifying the burden produced in terms of a single acidic substance – hydrogen ion or sulphur dioxide) and end-point methods (quantifying the actual impact of those burdens on the environment) are available. Using the latter involves assuming a model that relates the amount of acidic substance deposited to the extent of environmental damage caused (e.g. rate of plant species attrition per unit area per year per kg acid deposited). Acidification was known as acid rain during the 1970s and was mainly caused by sulphur and nitrogen oxides released during the combustion of coal and oil.

GWP includes emissions during combustion, albeit calculated in a slightly different way, so the above comments on thermal processing having a low impact are relevant also for acidification.

17.1.1.5 Eutrophication potential

Eutrophication is quantified in units of phosphate equivalents. Agricultural sources include nitrate (NO_3) and phosphate (PO_4) leaching to water and ammonia (NH_3) emissions to air. A consequence of phosphate leaching into watercourses is the prevalence of algal blooms in rivers and lakes. These clog up waterways are unsightly and deplete oxygen as the algae decompose. In addition, once minerals such as phosphates enter the river systems, they become so diluted that it is not possible to recover them economically. Phosphorous is one such element of which there is global concern over the long-term availability of deposits that can be mined.

Thermal processing is no different from other means of food preservation in that the crops and raw materials are grown using standard agricultural practices, which include fertiliser use. The impact of thermal processing on eutrophication is likely to be neutral and no different from other food preservation methods.

17.1.1.6 Land use

Assessing the impacts of land use for all parts of the system is problematic. Land occupancy is recorded by some practitioners for systems in which plant-derived resources play a significant part, as an indicator at least, of the extent to which this finite resource is utilised for the function being considered. Assigning impacts associated with changes in land use to particular products remain an area in which consensus is still being developed. It is a sensitive subject. For example, the destruction of rain forests to make way for cattle farming or crops such as soya or palm fruit is high on political agendas.

It could be argued that thermal processing enables raw materials to be grown in regions that are suited for this purpose and so land-use change does not need to be forced. However, this may be a biased argument and thermal processing is best considered neutral in its effects on land-use change.

17.1.1.7 Water use

No well-established lifecycle impact assessment (LCIA) method exists for the environmental impacts associated with water use. Two obstacles have hindered the emergence of an LCIA method relating water consumption to its environmental significance. The first is that the significance of any water consumption is highly location-dependent in terms of its implications for the water resource base; and LCA is a rather location-insensitive method in its current form. The second is that water consumption data is seldom included in LCA data sets in a manner that enables water-resource implications to be established. However, given the extent to which global warming is affecting rainfall patterns, the water use impact category is increasing in importance.

Thermal processing can be applied close to the growing region for a vegetable or fruit product, which allows the crops to be grown where water supply is guaranteed. Transport of a packaged and shelf-stable product is also more economical than for those that are perishable. Thus, thermal processing can help prevent water from being taken for agricultural purposes in parts of the world where its scarcity prevents it from being used for drinking.

17.2 GREENHOUSE GAS EMISSIONS

One of the most important LCA impact factors for today is the effect of an activity on global warming. This is assessed by calculating greenhouse gas (GHG) emissions in terms of their carbon dioxide equivalent values (kg CO_2e). The terms carbon footprint and GHG emissions have similar meanings and are interchangeable.

To calculate a carbon footprint for the production of a food product or a food business requires all of the stages involved in that activity to be considered. Figure 17.2 illustrates the linkage between raw materials production through to disposal of the waste and final product at the end of its life. Each stage must be assessed for the GHG emissions using a detailed series of calculations (see Example 17.1).

Calculation of GHG emissions is carried out relative to carbon dioxide that is given a value of unity (Table 17.1). PAS 2050 contains a more complete table that includes gases used in refrigeration, solvents, and packaging. It is apparent from Table 17.1 that a process releasing nitrous oxide has the potential for much greater global warming impact than one releasing carbon dioxide (Wiltshire et al. 2009).

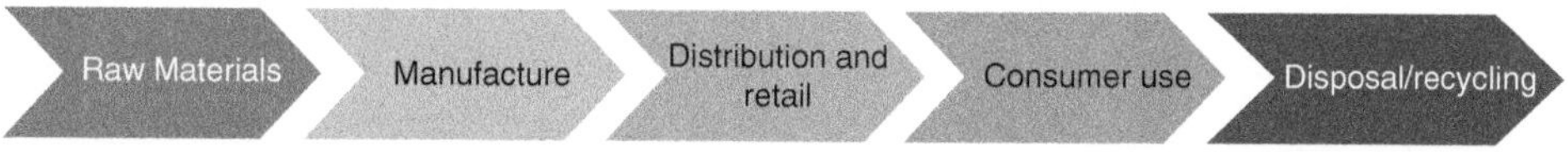

Fig. 17.2 Stages required in the GHG assessment of a food product.

Example 17.1 Beef, a high-carbon footprint ingredient

Beef cattle produce large quantities of methane as a result of enteric fermentation. Methane has a GWP of 25 times that of carbon dioxide, and so this can lead to cattle products such as beef and milk with high-carbon footprint values.

Over-use of nitrogenous fertilisers when growing crops results in the release of nitrous oxide into the atmosphere. This will result in a high-carbon footprint for that crop, because nitrous oxide has a GWP of 298 times that of carbon dioxide. To reduce the carbon footprint, the nitrogen-rich fertiliser should be applied at the correct dosing rate and at the correct time to maximise its absorption. Research is being undertaken to farm cattle in conditions that minimise the release of methane.

Table 17.1 Relative global warming potential (GWP) of the main greenhouse gasses of relevance to the food industry.

Greenhouse gas	Formula	GWP
Carbon dioxide	CO_2	1
Methane	CH_4	25
Nitrous oxide	N_2O	298

The next section presents a case study for a thermally processed product, bottled apple juice, which demonstrates the methods involved when calculating a carbon footprint value.

17.2.1 Case study: Bottled apple juice

This study estimates the embodied GHG emissions during the manufacture of a 75 cL bottle of Cox's apple juice. The bottled juice is in-pack pasteurised to destroy yeasts and moulds, which requires a process of at least two minutes at 70 °C. In addition to the primary packaging (glass bottle), the bottles were packaged into cardboard boxes of 12 bottles per box.

The apple juice carbon footprint (CF) includes all material emissions generated as a direct or indirect result of the product unit (PU) being produced. PAS 2050 allows immaterial emissions to be excluded, which are any single sources less than 1% of the total emissions, up to a maximum of 5%. In the apple juice example, ascorbic acid is added in very small quantities to minimise browning of the pressed juice; this is excluded from the calculations.

Figure 17.3 shows an outline flow diagram from the point at which apples are taken into the juicing room, up to when the bottled juice is ready for distribution from the farm. It shows the single raw material input (apples), the various packaging inputs and the two waste streams (pressed pulp and wastewater). The juicing operation takes place in a dedicated room, shown with a dotted line.

17.2.1.1 Raw materials (0.407 kg CO_2e/PU)

The chosen variety of apple is Cox's Orange Pippin; however, the data calculated for manufacture and packaging of apple juice is not specific to this variety. A CF value for Cox's apples of 0.075 kg CO_2/kg was used (Wiltshire et al. 2009), based primarily on assessments of fertiliser use and machinery outputs, with apples grown in a modern orchard. Allowing for a 70% juice recovery from milled apples, each 75 cL bottle contains 1.4 kg of apples, which is 0.105 kg CO_2e/PU.

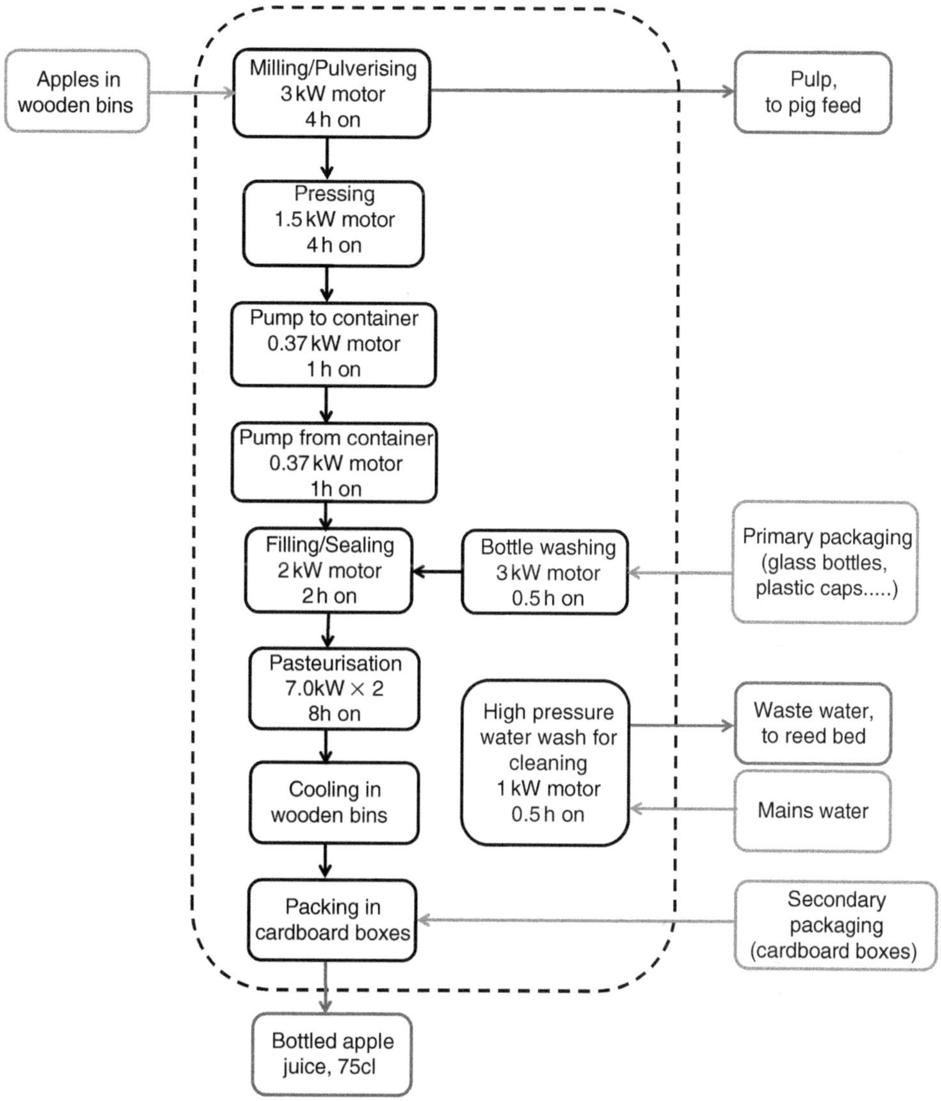

Fig. 17.3 Outline flow diagram for bottled apple juice manufacture.

This study used a product sold in green bottles, which have the advantage of allowing a higher proportion of recycled glass than for a clear bottle. Data for GHG emissions for glass were obtained from the British Glass Manufacturers Confederation (Enviros 2003), in which virgin glass and recycled glass have GHG values of 0.843 and 0.529 kg CO_2e/kg, respectively. A typical recycle rate for green glass is given as 81%; therefore, the GHG value for glass is 0.589 kg CO_2e/kg. The plastic cap is polypropylene, which has a GHG value of 4.4 kg CO_2e/kg

Table 17.2 GHG emissions (kg CO_2e) for the packaging materials used for 75 cL bottles.

Component	Weight (g/PU)	GHG value (kg CO_2e/kg)	GHG value (kg CO_2e/PU)
Glass bottle	400	0.589	0.2350
Plastic cap	3.0	4.4	0.0130
Paper label/ adhesive	0.5	1.03	0.0005
Plastic cap wrapping	0.5	2.4	0.0012
Cardboard boxes	47.5	1.03	0.0489
Total			**0.2989**

(Plastics Europe 2005). Minor components of the packaging are also considered, which are the paper label and adhesive, and the plastic wrapping over the cap (assumed to be low-density polyethylene). Table 17.2 details the mass and GHG values for each packaging component.

Bottles are packed into cardboard boxes (secondary packaging) in units of 12. A GHG value of 1.03 kg CO_2e/kg for cardboard was taken from the FEFCO LCA inventory (FEFCO 2006). Each box weighs 570 g and so the *pro-rata* weight for a one bottle PU is 47.5 g; therefore, the GHG value is 0.0489 kg CO_2e/PU.

17.2.1.2 Manufacture (0.061 kg CO_2e/PU)

Juice bottling includes the process steps within the dotted line in Fig. 17.3. Estimations were made for the duration of each stage that involves an electric motor, which is required because operation of these stages is intermittent. The thermal processing step is achieved in two water bath pasteurisers. Calculated data for each process step are given in Table 17.3.

The total electrical energy required to process 1200 bottles of juice is 136.74 kWh or 0.114 kWh per PU (bottle). Conversion of electrical energy to kg CO_2e uses the emission factor from Defra (2008), which results in the juice bottling process contributing 0.061 kg CO_2e/PU. Most of this is from the pasteurisers (thermal processing step).

17.2.1.3 Transportation (0.057 kg CO_2e/PU)

Apples are grown both on the farm and at a nearby farm 12 miles away. Transportation of apples from the trees to the juicing process is by tractor, with the apples packed in large wooden bins. These wooden bins are re-used.

Table 17.3 Electrical energy required for the milling, pressing and pasteurising stages of 1200 bottles of apple juice.

Unit operation	Motor power (kW)	Duration (hours)	Power used (kWh)
Two pasteurisers	7.0	8.0	112.0
Two pumps	0.37	1.0	0.74
Bottle sealer	2.0	2.0	4.0
Pulveriser/mill	3.0	4.0	12.0
Press	1.5	4.0	6.0
Bottle washing machine	3.0	0.5	1.5
High-pressure washer	r 1.0	0.5	0.5
Total			**136.74**

Table 17.4 GHG emissions (kg CO_2e) for transportation of the raw materials used for 75 cL bottles of apple juice.

Component	GHG value (kg CO_2e/PU)
Wooden bins of apples	0.0092
Glass bottles	0.0243
Plastic caps	0.0122
Cardboard boxes	0.0110
Total	**0.0566**

Estimations are made for the distances travelled in delivering the raw materials to the farm, and for the number of journeys made. Table 17.4 presents the breakdown of transportation GHG emissions for the various raw materials.

17.2.1.4 Waste (0 kg CO_2e/PU)

Two categories of waste are generated; wastewater from cleaning and washing operations, and apple pulp from milling and pressing.

Wastewater is gravity fed to a reed bed, which removes much of the organic matter, and fixes the carbon and nitrogen within the plant material. It is arguable that within the 100-year lifecycle suggested by PAS 2050, all of the organic carbon and nitrogen will end up as gases, and in doing so contribute to GHG emissions. This is not considered during this study because no data is available on the quantities of organic materials in the wastewater. No caustic or detergents are used for the cleaning operation.

Apple pulp is fed to the pigs that live on the farm, which is a carbon zero activity.

17.2.1.5 Overall carbon footprint (0.525 kg CO_2e/PU)

Table 17.5 compiles the GHG emissions for each of the categories detailed above. The CF for these 75 cL bottles of Cox's apple juice is 0.525 kg CO_2e/PU. The most significant component of this footprint is the glass bottles, which make up 45%, followed by the apples at 20% and the processing at 12%. The pasteurisation step only contributes 10%.

17.2.1.6 GHG emissions for other food products

There are now numerous publications freely available on the internet that quote carbon footprint values for food products (e.g. Tate & Lyle, British Sugar, Fat Tire Ale). Some of these have used the procedures defined in PAS 2050, although the source of the calculation procedures is not always clear. Data presented in Table 17.6 were calculated by a group of researchers whose aim was to advise Defra on applying PAS 2050 for food products and their raw materials (Wiltshire et al. 2009).

Food commodities with *low emissions* (<1 kg CO_2e/kg or L) tended to be crop commodities with high yields and low inputs, for example:

- apples (0.066–0.100) and bottled apple juice (0.71);
- potatoes (0.12–0.16);
- spring onions (0.23);
- animal feed crops (0.0043–0.7400);
- carrots (0.35);
- UK conventional tomatoes grown using 'waste' heat (0.39);
- wheat (0.40–0.74);
- onions (0.42–0.59).

Table 17.5 GHG emissions (kg CO_2e) for a 75 cL bottle of apple juice.

Category	**GHG emissions (kg CO_2e/PU)**
Raw materials	0.407
Manufacture	0.061
Transportation	0.057
Waste	0.000
Total	**0.525**

Source: Wiltshire et al. (2009).

Table 17.6 Summary of GHG emissions for manufactured food products, per kg product and product unit.

Food product	Kg CO_2e/ kg or L	Kg CO_2e/ PU	PU description
Beef cottage pie	7.6	3.3	Single 434.9 g chilled ready meal
White loaf of bread	0.73	0.60	827 g loaf in a plastic bag
Packed mild cheddar cheese	9.8	4.9	500 g plastic pack
Cox's apple juice	0.71	0.53	75 cL glass bottle
Chocolate coated cakes	2.5	0.42	165 g packet
Duck in Hoisin sauce	2.0	0.88	Single chilled ready meal (430 g inc. packaging)
Lamb shanks and roasted potatoes	19	25	Single chilled ready meal (1300 g inc. packaging)
Thai chicken pizza	3.5	1.6	1 pizza (460 g inc. packaging)

These GHG data were used to calculate GHG emissions for the manufacture of the food products in Table 17.6 (Wiltshire et al. 2009). Each of these products has a thermal processing step, either for preserving a raw material (e.g. milk for cheese) or as the final cook or process (e.g. bread or beef cottage pie).

Food commodities with *medium emissions* (1–5 kg CO_2e/kg or L) tended to be high-yielding livestock products or manufactured products, for example:

- milk (1.2–1.4 kg CO_2e/L);
- duck in Hoisin sauce ready meal (2.0);
- chocolate-coated cakes (2.5);
- Thai chicken pizza (3.5);
- chicken meat (3.1–4.4);
- duck meat (4.1);
- tea bags (4.1).

Food commodities with *high emissions* (>5 kg CO_2e/kg or L) tended to be livestock products and highly manufactured foods, for example:

- pig meat (5.5–9.9);
- beef cottage pie ready meal (7.6);
- packed mild cheddar cheese (9.8);
- beef (10–40);
- lamb shanks and roasted potatoes ready meal (19);
- lamb (27–39).

It is clear from the apple juice example, together with the data presented in Table 17.6 and the above examples, that the thermal processing step does not contribute significantly to the carbon footprint. The packaging materials influence because of the energy-intensive methods for their manufacture, although insignificant if the food product contains materials from a ruminant origin. However, glass and metal packaging, while having relatively high-carbon footprints, are environmentally friendly materials in terms of their minimal effect on abiotic resource depletion. This is not the case for plastic packaging that uses oil as its raw material.

As climate change continues to grab the headlines, thermal processing deserves to receive a more favourable press. This can only improve further as the true environmental impact of food manufacturing operations become important. Harvesting and processing of fruits and vegetables take place in close proximity so that deterioration of production is reduced. This offers many environmental and socio-economic benefits.

Perhaps the one environmental area in which thermally processed foods needs to improve is in its use of multi-layer plastics. Consumers seem to like semi-rigid plastic packages, such as pouches, cartons, and sachets, because they are lightweight and offer a large surface area for attractive graphics. However, few recycling facilities can manage these packages, resulting in the majority ending up in landfill sites. Compostable alternatives to plastics are available, however, their ability to withstand a wet thermal process is not yet proven. Multi-layer plastic packages will likely be used in thermal processing for some time. Re-use strategies can be employed to reduce the environmental burden with plastic packages. Mechanical recycling is the first and most environmentally economical option. Where this is not possible, chemical recycling can be used to break the plastic into its constituent parts. These can be used for fuel and as raw material for new plastics. This is a new technology but will experience widespread uptake as it develops. Another option is energy generation from plastic waste. Plastics are a highly efficient source of energy, making them attractive in waste heat recovery systems. Many European countries have well-developed systems for electricity and district heating via efficient energy from waste plants. Sweden is one such country. Data from The Swedish Institute in 2017 suggested 69% of all packaging was recycled, which included 85% for bottles and cans. Furthermore, 50% of household waste was turned into energy, which included plastic packages that could not be recycled or were put into the waste collection. Adoption of these types of energy recovery systems will help reduce the environmental impact of plastic packaging.

References

British sugar carbon footprint (n.d.). http://www.silverspoon.co.uk/home/about-us/carbon-footprint.

Defra (2008). *Guidelines to Defra's GHG Conversion Factors* – Annexes updated April 2008. http://www.defra.gov.uk/environment/business/envrp/pdf/ghg-cf-guidelines-annexes2008.pdf.

Enviros Consulting Limited (2003). *British Glass Manufacturers Confederation – Public Affairs Committee.* Glass Recycling – Life Cycle Carbon Dioxide Emissions http://www.britglass.org.uk/Files/Enviros_LCA.pdf.

FEFCO (2006). *European Database for Corrugated Board Life Cycle Studies.* European Federation of Corrugated Board Manufacturers www.fefco.org.

ISO 14044:2006 (2006). Environmental management – Lifecycle assessment – Requirements and guidelines.

ISO 14040:2006 (2006). Environmental management – Lifecycle assessment – Principles and framework.

PAS 2050 (2007). Publicly Available Specification for the assessment of the lifecycle greenhouse gas emissions of goods and services. www.bsigroup.com.

Plastics Europe (2005). Reports by I Boustead. *Eco-profiles of the European Plastics Industry LDPE FILM EXTRUSION, Polyethylene Terephthalate (PET) Bottle Grade*, PET film. www.plasticseurope.org.

Tate and Lyle carbon footprint (n.d.). http://www.tateandlyle.com/TateAndLyle/social_responsibility/environment/default.htm.

The carbon footprint of fat tire amber ale (n.d.). The climate conservancy. http://www.climateconservancy.org/cca_fattire.pdf

Tucker, G., Wiltshire, J., and Fendler, A. (2008). Carbon footprint of British food production. Food Science and Technology 22 (4): 23–26.

Wiltshire, J.J. et al. (2009). Scenario building to test and inform the development of a BSI method for assessing greenhouse gas emissions from food. Technical annex to the final report to Defra, Project Reference Number: FO0404. http://www.defra.gov.uk.

Index

Essentials of Thermal Processing, Second Edition. Gary Tucker and Susan Featherstone.